Immunology
of Receptors

IMMUNOLOGY SERIES

Edited by NOEL ROSE

Professor and Chairman
Department of Immunology and Microbiology
Wayne State University
Detroit, Michigan

Other Volumes in Preparation

11-77
524 pp
1977
$55

Immunology
of Receptors

Edited by

B. CINADER

Director, Institute of Immunology, and
Professor, Departments of Medical Biophysics,
Medical Genetics, and Clinical Biochemistry
University of Toronto
Ontario, Canada

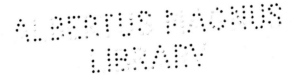

MARCEL DEKKER, INC.   New York and Basel

Library of Congress Cataloging in Publication Data

Main entry under title:

Immunology of receptors.

  (Immunology series ; 6)
  Includes bibliographical references and index.
  1. Autoimmune diseases--Etiology--Addresses, essays,
lectures. 2. Cell receptors--Addresses, essays, lectures. 3. Cellular control mechanisms--Addresses,
essays, lectures. I. Cinader, Bernhard. II. Series.
QR188.3.I47   616.07'9'3   77-14020
ISBN 0-8247-6674-1

These papers originally appeared in Immunological Communications,
Volume 5, Numbers 4, 5, 6, and 9, and Volume 6, Number 2, edited
by B. Cinader.

MARCEL DEKKER, INC.

270 Madison Avenue, New York, New York 10016

Current printing (last digit):
10 9 8 7 6 5 4 3 2 1

PRINTED IN THE UNITED STATES OF AMERICA

FOREWORD

The Editorial Committee of the Center for Immunology was privileged indeed to have found Dr. Bernhard Cinader willing to edit four Special Issues of *Immunological Communications* on the Immunology of Receptors. The 28 special contributions on that subject, after having been rearranged in a more logical sequence, have now been united in book form in the present volume, together with an introductory chapter by Dr. Cinader. We are convinced that a book on the "state of the art" of one of the most centrally important aspects of immunology is particularly timely and we hope that it will be well received.

Carel J. van Oss
Chairman, Editorial Committee
The Center for Immunology,
School of Medicine
State University of New York
at Buffalo

CONTENTS

CONTRIBUTORS

Louise T. Adler, Division of Immunology, St. Jude Children's Research Hospital, Memphis, Tennessee

K. Frank Austen, Departments of Medicine, Harvard Medical School and the Robert B. Brigham Hospital, Boston, Massachusetts

S. Avrameas, Département de Biologie Moléculaire, Institut Pasteur, Paris, France

M. Bornens, Département de Biologie Moléculaire, Institut Pasteur, Paris, France

R. Neal Boswell, Departments of Medicine, Harvard Medical School and the Robert B. Brigham Hospital, Boston, Massachusetts

B. Cinader, Director, Institute of Immunology, and Professor, Departments of Medical Biophysics, Medical Genetics, and Clinical Biochemistry, University of Toronto, Ontario, Canada

S. W. Craig, Department of Physiological Chemistry, Johns Hopkins University, School of Medicine, Baltimore, Maryland

P. Cuatrecasas, Wellcome Research Laboratories, Research Triangle Park, North Carolina

Janet M. Decker, Laboratory of Molecular Immunology, The Walter and Eliza Hall Institute of Medical Research, P.O., Royal Melbourne Hospital, Victoria, Australia

Charles DeLisi, Laboratory of Theoretical Biology/DCBD, National Cancer Institute, and Arthritis and Rheumatism Branch, National Institute of Arthritis, Metabolism, and Digestive Diseases, Bethesda, Maryland

Keith J. Dorrington, Department of Biochemistry and Institute of Immunology, University of Toronto, Toronto, Canada

Bonnie S. Dunbar, Department of Zoology, University of Tennessee, Knoxville, Tennessee

J. S. Flier, Diabetes Branch, NIAMDD, National Institutes of Health, Bethesda, Maryland

Luciana Forni, Basel Institute for Immunology, Basel, Switzerland

M. H. Freedman, Faculty of Pharmacy and Institute of Immunology, University of Toronto, Ontario, Canada

A. Froese, MRC Group, for Allergy Research, Department of Immunology, Faculty of Medicine, University of Manitoba, Winnipeg, Manitoba, Canada

Wm. R. Gallaher, Department of Microbiology, Louisiana State University Medical Center, New Orleans, Louisiana

E. W. Gelfand, Research Institute, Department of Immunology, Hospital for Sick Children, Toronto, Ontario, Canada

E. J. Goetzl, Departments of Medicine, Harvard Medical School and the Robert B. Brigham Hospital, Boston, Massachusetts

P. Halloran, Department of Medicine, Mt. Sinai Hospital, Toronto, Canada

P. A. Henkart, Immunology Branch, National Cancer Institute, National Institutes of Health, Bethesda, Maryland

Peter M. Henson, Department of Immunopathology, Scripps Clinic and Research Foundation, La Jolla, California

J. Holmgren, Institute of Medical Microbiology, University of Göteborg, Sweden

C. Howe, Department of Microbiology, Louisiana State University Medical Center, New Orleans, Louisiana

D. B. Jarrett, Diabetes Branch, NIAMDD, National Institutes of Health, Bethesda, Maryland

C. R. Kahn, Diabetes Branch, NIAMDD, National Institutes of Health Bethesda, Maryland

E. Karsenti, Département de Biologie Moléculaire, Institut Pasteur, Paris, France

P. J. Lachmann, MRC Group on Mechanisms in Tumour Immunity, Laboratory of Molecular Biology, The Medical School, Hills Road, Cambridge, England

P. Lake, ICRF Tumour Immunology Unit, Department of Zoology, University College London, Gower Street, London, England

S. T. Lee, Departments of Medicine and Immunology and the Manitoba Institute of Cell Biology, University of Manitoba, Winnipeg, Manitoba, Canada

Vanda A. Lennon, Molecular Neuropathology Laboratory, The Salk Institute for Biological Studies, San Diego, California

L. Lindholm, Institute of Medical Microbiology, University of Göteborg, Sweden

John J. Marchalonis, Laboratory of Molecular Immunology, The Walter and Eliza Hall Institute of Medical Research, Royal Melbourne Hospital, Victoria, Australia

Ian McConnell, MRC Group on Mechanisms in Tumour Immunity, Laboratory of Molecular Biology, The Medical School, Hills Road, Cambridge, England

K. L. Melmon, Division of Clinical Pharmacology, Departments of Medicine and Pharmacology, and the Çardiovascular Research Institute, University of California, San Francisco, California

Henry Metzger, Laboratory of Theoretical Biology/DCBD, National Cancer Institute, and Arthritis and Rheumatism Branch, National Institute of Arthritis, Metabolism, and Digestive Diseases, Bethesda, Maryland

N. A. Mitchison, ICRF Tumour Immunology Unit, Department of Zoology, University College London, Gower Street, London, England

E. Möller, Transplantation Immunology Laboratory, Huddinge Hospital and Division of Immunobiology, Karolinska Institutet Medical School, Stockholm, Sweden

K. B. Orr, Departments of Medicine and Immunology and the Manitoba Institute of Cell Biology, University of Manitoba, Winnipeg, Manitoba, Canada

F. Paraskevas, Departments of Medicine and Immunology and the Manitoba Institute of Cell Biology, University of Manitoba, Winnipeg, Manitoba, Canada

Benvenuto Pernis, Basel Institute for Immunology, Basel, Switzerland

U. Persson, Transplantation Immunology Laboratory, Huddinge Hospital and Division of Immunobiology, Karolinska Institutet Medical School, Stockholm, Sweden

H. Ramseier, Department of Experimental Microbiology, Institute for Medical Microbiology, University of Zurich, Zurich, Switzerland

O. Ringdén, Transplantation Immunology Laboratory, Huddinge Hospital and Division of Immunobiology, Karolinska Institutet Medical School, Stockholm, Sweden

J. Roth, Diabetes Branch, NIAMDD, National Institutes of Health, Bethesda, Maryland

J. L. Ryan, Immunology Branch, National Cancer Institute, National Institutes of Health, Bethesda, Maryland

V. Schirrmacher, Department of Cellular Immunology, Institute of Immunology and Genetics, Deutsches Krebsforschungszentrum, Heidelberg, West Germany

Max Schlamowitz, Department of Biochemistry, The University of Texas System Cancer Center, M. D. Anderson Hospital and Tumor Institute, Houston, Texas

M. Schlesinger, Department of Experimental Medicine and Cancer Research, The Hebrew University-Hadassah Medical School, Jerusalem, Israel

C. Alex Shivers, Department of Zoology, University of Tennessee, Knoxville, Tennessee

B. Rees Smith, Departments of Medicine and Clinical Biochemistry, Wellcome Research Laboratories, University of Newcastle upon Tyne, Newcastle upon Tyne, England

Tomio Tada, Laboratories for Immunology, School of Medicine, Chiba University, Chiba, Japan

Toshitada Takemori, Laboratories for Immunology, School of Medicine, Chiba University, Chiba, Japan

Masaru Taniguchi, Laboratories for Immunology, School of Medicine, Chiba University, Chiba, Japan

Gregory W. Warr, Laboratory of Molecular Immunology, The Walter and Eliza Hall Institute of Medical Research, Royal Melbourne Hospital, Victoria, Australia

Y. Weinstein, Department of Hormone Research, Weizmann Institute of Science, Rehovot, Israel

Immunology
of Receptors

Chapter 1

IMMUNOLOGY OF RECEPTORS - INTRODUCTION

B. Cinader
Institute of Immunology
Departments of Medical Biophysics,
Medical Genetics and Clinical Biochemistry
University of Toronto, Medical Sciences Building
Toronto, Canada M5S 1A8

The language of intracellular and intercellular communication
is the central theme of contemporary biology. This book is concerned with information transfer that results in the disposal of
foreign substances and in the antibody-dependent interference in the
interaction between a hormone and its receptor, i.e. with misinfor-
mation. Its first part deals with communication in terms of different receptors and of triggering molecules which are involved in
the defense mechanism and particularly in immune responsiveness.
Receptors which are affected by an autoimmune response are the
second theme of this book. Auto-antibodies, directed against
receptors, interfere with access of, or responsiveness to the normal chemical messenger, the hormone. Cells are triggered by antibodies to receptors, thus escaping control from other endocrine
organs and, ultimately, being turned off by continued stimulation.

Communications between cells involve chemical messengers which
bind to membrane receptors or to cytoplasmic and nuclear receptors.

Ligands bind with membrane "receptors" which are present at exceedingly high local concentrations on discrete particles. Such
distribution influences the experimentally measured binding constant.
Additional complexities are introduced when there is multipoint
attachment (1). Chemical messengers which become ligands for
membrane receptors induce the release of intracellular signals
which modify further cellular activities. As a consequence of
ligand binding, changes in membrane behaviour are initiated,
conformational changes of the membrane take place and, for ligands
with multivalent interaction, receptor aggregation and cap formation
occurs. Physico-chemical changes in the membrane, are followed by
altered membrane enzymes (e.g. adenyl cyclase), changes in the
permeability of the membrane to small molecules and, as a consequence, alteration of the intracellular concentration of monovalent and divalent ions (2,3). The rapid stimulation or inhibition
of many membrane-associated enzymes, within a few seconds after
ligand binding to the cell surface, may be attributable to mechanisms similar to those which operate in regulatory enzymes (4,5).
Receptors which can diffuse in the plane of the membrane may
reversibly associate with effectors to regulate their activity.
The affinity of the effectors would be increased when the receptor
is combined with the specific ligand (6). Some receptors are
glycoproteins or lipoproteins; co-operativity may not be exclusively
dependent on the protein moiety (7), but may also depend on the
lipid moiety. In fact, fatty acid compostion of phospholipid may

determine the magnitude of the reactivation of lipid-depleted

membrane $(Na^+ + K^+)$-ATPase (8).

After combination with a receptor, a triggering macromolecule

may, itself, undergo changes which result in interaction with a

second and different receptor site. There is good evidence that

this type of escalation may be involved in the biological effects

of cholera toxin (9). Such processes have also been suggested to

explain the mode of action of anaphylactic antibodies (10).

The chemical characterization of cell receptors is in its

infancy. It is perhaps most advanced where relatively simple

molecules occupy the receptor and receptor "shape" can be defined

in terms of the structural requirements for effective triggering

and for competitive interaction. This is the case of the chemo-

taxis controlling receptors for Val(Ala)-Gly-Ser-Glu on human

eosinophil polymorphonuclear leukocytes (11), but even with more

complicated receptor structures some advance has been made. This

is exemplified by the Fc receptors of different cells which have

different functional roles and differ in structure. Binding of

IgG to the Fc receptors on macrophanges (12,13,14) and heterologous

mast cells (15,16) involves the C3 domain, whereas binding to the

placenta involves a conformation, dependent on $C\gamma2$ and $C\gamma3$ domains

(17,18). It is not yet clear whether the Fc receptor of neutrophils

accomodates the $C\gamma2$ and $C\gamma3$ or only to the C3 domain (19). The

genetic control of Fc receptors has been a subject of considerable

recent interest but remains unresolved (20).

Interactions between macrophanges, B cells and T cells are essential to the full expression of immune responsiveness (21-23). Several non-antigen specific signals may be involved, but recognition of H-2 linked immune response with (I) region coded molecules; (see Fig. 1)(24) and interaction with receptor idiotypes appear to be of central importance for IgG antibody production (25).

B cell membranes contain several receptors: Ig molecules which are specific for antigen-antibody aggregates, complement receptors which combine with the third compound of complement (C3) and other membrane components (for instance, β2 microglobulin) through which   . the B cell can be stimulated by antibody, but for which a physiological function is not known. It is not clear whether membrane Ig can trigger the cell, i.e. whether combination with antigen has a direct effect on transductional devices (such as enzymes or calcium gating protein), or whether the receptor functions solely by enabling antigen to interact with some other component of the membrane (26). In fact, there are two different types of Ig molecules on B cell membranes. The majority of B cells in spleen, lymph nodes and Peyer's patches, but not in bone marrow, are IgD and IgM-bearing. During ontogeny and evolution, IgM appears before IgD. IgD receptors disappear from B cells during maturation to secretion and are never found on the membrane of IgG-containing cells. The signal delivered by the IgD receptor is unknown. If the signal were related to stimulation, it would affect proliferation rather than maturation (27).

The Fc receptor of B cells appears to have a regulatory
function and probably can be triggered so as to interfere with
B cell operation (28). The functional role of complement receptors is not established. Initially, it was thought to be a
receptor by which the cell could be triggered (29). More
recently, a less central role was assigned to this receptor and
it was thought to increase the binding of antigen to the B cell
(29-33,35). There is an association, perhaps identity between
C3d and the receptor for Epstein Barr virus; this may be responsible for target specificity of the virus (34). Modification
of immune responsiveness by viruses and other features of
infectivity may be better understood when our knowledge of virus
receptor is further advanced (36).

T cells have receptors for antigen which show high specificity
and appear to contain the hypervariable region of immunoglobulin.
One subpopulation of T cells has Fc receptors (37,38). Soluble
antigen-antibody complexes may combine with such receptors and
prevent interaction between T cells and other cell types (39)
and interfere with the cell's response to Concanavalin A stimulation (28,38,40,41). T cells also appear to have receptors for
cytophilic Ig or Ig complexes which amplify helper cell function
(42).

Passively administered anti-idiotypic antibody of a
particular subclass can sensitize mice to the antigen which is
recognized by the corresponding idiotype. This sensitization
affects T helper cells as well as precursor B lymphocytes. It,

therefore, appears that both types of cells possess the same
idiotype as part of some receptors for antigen. Interaction of
these receptors with anti-idiotypic antibody can be functionally
equivalent to interaction with antigen (43). In fact, an antigen
binding and idiotype-containing two-chain molecule (molecular
weight 150,000) may be produced by T lymphocytes (44). In short,
T cell membranes contain idiotypic determinants which confer on the
T cells capacity for antigen-recognition. Immune response gene
products (Fig. 1) are involved in various cell-interactions which
result in the final expression of immune responsiveness. If
antigen interaction with the Ig receptor is the first signal, the
interaction between the receptor and the T cell factor may be the
second signal, and cause an increase in cyclic AMP (45). The
interaction of the antigens and T cells can be blocked by sera
directed against I region determinants (46-49; Fig. 1). Furthermore,
the T cell helper factor (50) and the T cell suppressive factor (51)
can react with antibodies directed against I region products (52).
There are receptor or acceptor sites for various T cell factors.
Probably the helper factor and the acceptor is encoded by genes
present in the I-A subregion; the I-J subregion codes for the
suppressor factor and acceptor (Fig. 1).

A subset of T cells, characterized by passage over nylon-
wool columns, possesses receptors for suppressive T cell factors
(53). Cells which have this receptor are not helper cells but
appear to be a third T cell type which is involved in the trans-

THE MAJOR HISTOCOMPATABILITY COMPLEX(H-2) OF THE MOUSE

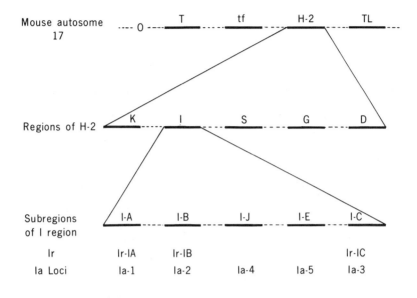

FIGURE 1:   0 = centromere; T = brachyury ("short" tail), controls development of notochord, neural tube and spinal column; tf = "tufted," controls
pattern of abnormal hair growth; H-2 = major histocompatibility complex;
TL = thymus leukemia; K and D = regions which code for H-2 antigens; G =
region which codes for erythrocyte determinants; I = immune response region;
immune response Ir genes are localized to the I-A, I-B and I-C subregions.
Serologically detectable I-region associated (Ia) antigens are determined
by Ia loci in the I-A, I-J, I-E and I-C subregions. The mixed lymphocyte
reaction (MLR stimulation) is controlled by Ia loci in the I-A and I-C
subregions; I-J subregion codes for Ia determinants (Ia-4 locus) on
a soluble suppressor factor.

mission of suppressor activity. It is possible that the acceptor
site is an *I-J* subregion product (54,55). Nylon adherent cells also
possess receptors for enhancing factor and *I-A* subregion genes may
code for both the factors and the corresponding acceptor sites.
The expression of these two types of acceptor sites may be
dependent on the time after antigenic stimulation (56). Another
receptor site appears to be present on responding T cells and is
involved in the suppression of the mixed lymphocyte reaction. It
is probable that this receptor molecule is coded by genes in the
I-A subregion of the H-2 complex (57; Fig. 1)

Receptors for histamine, beta mimetic catecholamines and
prostaglandins are found on lymphocytes. The corresponding drugs
increase intracellular cyclic AMP concentrations. Subpopulations
of suppressor T cells can be separated by their differential
adherence to insolubilized conjugates of histamine with serum
albumin. These cells are Ia positive and when stimulated with
histamine, beta adrenergic amines and prostaglandins of the E
series, lose their suppressor activity (58). Not only suppressor
cells, but also T cell precursors of cell mediated immunity possess
receptors for amines and prostaglandins (58).

The known receptors and the postulated modes of interactions
do not yet encompass all the parameters involved in the regulation
of the immune system. Various properties of the immunogen influence
the activating process and the cell types which are activated. The
relevant properties of the antigens are not simply the summation
of the reactivity of individual determinants. In particular,

repeating identical determinants affect the quality of response to
an immunogen. This could be the consequence of the high avidity
which results from multipoint binding, but it is quite possible
that membrane fluidity (59) and aggregation of cell surface
components may be involved (60, 61). This would trigger the cell
to differentiate and to increase the density of antigen receptors
(62). In fact, increase in receptor density has been observed on
exposure to mitogens(2).

At relatively high antigen concentrations, which favour high
zone tolerance in an intact animal, receptors may become inter-
linked, but may be prevented from aggregation. Such a frozen state
of the antigen recognition system may render the cell unresponsive
to antigen stimuli (62). Other chemical properties of antigen
affect the nature of immune responsiveness, presumably by selective
effects on different receptor sites. This has been demonstrated by
chemical modification of antigens, which results in a diversion of
the immune response from antibody formation into delayed type
hypersensitivity and, in some cases, into antibody-tolerance. This
has been shown with flagellin (63), serum proteins (64), carcino-
embryonic antigen (65) and lysozyme (66).

The effect of antibodies, which are directed against receptors,
is an intriguing chapter in the exploration of auto-immune disease.
Receptors appear to relay similar signals whether they are occupied
by their normally triggering molecules (hormones or antigens) or by
antibodies to receptors.

We shall now turn to this aspect of receptor-immunology. Suppression of immunoglobulin formation with antibody against immunoglobulin may be regarded as a model system for receptor-antibody interaction (67-68). In a given antiserum, different antibodies of different subclasses against the same lymphocyte receptors may either inhibit or initiate immunoglobulin formation (43,67). Depending on the specificity of the suppressive antibody, formation of allotype, idiotype or isotype can be inhibited. Neonatally injected or maternally transferred antibody prevents the appearance of circulating Ig (69-74), circulating levels of experimentally induced antibodies (75,76), plaque forming cells (77,78), appearance of cell associated Ig, detectable by fluorescent staining (79), and prevents the *in vivo* induction of antibody-mediated thymidine uptake and blast transformation of lymphoid cells (80,81). Absence of rabbit cells with a particular type of receptor is the result of the administration of suppressive antibody against this type of Ig within the preceding 24 hours (79). For operational reasons, suppression of circulating Ig can be demonstrated only at a much later time, i.e. when untreated controls develop significant levels of the parental allotype. Complete suppression results in loss of natural tolerance to the suppressed allotype. As a consequence, subsequent immunization may elicit antibody to the genetically determined, but phenotypically unexpressed, allotypic specificity (82). The induction and duration of suppression depends on the properties of suppressive antibody and on the regenerative capacity of the immune system for the suppressed cell. Thus, two factors are involved: one a property of the suppresive antibody and connected with the initial suppressive event, the other connected with the recipient's capacity to replace the suppressed component of the immune system.

The *in vivo* suppressive capacity of antibody depends on its Fc portion, on its affinity and on its specificity. Irrespective of specificity, antibody must possess an intact Fc portion in order to be effective in inducing suppression. *In vivo* suppression cannot be induced with antibody fragments $F(ab)_2$, although they combine with the target cell and trigger a synthetic burst of Ig formation (67,78,83). In contrast, both Fab and $F(ab)_2$ bring about *in vitro* inhibition of allotype synthesis by adult cells in short-term tissue culture (84). Thus, neither bivalency nor the Fc portion plays a role in *in vitro* inhibition, which may depend only on blocking of receptors. Additional processes must be involved in long-lasting *in vivo* suppression, since both Fc and bivalency are necessary. Bivalency, but not the Fc portion, is required to induce blast transformation (85). It follows that a process other than, or in addition to, blast transformation is involved in the *in vivo* induction of suppression in the newborn.

Attempts to induce suppression with specifically purified antibody have been unsuccessful (86). This may indicate that affinity of antibody molecules is crucial to suppressive capacity, since specific purification invariably eliminates the most avid antibody molecules.

The relation between antibody specificity and suppressive capacity may be illustrated by comparing the effect of neonatally administered antibody to a Fab determinant with that of an antibody directed against an Fc determinant of the Ig receptors. Rabbit allotypes of the Fab portion are located on the heavy γ, μ , α and ε chains (a locus) and on the light κ chain (b locus) which is associated with heavy chains of all classes (87). These allotypes can be suppressed by neonatal administration of an appropriate allogeneic antiserum. In contrast, attempts to suppress allotypes of the Fcγ portion (d and e loci) have, so far, failed (83,88,89), and Ig appear in the circulation as soon as the passively adminis-tered antibody disappears from it. In fact, neither suppression nor a synthetic burst can be induced with antibody against Fcγ determinants. There are very few cells with such determinants. The majority of Ig-receptor-carrying rabbit cells with Fab deter-minants appear to have also Fcμ determinants and only fewer than 5% have Fcγ determinants (90). This minority of cells is clearly not a suitable target for suppression; target cells for long-lasting suppression are lymphoid cells in early differentiation stages. In adult animals, this target cell may no longer play a crucial role in the generation of B cells. In the adult, B cells may primarily arise from differentiation products of these early target cells. These later B-cell precursors can be blocked at the level of their receptors, but are not readily suppressed (68). Do antibodies to other receptors and on other cell types have compar-able triggering and inhibitory effects? Auto-antibodies against

receptors appear in a variety of diseases and some forms of in-
fertility may be attributable to auto-antibodies against zona
pellucida antigens which block sperm attachment to the zona
pellucida (91). *In vitro* stimulation and increased thymidine
incorporation is not dependent on the Fc portion of the receptor-
antibodies. Upon acute exposure of fat cells to insulin-antibody,
stimulation of basal glucose oxidation, i.e. insulin-like effects,
were produced. In this case, as in the stimulation of thymidine
uptake by antibodies directed against Ig receptor of lymphocytes,
in vitro stimulation does not involve the Fc portion of the antibody molecules and stimulation is as effective with Fab_2 fragments
as with the intact antibody molecule (92). Stimulation of the
thyrotropine receptor with antibody results in increased cyclic
AMP production and the monovalent Fab antibody fragment is as
effective in promoting this increase as the intact bivalent
molecule (93). Slight contaminent with Fab aggregates or with
bivalent molecules could be responsible for this observation. If
this should not prove to be the case, the mechanism of thyrotropine receptor stimulation would be different from that of Ig
receptors of lymphocytes, which is dependent on the bivalency of
the antibody (94). We do not know whether divalency and Fc
fragments are involved in the long lasting effects upon chronic *in
vivo* exposure to antibody directed against receptors. Patients
with antibody against receptors are insensitive to the normally
stimulating hormones and it is not clear whether this is due to

blockade of receptors or to the loss of receptor sites. It will
be of some interest to determine whether allotype suppression and
autoimmune receptor interference are subject to similar reaction
mechanisms.

There is persuasive evidence that the hyperthyroidism
associated with Graves' Disease may be caused by antibodies
directed against the thyrotropin receptor; that one group of
patients with insulin resistance, acanothosis nigricans and a
variety of symptoms which indicate generalized autoimmunity have
circulating inhibitors which interfere with insulin receptor
function (95) and finally, that antibodies and lymphocytes
specifically reactive with nicotine acetylcholine receptors, may
be a factor in the etiology of myasthenia gravis (92,93,96,97).
It remains to be seen whether or not cell mediated immunity may be
involved in autoimmune damage to receptor structures. Significant
correlation between severity of myasthenia gravis and *in vitro*
responsiveness of lymphocytes to cell receptors have been reported
(98). Claims of direct stimulation of isolated bovine thyroid cells
by peripheral blood lymphocytes from patients with Graves' Disease
(99) have not been confirmed (93). In experimentally induced
disease, pathological effects can be induced by cell mediated
immunity and important observations on the specificity of humoral
and cell mediated immunity have been made. Distinct antigen
determinants induce antibody and cell mediated immunity in
experimental autoimmune encephalomyelitis (100). It appears that

some factors which regulate susceptibility to autoimmune disease,

as well as other diseases, are controlled by the HL-A major

histocompatibility complex of man (101). The identification of

these regulatory mechanisms will represent a major advance in our

understanding of disease induction and progression and in the

development of medical practice which is based on the individual-

ity of disease susceptibility and progression (102).

Acknowledgments

Thanks are due to Dr. Terry Delovitch for critical reading of the

manuscript, to Mr. Ladislav Horvath for checking the references, and

to MRC and NCIC for support of the authors' research work.

References

1. DeLisi, C. and Metzger, H., this Volume, pp. 21-40.

2. Bornens, M., Karsenti, E. and Avrameas, S., this Volume, pp. 41-61.

3. Freedman, M.H. and Gelfand, E.W., this Volume, pp. 97-114.

4. Changeux, J.P., Thiery, J., Tung, Y. and Kittell, C., Proc. Nat.
    Sci. US, 57: 335, 1967.

5. Monod, J., Wyman, J. and Changeux, J.P., J. Mol. Biol., 12: 88,
    1965.

6. Jacobs, S. and Cuatrecasas, P., Biochim. Biophys. Acta, 433:
    482, 1976.

7. Edelman, G.M., Yahara, I. and Wang, J.L., Proc. Nat. Acad. Sci.
    US, 70: 1442, 1973.

8. Kimelberg, H.K. and Paphadjopoulos, D., J. Biol. Chem., 249:
    1071, 1974.

9. Craig, S.W. and Cuatrecasas, P., this Volume, pp. 63-76.

10. Stanworth, D.R., Haematologia, 8: 299, 1974.

11. Boswell, R.N., Austen, K.F. and Goetzl, E.J., this Volume, pp. 149-159.

12. Yasmeen, D., Ellerson, J.R., Dorrington, K.J. and
     Painter, R.H., J. Immunol., 116: 518, 1976.

13. Yasmeen, D., Ellerson, J.R., Dorrington, K.J. and
     Painter, R.H., J. Immunol., 110: 1706, 1973.

14. Dorrington, K.J., this Volume, pp. 183-200.

15. Minta, J.O. and Painter, R.H., Immunochemistry, 9: 1041, 1972.

16. Froese, A., this Volume, pp. 235-251.

17. McNabb, T., Koh, T.Y., Dorrington, K.J. and Painter, R.H.,
     J. Immunol., 117: 882, 1976.

18. Schlamowitz, M., this Volume, pp. 253-272.

19. Henson, P., this Volume, pp. 131-147.

20. Halloran, P. and Schirrmacher, V., this Volume, pp. 201-219.

21. Greaves, M.F., Owen, J.T. and Raff, M.C., in T and B
     Lymphocytes: Origins, Properties and Roles in Immune Responses.
     Excerpta Medica, Amsterdam, 1973.

22. Katz, D.H. and Benacerraf, B., in The Immune System, Genes,
     Receptors, Signals, edited by E. Sercarz, A.R. Williamson and
     C.F. Fox, p. 569, Academic Press, New York, 1974.

23. Cooper, H.L., Transpl. Rev., 13: 3, 1972.

24. Katz, D.H., Graves, M., Dorf, M.E., Dimuzio, H. and
     Benacerraf, B., J. Exp. Med., 141: 263, 1975.

25. Lake, P. and Mitchison, N.A., this Volume, pp. 407-417.

26. Karlin, A., Cowburn, D.A. and Reiter, M.J., in Drug Receptors,
     edited by H.P. Rang, p. 193, MacMillan, London, 1972.

27. Pernis, B. and Forni, L., this Volume, pp. 371-390.

28. Ryan, J.L. and Henkart, P.A., this Volume, pp. 221-234.

29. Dukor, P., Schumann, G., Gisler, R.H., Dierich, M., König, W.,
     Hadding, U., and Bitter-Suermann, D., J. Exp. Med., 139: 337,
     1974.

30. Dukor, P., Dietrich, F.M., Gisler, R.H., Schumann, G. and Bitter-Suermann, D., in Progress in Immunology II, edited by L. Brent and J. Holborow, Vol. 3, p. 99, North Holland Publ., 1974.

31. Feldmann, M. and Pepys, M.B., Nature, 249: 159, 1974.

32. Janossy, G., Humphrey, J.H., Pepys, M.B. and Greaves, M.F., Nature, New Biol., 245: 108, 1973.

33. Pepys, M.B., J. Exp. Med., 140: 126, 1974.

34. Jondal, M. Klein, G., Oldstone, M.B.A., Bokish, V. Yefenof, E., Scand. J. Immunol., 5: 401, 1976.

35. McConnell, I. and Lachmann, P.J., this Volume, pp. 161-182.

36. Gallaher, W.R. and Howe, C., this Volume, pp. 491-508.

37. Grey, H.M., Kubo, R.T. and Cerottini, J.C., J. Exp. Med., 136: 1323, 1972.

38. Stout, R.D. and Herzenberg, L.A., J. Exp. Med., 142: 611, 1975.

39. Stout, R.D. and Herzenberg, L.A., J. Exp. Med., 142: 1041, 1975.

40. Ryan, J.L., Arbeit, R.D., Dickler, H.B. and Henkart, P.A., J. Exp. Med., 142: 814, 1975.

41. Ramseier, H., this Volume, pp. 293-312.

42. Paraskevas, F., Lee, S.T. and Orr, K.B., this Volume, pp. 313-328.

43. Eichmann, K. and Rajewsky, K., Eur. J. Immunol., 5: 661, 1975.

44. Binz, H. and Wigzell, H., Scand. J. Immunol., 5: 559, 1976.

45. Mozes, E., in The Role of Products of the Histo-compatibility Gene Complex in Immune Responses, edited by D.H. Katz and B. Benacerraf, p. 485, Academic Press, New York, 1976.

46. Crone, M., Koch, C. and Simonsen, M., Transpl. Rev., 10: 36, 1972.

47. Hammerlind, G.J. and McDevitt, H.O., J. Immunol, 112: 1734, 1974.

48. Basten, A., Miller, J.F.P. and Abraham, R., J. Exp. Med., 141: 547, 1975.

49. Wekerle, H., Eshhar, Z., Lonai, P. and Feldmann, M., Proc. Nat. Acad. Sci. USA, 72: 1147, 1975.

50. Taussig, M.J., Munro, A.J., Campbell, R., David, G.S. and Staines, N.A., J. Exp. Med., 142: 694, 1975.

51. Tada, T., Okumura, K. and Taniguchi, M., J. Immunol., 111: 952, 1973.

52. Munro, A.J. and Taussig, M.J., in Role of Products of the I-Region of the H-2 Complex in Cell Cooperation in Mitogens in Immunobiology, edited by J.J. Oppenheim and D.L. Rosenstreich, p. 261, Academic Press, 1976.

53. Tada, T., Taniguchi, M. and Davis, C.S., in Cold Spring Harbor Symposium on Quantitative Biology, Vol. XLI, in press, 1976.

54. Murphy, D.B., Herzenberg, L.A., Okumura, K., Herzenberg, L.A. and McDevitt, H.O., J. Exp. Med., 144: 699, 1976.

55. Okumura, K., Herzenberg, L.A., Murphy, D.B., McDevitt, H.O. and Herzenberg, L.A., J. Exp. Med., 144: 685, 1976.

56. Tada, T., Taniguchi, M. and Takemori, T., this Volume, pp. 351-370.

57. Rich, S.S. and Rich, R.R., J. Exp. Med., 142: 1391, 1975.

58. Weinstein, Y. and Melmon, K.L., this Volume, pp. 115-130.

59. Singer, S.J. and Nicolson, G.L., Science, 175: 720, 1972.

60. Raff, M.C. and De Petris, S., Fed. Proc., 32: 48, 1973.

61. Unanue, E.R., Perkins, W.D. and Karnovsky, M.J., J. Exp. Med., 136: 885, 1972.

62. Diener, E. and Paetkau, V.H., Proc. Nat. Acad. Sci., USA, 69: 2364, 1972.

63. Parish, C.R., in Progress in Immunology II, edited by L. Brent and J. Holborow, Vol. 2, p. 39, North Holland Publishing Co., Amsterdam, 1974.

64. Benacerraf, B. and Gell, P.G.H., Immunology, 2: 53, 1959.

65. Chao, H-F., Peiper, S.C., Aach, R.D. and Parker, C.W., J. Immunol., 111: 1800, 1973.

66. Thompson, K., Harris, M., Benjamini, E., Mitchell, G. and Noble, M., Nature New Biol., 238: 20, 1972.

67. Cinader, B. and Dubiski, S., Brit. Med. Bull., 32: 171, 1976.

68. Adler, L.T., this Volume, pp. 419-438.

69. Dray, S., Nature (London), 195: 677, 1962.

70. Dubiski, S., Quant. Biol., 32: 311, 1967.

71. Dubiski, S. and Swierczynska, Z., in Human Anti-Human Gamma-globulins, edited by R. Grubb and G. Samuelsson, p. 39, Pergamon Press, Oxford, 1971.

72. Dubiski, S. and Swierczynska, Z., Int. Arch. Allergy Appl. Immunol., 40: 1, 1971.

73. Cinader, B., Koh, S.W. and Kuksin, P., Cell. Immunol., 11: 170, 1974.

74. Herzenberg, L.A. and Herzenberg, L.A., Contemp. Top. Immunobiol., 3: 41, 1974.

75. Mage, R.G. and Dray, S., Nature (London), 212: 699, 1966.

76. Hart, D.A., Pawlak, L.L. and Nisonoff, A., Eur. J. Immunol., 3: 44, 1973.

77. Chou, C.T., Cinader, B. and Dubiski, S., Int. Arch. Allergy Appl. Immunol., 32: 583, 1967.

78. Strayer, D.S., Cosenza, H., Lee, W.M.F., Rowley, D.A. and Kohler, H., Science, 186: 640, 1974.

79. Harrison, M.R. and Mage, R.G., J. Exp. Med., 138: 764, 1973.

80. Sell, S., J. Exp. Med., 128: 341, 1968.

81. Harrison, M.R., Elfenbein, G.J. and Mage, R.G., Cell. Immunol., 11: 231, 1974.

82. Lowe, J.A., Cross, L.M. and Catty, D., Immunology, 28: 46, 1975.

83. Shek, P.N. and Dubiski, S., J. Immunol., 114: 621, 1975.

84. Schuffler, C. and Dray, S., Cell. Immunol., 11: 377, 1974.

85. Fanger, M.W., Hart, D.A., Wells, J.V. and Nisonoff, A., J. Immunol., 105: 1484, 1970.

86. Goldman, M.B. and Mage, R.G., Immunochemistry, 9: 513, 1972.

87. Dubiski, S., Med. Clin. North Am., 56: 557, 1972.

88. Mage, R.G., Ann. New York Acad. Sci., 190: 203, 1971.

89. Lowe, J.A., Immunology, 23: 591, 1972.

90. Pernis, B., Forni, L. and Amante, L., J. Exp. Med. 132: 1001, 1970.

91. Dunbar, B.S. and Shivers, C.A., this Volume, pp. 509-519.

92. Flier, J.S., Kahn, C.R., Jarrett, D.B. and Roth, J., this Volume, pp. 477-489.

93. Smith, B.R., this Volume, pp. 461-476.

94. Weiner, H.L., Moorhead, J.W., Yamage, K. and Kubo, R.T., J. Immunol., 117: 1527, 1976.

95. Flier, J.S., Kahn, C.R., Roth, J. and Bar, R.S., Science, 190: 63, 1975.

96. Toyka, K.V., Drachman, D.B., Griffin, D.E., Pestronk, A., Winkelstein, J.A., Fischbeck, K.H. Jr., and Kao, I., New Engl. J. Med., 296: 125, 1977.

97. Lennon, V.A., this Volume, pp. 439-460.

98. Richman, D.P., Patrick, J. and Arnanson, B.G.W., New England J. Med., 294: 694, 1976.

99. Edmonds, M.W., Row, V.V. and Volpe, R., J. Clin. Endocr., 31: 480, 1970.

100. Lennon, V.A., Wilks, A.V. and Carnegie, P.R., J. Immunol., 105: 1223, 1970.

101. Dausset, J. and Hors, J., Transplant. Rev. 22: 44, 1975.

102. Cinader, B., Canadian Medical Association Journal, 113: 11, 1975.

Chapter 2

SOME PHYSICAL CHEMICAL ASPECTS
OF RECEPTOR-LIGAND INTERACTIONS

Charles DeLisi and Henry Metzger
Laboratory of Theoretical Biology/DCBD,
National Cancer Institute, and
Arthritis and Rheumatism Branch,
National Institute of Arthritis, Metabolism,
and Digestive Diseases, Bethesda,
Maryland 20014

Abstract

Quantitative evaluations of a variety of binding reactions of
interest to immunologists have usually assumed that the reactants are homo-
geneously dispersed in solution for purposes of calculation. In fact, many
of these reactions involve cell-bound "receptors" which are present at
exceedingly high local concentrations on discrete particles. We describe
how such a distribution can influence the experimentally measured binding
constants. We also briefly consider the additional complexities introduced
when multipoint attachment between the ligand and cell-bound receptors is
possible, and discuss the possible biological implications.

I. Introduction

There has been increasing interest in immunologic phenomena which

occur on the surfaces of cells. Experimental techniques now make possible

the quantitative study of such reactions, but the interpretation of the data

is not immediately obvious. For example, it has been found that the reaction

of IgE with its cell surface receptor on basophils, and of mouse IgG_{2a} with

a cell receptor on macrophages, is consistent with a simple reversible re-

action

$$Ig + R \xrightarrow[\quad k_r \quad]{k_f} Ig \ldots R$$

21

These reactions have been characterized with regard to rate and equilibrium

constants (1, 2), calculation of which has been based on formulae which

assume that the reactants are randomly distributed in the reaction volume.

Thus, if the incubation mixture contained 1×10^9 cells/L, each having 10^6

receptors/cell, the molarity of receptors was taken as $1 \times 10^{15}/6 \times 10^{23}$ or

1.67×10^{-9} M. In fact, however, the receptors are situated on the surface

membranes of the cells where their local concentration is 10^6 -fold higher,

i.e., $\sim 10^{-3}$M! One of the aims of this paper is to consider what the re-

lationship is between the effective (measured) rate constants, k_f and k_r,

and the intrinsic rate constants, k_1 and k_{-1}, under these conditions. In

which situations will there be substantial deviations and, if so, what will

be their direction?

In many reactions of interest the potential for multipoint attach-

ment exists. Its likelihood may depend upon the relative rates of diffusion

of receptors in the plane of the membrane vs the effective dissociation rate

of the initial monovalent ligand-receptor interaction. Here again the ques-

tion of whether the local high concentration of receptors on the cell surface

will cause k_r to be substantially different than k_{-1} becomes important. The

third section of this paper considers such reactions.

We recognize that a variety of factors, which may be difficult to

quantify, may importantly influence the reactions of ligands with cell

surface receptors: the complicated topology of the membrane itself (cells

are not smooth spheres), the proximity of "irrelevant" surface structures,

receptor turnover, and so on. Still, it seems reasonable to neglect these

uncertainties temporarily and see where theory leads us. Only be assessing

the extent to which the theory fails to conform to experiment can we gauge

the significance of these other factors.

II. Monofunctional Reactions

II.1.   General Remarks

Many reactions of immunological interest appear to be representable
by the single step bimolecular process

$$L + R \underset{k_r}{\overset{k_f}{\rightleftharpoons}} R_B \tag{1}$$

where L and R are the concentrations of the free reacting units, R_B the
concentration of the complex, and k_f and k_r the effective forward and reverse
rate constants. Just as the reaction of antibody in solution with simple
ligands (3-6) and more complicated ligands (7, 8) is representable in this
way, we could anticipate that qualitatively the reaction of soluble ligand
with cell-bound antibody would behave similarly, as would the reaction between the Fc region of IgE and its receptor sites (1) and mouse IgG_{2a} with
receptors on macrophages (2). The evidence in these cases is based on the
dose dependence of the reaction rate as well as on an analysis of the kinetic
binding curves. For such a reaction the concentration of bound receptors at
time t can be expressed in several ways (4, 9); the formulation of Bell and
DeLisi being

$$R_B = a + \left\{ \frac{\exp(-k_f bt)}{R_0 - a} - \frac{1}{b} \left(1 - \exp(-k_f bt) \right) \right\}^{-1} \tag{2}$$

where

$$a = \frac{1}{2} \left[R_0 + L_0 + \frac{1}{K} + b \right] \tag{3}$$

$$b = \left[(R_0 + L_0 + \frac{1}{K})^2 - 4R_0 L_0 \right]^{\frac{1}{2}} \tag{4}$$

$$K = k_f/k_r \tag{5}$$

and R_0 and L_0 refer to the initial concentrations of reactants.

It is often experimentally useful to set conditions so that equation 2 can be approximated by

$$R_B = \frac{KL_0}{1 + KL_0} \left[1 - \exp\left(-t/\tau\right)\right] \tag{6}$$

where
$$\frac{1}{\tau} \equiv k_r + k_f \left(\bar{L} + \bar{R}\right) \tag{7a}$$

$\bar{L}$ and $\bar{R}$ being the concentrations of free ligand and receptor at equilibrium. Equation 6 is true for all concentrations if the reaction is close to equilibrium. If one of the reactants is in excess (for example, ligand), then

$$\frac{1}{\tau} = k_r + k_f L_0 \tag{7b}$$

and equation 6 holds throughout the entire reaction. The relaxation time of the process, τ, is a useful measure of the time required to reach equilibrium.

When both receptor and ligand are homogeneous, monofunctional and randomly dispersed, k_f and k_r can reasonable be interpreted as the rate constants k_1 and k_{-1} for the site-site interaction. However, when receptors are concentrated on a surface, a number of factors may affect the effective reaction rates so that k_f and k_r may not be identical to k_1 and k_{-1}.

II.2.   The Forward Rate Constant

II.2.1.   The effect of diffusion coefficient differences. Specific bond formation requires that the centers of the reacting units be separated by some distance, r_0. Assuming the units are spherical, r_0 is approximately equal to the sum of their radii. Since the frequency of such encounters is clearly dependent upon the diffusion rates as well as the concentration of reactants, we must ask whether the decreased diffusion coefficient of cell-bound as compared to free receptors affects the rate of reaction with ligand.

To begin with, since both reactants diffuse, it is the sum of the diffusion coefficients which enters the description of the process. Thus, if D_L, D_R and D_C represent coefficients of ligand, isolated receptor and cell (or cell-bound receptor), respectively, what is important is the difference between $D_1 = D_L + D_R$ and $D_2 = D_L + D_C$. Clearly, when D_L is much larger than D_R and therefore much larger than D_C, as would be the case if L were a small molecule and R an immunoglobulin, the difference between D_1 and D_2 would be insignificant.

More generally, where the ligand diffusion does not dominate and the encounter frequency does change, there may still be little effect on the reaction rate. The reason is that attaining the encounter distance r_0 usually does not guarantee bond formation (10, 11). For ligands at r_0 there will be many collisions before a successful reaction occurs. Thus, although a smaller diffusion coefficient reduces the encounter frequency, it also increases the number of collisions per encounter. It can be shown formally (11) that these two effects may cancel under a wide range of conditions and that diffusion is not rate controlling unless the reaction rate or viscosity is exceptionally high. The argument finds considerable support from polymerization studies: as polymers grow there is a continuous change in the diffusion coefficient of the reacting species, but this causes no observable deviation from predicted growth kinetics based on assuming constant reaction rates (11).

II.2.2.   The effect of concentration gradients. Consider the reaction between ligand and cell-bound receptor. Since there is net binding (i.e., the association rate is greater than the dissociation rate), the concentration of ligand in the immediate vicinity of the cell surface will be depleted somewhat. Consequently, there will be a concentration gradient near the cell, and the ligand concentration, $L(r,t)$, at some time t and distance r from the cell center will satisfy the diffusion equation

$$\frac{\partial L(r,t)}{\partial t} = D\nabla^2 L(r,t)$$

where $V^2 L$ represents the spatial change in the concentration gradient, i.e.,
the second derivative of $L(r,t)$ with respect to position and $D = D_L + D_C$.

Two conditions must be specified in order to solve equation 8. First,
far from the cell the ligand concentration should be undisturbed by the binding
reaction and therefore be equal to L_0. Second, Ficks first law must hold near
the reaction surface at r_0 i.e., the rate at which ligand crosses a unit area
at distance r_0 must be proportional to the concentration gradient at r_0 (12).
This second condition can be expressed as:

$$\frac{Nk_1 L(r_0 t)}{4\pi r_0^2} = -D \frac{\partial L(r_0 t)}{\partial r} \tag{9}$$

where k_1 is the intrinsic forward rate constant for receptor site-ligand binding
and N = number of receptors per cell. The left side of equation 9 is the rate
at which ligand is crossing a unit area of a spherical surface with radius r_0,
and the right side is proportional to the concentration gradient in a direction
perpendicular to the surface. Then the concentration of ligand at r_0 and time
t is (13)

$$L(r_0,t) = \left\{ 1 - \frac{r_0 - \beta}{r_0} \left[1 + \exp\left(\frac{Dt}{\beta^2}\right) \text{erfc} \frac{\sqrt{Dt}}{\beta} \right] \right\} L_0 \tag{10}$$

where
$$\beta \equiv \frac{r_0}{1 + Nk_1/4\pi D r_0} \tag{11}$$

and erfc $(\sqrt{Dt}/\beta)$, called the co-error function, is tabulated (14).

After a rapid transient that lasts for a time

$$t \stackrel{\sim}{=} \beta^2/D \tag{12}$$

the concentration of ligand at r_0 approaches a time-independent (steady state)
value and equation 10 can be approximated by

$$L(r_0) = \frac{L_0}{1 + Nk_1/4\pi D r_0} \tag{13}$$

The main point of this development is that, if the binding reaction
has not come to equilibrium, neither ligands nor cells can be considered

randomly distributed throughout the entire volume as would be the case in an

ideal gas. Consequently, there will be a net flux of unreacted ligand to-

ward free receptors, and as a result, the reaction rate will be given by

$$Nk_1 L(r_0) = \frac{Nk_1 L_0}{1 + Nk_1/4\pi Dr_0} \tag{14}$$

In an experimental determination of the forward rate constant, the

usual assumption is that the ideal gas distributions hold close to the cell.

For example, with ligand in excess, L_0 would be used rather than $L(r_0)$ [see

equation 7b]. Consequently, according to equation 14, the measured rate

constant k_f should be interpreted as

$$k_f = \frac{Nk_1}{1 + Nk_1/4\pi Dr_0} \tag{15}$$

Thus, if $Nk_1 \lesssim 4\pi Dr_0$, then

$$k_f \simeq Nk_1, \tag{16}$$

the role of the concentration gradient is insignificant, and there is no

error in using L_0 rather than $L(r_0)$. On the other hand, if

$$Nk_1 > 4\pi Dr_0 \tag{17}$$

then $$k_f \simeq 4\pi Dr_0 \tag{18}$$

and the reaction rate is diffusion controlled. This last result is the

Smoluchowski diffusion-limited rate constant (15). Since the diffusion co-

efficient is $kT/f(T)$, where k is Boltzmann's constant, T the temperature in

$^\circ K$, and $f(T)$ the temperature-dependent) frictional coefficient, equation 18

can also be written as

$$k_f \simeq 4\pi DkTr_0 \left(\frac{1}{f_L(T)} + \frac{1}{f_R(T)} \right) \tag{19}$$

The frictional coefficient is linearly proportional to viscosity so

equation 19 predicts that the diffusion-limited rate constant is inversely

proportional to viscosity. This is characteristic of diffusion-controlled
reactions. In addition, since the viscosity is temperature dependent, the
forward rate constant will be characterized by a positive activation energy,
typically about one-third the latent heat of vaporization of the solvent.
For most solvents, this is about 2-5 Kcal/mole.

The above development assumes the existence of a steady-state gradient. However, as indicated by G.I. Bell (personal communication), if
chemical equilibrium is reached rapidly, such a gradient may never be established. Thus the length of the experiment must be long compared to the time
required to establish the gradient, and short compared to the time required
to reach equilibrium, i.e., according to equations 7b and 12

$$\frac{\beta^2}{D} \leq t \leq \frac{1}{k_1 L(r_0) + k_{-1}} \tag{20}$$

Steady-state concentration gradients will be important only when equations
17 and 20 are both satisfied.

Although these results have been presented explicitly in terms of
ligand interacting with cells having a large number of receptors, they
also apply to reactions between ligand and receptors that are not cell bound.
In such cases, units which have reacted cannot continue to react, but the
development can be carried through in terms of a net average flux. The
result is equation 15 with N = 1. This means that the Smoluchowski diffusion-limited rate constant for the free receptor-ligand interaction is

$$k_{D1} = 4\pi D_1 r_1 \tag{21}$$

where $r_1 = r_L + r_R$

whereas for the cell-bound receptor-ligand interaction

$$k_{D2} = 4\pi D_2 r_2 / N \tag{22}$$

where $r_2 = r_L + r_C$

r_L, r_R and r_C denoting the radii of ligand, receptor and cell, respectively. The numerical values of k_{D1} and k_{D2} will generally differ substantially, so that when they determine reaction rates, one must expect differences between cell-bound and dispersed systems. Alternatively, with k_{D1} and k_{D2} greater than the intrinsic site-site forward rate constant, measurements on the two systems should yield the same results.

As a numerical example, consider the reaction between IgE and Fc receptors that are uniformly distributed over a cell surface. Then r_2 = $(5 \times 10^{-4} + 10^{-6})$ cm $\stackrel{\sim}{=}$ 5×10^{-4} cm, and $D_2 = D_C + D_L \stackrel{\sim}{=} D_L = 2 \times 10^{-7}$ cm^2/sec. Hence, $K_{D2} = 7.5 \times 10^{11}$ (M-sec)$^{-1}$/N, which becomes 7.5×10^6 (M-sec)$^{-1}$ and 7.5×10^5 (M-sec)$^{-1}$ with N = 10^5 and 10^6, respectively. Since the experimental result is $\stackrel{<}{\sim} 10^5$ (M-sec)$^{-1}$ (1), it apparently does not represent the Smoluchowski diffusion-limited value. (Incidentally, for this particular case, the reverse rate constant is $\stackrel{<}{\sim} 10^{-5}$ sec^{-1} and the relaxation time for chemical reaction is relatively slow under the usual concentration conditions.) This situation is to be compared to reaction between immunoglobulin and dispersed receptor. Assuming the receptor is comparable in size to IgE, r_1 = $r_L + r_R \stackrel{\sim}{=} 2 \times 10^{-6}$ cm and $D_1 = D_L + D_R \stackrel{\sim}{=} 4 \times 10^{-7}$ cm^2/sec. Therefore, k_{D1} = 6×10^9 (M-sec)$^{-1}$. The preliminary experimental value of the forward rate constant for dispersed receptors is 8×10^5 (M-sec)$^{-1}$. Both experimental results are well below their respective Smoluchowski diffusion-limited values, but they nevertheless differ; and although the variation is not large, it suggests that other factors not yet considered play some role.

II.2.3.   <u>Other effects.</u> Ligand-receptor interactions, whether the receptor be cell bound or not, invariably proceed more slowly than one would predict by considering only translation of the center of mass of re-actants, as is done in deriving the Smoluchowski result. There are several sources giving rise to this difference, the most important probably being the invalid assumption that all portions of ligand and receptor are equally reactive. It is evident that for units to react, not only must their centers

be separated by r_0, but they must also be appropriately oriented. Inclusion

of this orientational requirement leads to the introduction of a multiplica-

tive factor, f (<1), on the right side of equation 18 (16), thus reducing

the diffusion-limited rate constant. The actual magnitude of f will depend

upon the precision required to form a bond, but it may easily be $\lesssim 10^{-2}$ (17,

18). However, just how this varies as the result of receptors being cell

bound rather than dispersed is difficult to say at present; in part because

of the complicated nature of the cell surface, and in part because orienta-

tional theories are in the early stages of development. It would seem

reasonable to suppose, however, that for a spherical cell with receptors

uniformly distributed over its surface, only the rotational diffusion of li-

gand should matter (i.e., only it is considered asymmetric) and f should be

substantially larger than when both interacting units are asymmetric, as they

are when ligand interacts with dispersed receptor.

In addition to site asymmetry (i.e., confining the reactivity of the

unit to a small portion of its surface), there is the related problem of

geometrical asymmetry, i.e., the effect which deviation from spherical shape

has on reaction kinetics. Hill (18) has studied this question for the re-

action of ellipsoidal particles with a site localized on a stationary plane.

An interesting result is that for ligands shaped as oblate ellipsoids of

revolution, the orientational factor will include an activation energy term,

i.e., it will show a temperature dependence which arises because of hindered

rotation. Although the model is highly idealized, the result is significant

since it illustrates how a temperature-dependent forward rate constant can

arise from geometrical asymmetries of the reacting units.

The diffusion-limited rate constant may also differ from what is pre-

dicted by the Smoluchowski equation, i.e., equation 18, as the result of the

presence of any type of long-range interaction, for example, a coulombic repulsion. It can be shown that the effect can, to a first approximation,

be taken into account by replacing r_0 with

$$r^* \equiv \frac{Z_L Z_R e_0^2}{4\pi\epsilon kT} \quad \frac{1}{[\exp\left(\frac{Z_L Z_R e_0^2}{4\pi\epsilon r_0 kT}\right) - 1]} \tag{23}$$

where e_0 is the electronic charge, ϵ the dielectric permitivity of the sol-

vent, and Z_L and Z_R the effective charge valences of ligand and receptor,

respectively. Z_L has the sign of the net charge on ligand, and Z_R the sign

.of the net charge on receptor.

When $\frac{Z_L Z_R}{4\pi\epsilon r_0 kT} > 1$, replacing r_0 by r^* has two main effects. First,

the value of k_1 required for equation 17 to hold is reduced, thus providing

another source of reduction of the forward rate constant below the value

given by the Smoluchowski equation. In addition, the forward rate constant

will be more temperature dependent than in the absence of electrostatics.

In particular, the free energy of activation $G^{\ddagger}$, will be given by

$$G^{\ddagger} = kT^2 \, \partial \ln k_f / \partial T = RT^2 \, \frac{\partial}{\partial T} \, \ln\left(\frac{1}{f_L} + \frac{1}{f_R}\right) + \frac{Z_L Z_R e_0^2}{4\pi\epsilon r_0} \tag{24}$$

where the temperature dependences of thermal expansion and dielectric con-

stant have been neglected. The first term in equation 24 is the activation

energy associated with solvent viscosity and, as already indicated, it is in

the range of 2-5 Kcal/mole for many liquids. The second is the electrostatic

energy.

As a numerical example, with an electrostatic energy about five

times larger than the thermal energy, the electrostatic term can contribute

another three Kcal/mole to $\Delta G^{\ddagger}$. In such a case, the magnitude of the for-

ward rate constant will also be affected. In particular, with an electro-

static energy of 3 Kcal/mole, $Nk_1 = 2.5 \times 10^{10}$ $(M\text{-sec})^{-1}$. This puts k_1 in

the range of $(10^4 - 10^5)$ $(M\text{-sec})^{-1}$.

To summarize, with orientational and long-range forces included, equation 15 can be written as

$$k_f = \frac{k_i}{1 + k_i/4\pi D_i r_i {}^* f_i} \qquad i = 1, 2 \qquad (25)$$

where $i = 1, 2$ refers to systems involving dispersed and cell-bound receptors, respectively (therefore $k_2 \equiv N k_1$), and f_i may include an activation energy. If it is still true that

$$k_i < 4\pi D_i r_i {}^* f_i \qquad (26)$$

for both systems, the theory predicts that the effective forward rate constants per receptor site should be the same in both cases. Because of lack of adequate knowledge of the parameters, it is difficult to estimate whether this inequality holds for any particular system. However, it is easy to gain insight experimentally, since if it does not hold, an inverse dependence of reaction rate on viscosity is expected, as explained above. In fact, for the IgE receptor system, a positive result would provide a rational basis for interpreting the finding that the forward rate constant depends on whether the receptors are cell bound or dispersed. If, on the other hand, a viscosity dependence is not found, other factors not yet considered may be responsible for the difference; for example, a conformational change in the receptor as the result of interaction with other membrane constituents. As a working hypothesis, we will assume for the moment that the reaction is diffusion limited.

In connection with this possibility, Kulczycki and Metzger (1) find a forward rate constant which decreases from 8.8×10^4 (M-sec)$^{-1}$ at 37°C to 2.1×10^4 (M-sec)$^{-1}$ at 4°C, a temperature dependence which implies an activation energy of 7.8 Kcal/mole. This is about 3-6 Kcal/mole larger than one would expect from viscosity alone. In addition, they find very little change in the results over a wide range in pH, so it seems unlikely that electrostatics is making a significant contribution to this difference. We therefore speculate that the remainder of the activation energy is contributed by

hindered rotation of the IgE when it is in the vicinity of the cell surface,

and this is probably due to the geometrical complexities of the basophil

membrane as well as the irregular geometry of the IgE. This explanation is

consistent with preliminary observations on dispersed receptor - IgE inter-

actions, which appear to show a smaller temperature dependence.

Finally, we note that there are rather large differences found in

forward rate constants for antibodies binding to free as compared with cell-

bound epitopes; the former values being in the range of $(10^7 - 10^8)$ (M-sec)$^{-1}$

and the latter in the range of $(10^4 - 10^5)$ (M-sec)$^{-1}$ (19, 20). Such differ-

ences are easily accounted for if the antibody cell-bound epitope reaction is

diffusion controlled, i.e., if $k_2 > 4\pi D_2 r_2^* f_2$; for in such a case the collision

frequency in the cellular system is expected to be substantially smaller than

that in the dispersed system. However, there are other explanations (see, for

example, Section III), and a useful test of this possibility would be provided

by viscosity-dependent measurements.

II.3.   The Reverse Rate Constant

The dissociation of interacting units is a two-step process involving

bond breakage followed by diffusion-mediated separation. In this case, after

a steady-state gradient has been established, and with the diffusion step

rate limiting, Ficks first law implies that

$$k_r = 4\pi r^2 D \frac{dL(r)}{dr} \tag{27}$$

With the boundary conditions $L(r_0) = \frac{3}{4\pi r_0^3}$ and $L(\infty) = 0$, equation 27 inte-

grates to

$$k_r = \frac{3D}{r_0^2} \tag{28}$$

Equation 28 is the diffusion-limited reverse rate constant. A rough esti-

mate for diffusion-limited dissociation of hapten from antibody (where $D \stackrel{\sim}{=}$

$D_L = 10^{-6}$ cm^2/sec and $r_0 \stackrel{\sim}{=} r_R = 1.2 \times 10^{-6}$ cm) is 2×10^6 sec^{-1}. This is

well above the observed reverse rate constants and it therefore seems unlikely that gradient effects are important in this case. However, for the
dissociation of a high molecular weight antigen from an antibody receptor or
antibody from an Fc receptor on a cell of radius 5×10^{-4} cm and where $D =$
2×10^{-7} cm^2/sec, equation 28 gives 2-4 sec^{-1}. This number is well within
the observed range of antibody-hapten dissociation rates and suggests that
the dissociation of singly bound antibody from a cell may be smaller than the
corresponding dissociation from free hapten because of diffusion limitations.

Whether or not diffusion plays a role in _experimental_ measurements
will depend upon protocol. For example, in an experiment in which there is
an excess of unlabeled ligand which competes with labeled ligand for receptor,
the concentration of the competitor in the vicinity of cell surfaces may be
sufficiently high to prevent recombination of labeled ligand. Under such
circumstances, diffusion should be unimportant since re-association of labeled
ligand with receptors will be blocked.

III. Bivalent Ligands

III.1.   Basic Concepts

The interaction of bivalent ligand with cell-bound receptors occurs
in two steps: a bimolecular site-site interaction with intrinsic forward and
reverse rate constants k_1 and k_{-1}, followed by an intramolecular (intracom-
plex) reaction with rate constants k_2 and k_{-2} (Fig. 1). If receptors are
also bivalent, then the second step may occur with a site on the same receptor molecule or on another molecule. We wish to outline some of the general features of the model shown in Fig. 1, and their relation to the effects
discussed in Section II.

In general, if neither reactant is in great excess, a complete
description of the process requires two coupled nonlinear differential
equations which are difficult to solve. This should be contrasted to the

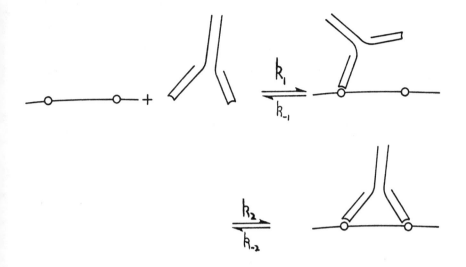

Figure 1

single-step process which leads only to a single nonlinear equation for which
a solution is readily obtained [equation 2]. As noted, if linearization
is valid (that is, at times close to equilibrium or when either L_0 or R_0 is
in excess), the single-step process leads to a particularly simple result:
a single exponential decay with relaxation time τ given by equation 7. In
the two-step reaction, when the linearized equations are valid, the kinetic
curve is a sum of two exponentials with relaxation times (21)

$$\frac{1}{\tau_+} = -\frac{B}{2} + \frac{1}{2}\sqrt{B^2 - 4C} \qquad\qquad (29a)$$

$$\frac{1}{\tau_-} = -\frac{B}{2} - \frac{1}{2}\sqrt{B^2 - 4C} \qquad\qquad (29b)$$

and where $\quad B \equiv k_1{}^* + k_2 + k_{-1} + k_{-2}$ $\qquad\qquad (30)$

$$C \equiv k_1{}^*(k_2 + k_{-2}) + k_{-1}k_{-2} \qquad\qquad (31)$$

and $k_1{}^*$ is either k_1 $(\bar{R} + \bar{L})$, $k_1 L_0$ or $k_1 R_0$, depending upon which conditions are used to satisfy linearization. For bivalent ligand interacting with cell-bound receptors the last (R_0 in excess) is (mathematically) the most appropriate since only then will k_2 be independent of the extent of the reaction. Alternatively the second will be realistic provided the reaction has not proceeded too far while the first is valid close to equilibrium. It is evident that even with linearization, the observed binding curve is a complicated function of all elementary rate constants.

The experimental binding curve need not be so complicated as the above remarks indicate. In particular, with $B^2 >> 4C$ and using the second linearization condition, equations 29 become

$$\frac{1}{\tau_+} = - \frac{k_1{}^*(k_2 + k_{-2}) + k_{-1}k_{-2}}{k_1{}^* + k_2 + k_{-1} + k_{-2}} \equiv - (k_f L_0 + k_r) \tag{32}$$

and

$$\frac{1}{\tau_-} = - (k_1{}^* + k_2 + k_{-1} + k_{-2}) \tag{33}$$

where

$$k_f = \frac{2k_1(k_2 + k_{-2})}{k_1 L_0 + k_2 + k_{-1} + k_{-2}} \tag{34}$$

$$k_r = \frac{k_{-1}k_{-2}}{k_1 L_0 + k_2 + k_{-1} + k_{-2}} \tag{35}$$

and

$$\frac{k_f}{k_r} = 2K_1(1 + K_2) \tag{36}$$

Because of the assumption $B^2 >> 4C$, the two time constants are well separated, the second being much shorter than the first. In addition, since $k_{-1} \gtrsim \sec^{-1}$, the exponential and τ_- represents a rapid transient and the observed kinetics is determined by τ_+, and is a single exponential decay with k_f and k_r given by equations 34 and 35. Under these conditions, the analysis of Section II is unchanged. Experiments involving the binding of multifunctional ligands to surfaces with an array of sites, and for which single-step kinetics are observed (for example, the Hornick-Karush experiments and, perhaps, also those

of Hughes-Jones and his colleagues) apparently operate in this limit. In
such cases, the observed rate constants are generally complicated functions
of those for the elementary steps.

If k_2 is substantially larger than all other rate constants, the result becomes especially simple. In particular,

$$k_f = 2k_1 \tag{37}$$

and $$k_r = \frac{k_{-1}}{K_2} \tag{38}$$

Under these very special conditions, the forward rate constant can be unambiguously identified as the intrinsic ligand-receptor site rate constant that
would be measured using dispersed receptor, provided diffusion is not controlling.

III.2.   Biological Implications

Equation 38 predicts that the reverse rate constant is reduced by
K_2 (>>1), and this, along with "retention" effects (22) may have important
consequences on the interpretation of certain experiments. For example, in
vitro tolerization studies require that the tolerizing antigen be washed from
the cells before a subsequent, normally immunogenic dose is administered.
However, the actual washing time is not the period of centrifugation, but
only the time during which cells are more or less suspended. This is because
once they have sedimented to the bottom, any ligand still bound will have to
diffuse through densely packed cells to escape to the supernatant. Because
of retention, the rate at which this happens is reduced by ($\frac{1}{1 + KR_0}$). More
correctly, it can be shown that the diffusion coefficient is reduced by this
amount (23). To be more specific, the fraction of ligand remaining bound to
cells that have been dispersed (sedimenting) for time t is $\exp(-k_r t)$.
Thus, if t = 10 minutes and $k_r = 10^{-4}$ $\sec^{-1}$, over 94% will remain bound after
a single washing cycle, and over 83% after three cycles. What is important

to notice is that (over a very long time range) the result is independent of
centrifugation time. If the cells are pelleted within the first ten minutes,
it makes little difference if centrifugation is for ten minutes or an hour.
What is required is that the washing time of dispersed cells be comparable to
or longer than the reciprocal of the reverse rate constant, and this may
require using slower rotor speeds as well as longer washing cycles.

The retention factor $\frac{1}{1 + KR_0}$ may also be important in vivo. Where
cell concentrations are ordinarily high, ligand may dissociate and reassociate
many times before being removed. Under such circumstances, the rate of removal may be many times slower than would be expected on the basis of in
vitro studies. As noted elsewhere (1) this may account for the apparent discrepancy between the decay of IgE-mediated local cell sensitization in vivo
and the half-life of IgE-cell dissociation in vitro.

Equations 37 and 38 represent only one of several possible sets of
limiting conditions for two-step processes. An alternative limit arises
from equations 34 and 35 if k_{-1} is the dominant rate constant. This becomes
relevant when receptors are widely spaced compared to the distance between
determinants on the ligand, so that diffusion of receptors in the plane of
the membrane is required for cross linking. For example, the rate constant
for single-site dissociation (k_{-1}) is in the range of (1-100) sec^{-1} whereas
with 10^5 receptors per cell, Bell (24) has estimated that the diffusion-
limited cross-linking rate constant is about 0.1 sec^{-1}. Under these conditions

$$k_f = 2K_2K_1 \tag{39}$$

$$k_r = k_{-2} \tag{40}$$

The interpretation of equation 50 is that the effective forward rate constant
is proportional to the concentration of singly bound ligand ($2K_1R_0$) multi-
plied by the rate constant for forming the second band.

Although k_2 has been introduced as an intramolecular rate constant, it will depend upon the density of free receptor sites on the cell surface. Thus $k_2 = \gamma R_0 k_2'$, where k_2' is the forward rate constant for the bimolecular reaction in the plane of the membrane and γ is a proportionality constant. The effective forward rate constant for multisite attachment

$$k_f \overset{\sim}{=} 2\gamma k_2' KR_0 \tag{41}$$

depends on both affinity and receptor density. In addition, if k_2' is diffusion limited, it will vary inversely as the membrane viscosity, so that k_f should be significantly temperature dependent. This possibility is in accord with the observed temperature dependence of patch formation(25). In addition, equation 41 predicts that the effective forward rate constant is affinity dependent. As noted elsewhere (26), this provides a mechanism for affinity-dependent selection and therefore maturation even when multivalent bonding is irreversible within the time required to deliver an activating signal. Finally, polymeric mitogens which do not lead to maturation may, because of their size, bind many receptors more or less simultaneously. Their binding kinetics would therefore more closely correspond to equations 37 and 38, and with K_2 very large (and therefore $k_r \overset{\sim}{=} 0$ within the time required to deliver a signal) binding would be controlled only by k_1 (equation 37). It would therefore be affinity independent and the response would not mature.

References

1.   Kulczycki, A., and Metzger, H., J. Exp. Med. 140:1676, 1974.

2.   Unkeless, J.L., and Eisen, H.N., J. Exp. Med., 142:1520, 1975.

3.   Froese, A., Sehon, A.H., and Eigen, M., Canad. J. Chem., 40:1786, 1962.

4.   Day, L.A., Sturtevandt, J.M., and Singer, S.J., Ann. N.Y. Acad. Sci., 103:619, 1963.

5.   Froese, A., Immunochem, 5:253, 1968.

6. Pecht, I., Givol, D., and Sela, M., J. Mol. Biol., 67:421, 1972.

7. Levison, S.A., and Dandliker, W.B., Immunochem., 6:253, 1969.

8. Hornick, C., and Karush, F., Immunochem., 9:325, 1972.

9. Bell, G.I., and DeLisi, C., Cellular Immunol., 10:415, 1974.

10. Benson, S.W., The Foundations of Chemical Kinetics. McGraw-Hill, Inc., New York, 1960.

11. Flory, P.J., Principles of Polymer Chemistry. Cornell University, Ithaca, N.Y., 1953.

12. Tanford, C., The Physical Chemistry of Macromolecules. John Wiley and Sons, New York, 1961.

13. Noyes, R.M., Progress in Reaction Kinetics, 1:129, 1961.

14. Abramowitz, M., and Segun, I.A., Handbook of Mathematical Functions. Dover Publications, New York, 1965.

15. Smoluchowski, M., Physik Z., 17:557, 1916.

16. Sloc, K., and Stockmayer, W.H., J. Chem. Phys., 54:2981, 1971.

17. Schmitz, K.S., and Schurr, J.M. J. Phys. Chem., 76:534, 1972.

18. Hill, T.L., Proc. Nat. Acad. Sci. U.S.A., 72:4918, 1975.

19. Hughes-Jones, N.C., Gardner, B., and Telford, R., Biochem. J., 85:466, 1962.

20. Hughes-Jones, N.C., Gardner, B., and Telford, R., Biochem. J., 88:435, 1963.

21. DeLisi, C., Antigen-Antibody Interactions. Springer-Verlag, Heidelberg, 1976 (in press).

22. Selhary, T.J., Sevec, S., Boos, W., and Schwartz, M., Proc. Nat. Acad. Sci. U.S.A., 72:2120, 1975.

23. DeLisi, C., and Goldstein, B., J. Theor. Biol., 51:313, 1975.

24. Bell, G.I., in Proc. First Annual Los Alamos Life Sciences Symposium, edited by D.F. Petersen, ERDA Technical Center, Oak Ridge, in press.

25. de Petris, S., and Raff, M.C., Nature New Biol., 241:257, 1973.

26. Bell, G.I., Nature, 248:430, 1974.

Chapter 3

RECEPTOR MOBILITY AND ITS COOPERATIVE RESTRICTION BY LIGANDS

M. Bornens, E. Karsenti and S. Avrameas
Département de Biologie Moléculaire,
Institut Pasteur, Paris

Abstract

The functional significance of membrane fluidity and receptor mobility in lymphoid cells has been studied in the recent literature. Although far from clarified, the role of membrane fluidity in achieving control over cell activity is probably important; it allows cooperative interactions over long distances.

Here, the emphasis is put on the phenomenon of restriction of receptor mobility by ligands such as Concanavalin A, a phenomenon discovered in recent years using morphological techniques. We discuss in some detail our own approach for studying this phenomenon. This consists of quantitatively measuring the active sites on cell-bound lectin molecules by subsequent fixation of horse-radish peroxidase. This study has shown a cooperative binding of Concanavalin A to cells which corresponds to a modification of the membrane, leading to the recruitment of new receptors. The existence of a post-binding event, that we have called micro-redistribution, has been shown at 4°C, through the use of peroxidase binding to cell-bound lectin.

A cooperative restriction of receptor microredistribution is observed when the cooperative recruitment of receptors induced by increasing concentrations of Concanavalin A occurs. Both phenomena were shown to be modulated by drugs such as colchicine and cytochalasin B. The characteristics of this modulation suggest that density and distribution of receptors are dependent upon the state of a multimeric submembrane structure which is still functional at 4°C.

The Fluid Mosaic Membrane Model of Singer and Nicolson (1) proposed that membranes are composed of a fluid lipid bilayer matrix which allows lateral movements of membrane components. The success of this model is well known, due to the overwhelming experimental data fitting it. There is now strong evidence indicating that plasma membrane components, either lipids or proteins, are capable of lateral mobility in the membrane plane; several reviews on this topic have been published recently (2-3).

41

The functional significance of such a mobility of membrane components has been considered and the possibility of transmembrane (or cortical) control on cellular activity through cell-surface modulation has been proposed. As an example, one can note that mobility of membrane proteins plays a key role in recent models for hormone action (4) or triggerring of B lymphocytes (5). But, rather than a part of a particular mechanism of a specific function, lateral mobility of surface proteins could be just the necessary condition for a coordinated activity of the whole surface of a living cell.

Such an idea has appeared quite supported in the recent years by studies on the effects of lectins on cells, particularly on lymphoid cells. In a lesser extent, experiments with anti-immunoglobulins of varying valences have also supported this idea. To study receptor mobility, cytological approaches such as immunofluorescence or immuno-autoradiography have been extensively used and reviewed recently (6); there exists also possibility of following cell surface events through binding parameters of ligands to cell surface receptors. In this short communication we will mainly review works which used the latter approach with lectins and anti-immunoglobulins in order to understand the modification of the membrane that such ligands can induce. For comparison, we will see very briefly, in a very specific hormonal system, what can be learned from binding studies for the problem of cell receptor mobility.

I. Ligand Binding and Cell Receptor Mobility

A. Insulin

Negative cooperativity has been shown for the binding of insulin to cells or purified membranes (7). The data show that a mechanism exists which regulates the affinity of the receptor sites according to the concentration of insulin available to the cell. The receptor sites are able to undergo a transition from a slow dissociating state to a fast dissociating state when an increased fraction of them is occupied by insulin.

These data can be explained in terms of change in configuration of individual oligomeric receptors or in terms of clustering of receptors through ligand-induced movements
in the plane of the membrane or in both terms.   There is good experimental evidence by
electron microscopy of clustering of receptors upon binding of biologically active fer-
ritin-insulin to fat cells or liver membranes (8).   Other data such as the effect of temperature and the effect of Con A (see below) on the binding of insulin also support the
idea that receptor clustering occurs (7).

   The coupling between hormone binding and biological activation is another step in
which receptor mobility is probably important.   Actually, such a coupling is formalized
in the "mobile receptor hypothesis" of Cuatrecasas (4); this model states that the receptor, which can diffuse independently in the plane of the membrane, reversibly associates with effectors (catalytic units) to regulate their activity.   Recently, Jacobs et al
(9) have shown that negative cooperativity of insulin receptor binding could be explained by the mobile receptor hypothesis.   A variety of mechanisms can probably be used
for the coupling between binding and biological activation: negative cooperativity in
binding can be associated with negative cooperativity in biological effects (10-11), or
with positive cooperativity (12).   Recently, Levitzki et al (13) have proposed a model
based on clustering of oligomeric receptors to generate cooperative membrane response
from a non-cooperative ligand binding.   Obviously, such mechanisms could be of great
physiological significance.

   B. Specific Receptors on Lymphocytes

   A very abundant literature exists on receptor mobility in the membranes of lymphoid
cells, as followed by morphological techniques; recent reviews have been published (6,
14).   Much less has been reported on ligand binding properties as compared to those
very large surface events.

Authors who have studied the binding of purified and labeled anti-immunoglobulins on splenic lymphocytes have all found a classical hyperbolic curve of binding as a function of ligand concentration, with a maximum around 2.10^5 molecules per cell (15-18). Is such a simple binding able, however, to modify mobility parameters of the cell surface receptors? This is indeed suggested by the well known induction of patching and capping (6, 19-26). This induction seems clearly to be related to the valence of the anti-immunoglobulins; Fab fragments are unable to do it (6, 19, 21, 24). That cross-linking is needed is further shown by the formation of patches when cells treated with Fab fragments are further treated with anti-Fab (21).

Induction of thymidine incorporation in B lymphocytes by anti-immunoglobulins can be observed, depending on the origin of the cells; rabbit B cells do respond to such antibodies (27), while mouse and human B cells do not (28). Mitogenic properties of soluble polymers of Fab fragments which had been cross-linked by glutaraldehyde were tested in our laboratory (29); such compounds do stimulate thymidine incorporation in rat B cells with a dose response curve which is reminiscent of thymocyte stimulation by Con A (narrow dose response curve), while native anti-immunoglobulins showed a dose-response curve which is reminiscent of thymocyte stimulation by Succinyl-Con A (stimulation is obtained for a large concentration range).

A recent model for B lymphocyte triggering has been proposed (5) which involves lateral mobility of surface receptors (IgD); it is close to the mobile receptor-hypothesis for hormone action (4) except that it involves proteolysis at a specific site of the cell receptor, possible only for binding of the antigen; such a proteolysis results in exposure of a site which can interact, through lateral mobility, with an effector responsible for triggering.

C. Lectin Binding to Cells

Here one deals with a different situation from hormones or antigens and the concept
of "lectin receptor" has to be taken with caution.

The so-called receptors are probably, in each case, a broad spectrum of cell surface
glycoproteins, possibly all of them for some lectins. The maximum number of binding
sites per cell is usually quite higher than corresponding values for hormones or antigens. The equilibrium constants also indicate lower specificity. Another reason for
not taking lectin-cell membrane interaction as a general model for ligand-specific receptor interaction is the great variety of biological effects that can be induced by the
same lectin on different cellular systems or on the same cell at different lectin concentrations.

The great interest of lectins, Concanavalin A[1] being the best example, is that
these molecules are interacting with some basic constitutive element or property of
the cell surface which controls or allows any other specific activity; lateral mobility
in the plane of the membrane, for example. Before discussing this aspect in the next
section, we will review binding studies of lectins to cells and see how they can be
correlated with receptor mobility.

The effective equilibrium constant of binding of a lectin to membrane receptors is
often quite higher than it is with the hapten sugar (30-32). Spontaneous dissociation
of the Con A membrane complex is very slow (33-34). Exchange of bound labeled
Con A with unlabeled Con A is very slow. A small but significant part of the cell-
bound lectin cannot be dissociated from the membrane by the hapten sugar (34-36),
although the left-over lectin is still on the external side of the membrane (36).

All these data strongly suggest already that lectin binding is not a simple bimolecular reaction. The first reports on binding studies of lectins to cells pointed to homo-

geneity of the receptor sites on the cell (37-38). More recently, the binding of Con A

and Wheat Germ Agglutinin to human lymphocytes was studied and shown to be com-

plex in the case of Con A (39); this result suggested the presence of multiple binding

sites for Con A. Binding of the same two lectins was measured on isolated fat cells

(40); Wheat Germ Agglutinin was shown to bind fat cells in quite a complex manner,

suggesting the existence of multiple and interacting classes of binding sites; the same

cells were shown to possess heterogeneous classes of binding sites for Con A and there

was evidence suggesting interaction between these binding sites.

The binding of Con A to rat thymocytes was studied in some detail in our labora-

tory (34-35). We compared the binding at $4°C$ of Con A and Succinyl-Con A, which

is a dimer and which cannot reassociate as a tetramer (41). The data of Con A bind-

ing to rat thymocytes at $4°C$ suggested positive cooperativity in the binding of the

lectin to the cell membrane. This was not the case for Succinyl-Con A (Figure 1).

This suggested that multivalency was necessary for cooperative binding to occur. We

do not know the actual mechanism responsible for such a cooperative binding of Con A

nor how it affects binding parameters of individual receptors to lectin molecules.

What seemed to occur in this cooperative binding of Con A was the appearance of new

receptors: the maximum number of Con A molecules which could be bound per cell

was more than twice the number of Succinyl-Con A molecules (3.8×10^6 versus 1.6

$\times 10^6$). Moreover, since we could assume that multivalent binding of Con A occurred,

this meant that many more than 3.8×10^6 cell receptors were involved in the binding

of Con A when the cell was saturated. When cells were pre-fixed with glutaralde-

hyde, there was no cooperative binding of Con A to the cells. Significantly fewer

receptors were available. We, therefore, interpreted our results on living cells as

showing that a cooperative membrane modification was brought about by the binding

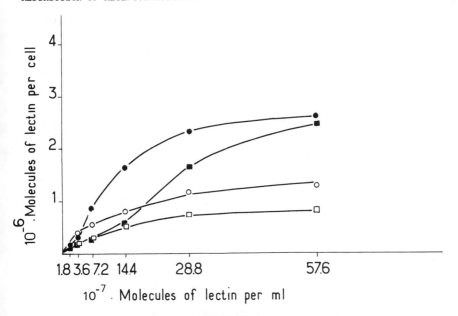

FIGURE 1.
Binding of ^{125}I-labelled Con A and I[125]-labelled Succinyl-
Con A to rat thymocytes at 4°C. (●) Con A binding to normal
cells. (O) Succinyl-Con A binding to normal cells. (■)
Con A binding to cells which were pretreated with colchicine
(10[-4]M) plus cytochalasin B (10 µg/ml) for 30 min at 37°C
and then incubated at 4°C for binding experiment. (□)
Succinyl-Con A binding to drugs-treated cells.
For technical details, see references 35 and 63.

of the first molecules of Con A; appearance of new receptors would be the effect of

this membrane modification. Such an interpretation fits well with a previous observa-

tion from our laboratory (42). It is also supported by a recent report on the effect of

Con A on the expression of cell surface sialytransferase activity of mouse thymocytes

(43); the results are consistent with a rapid Con A-induced exposure of enzyme mole-

cules that were previously inaccessible to substrate. The process does not require

active protein synthesis.

A recent report on the interaction of Con A with isolated thymocyte membranes

confirmed that positive cooperativity in the binding occurred (44). Moreover, as in

our case, the binding of Con A to glutaraldehyde-treated membranes was non-cooperative, with high affinity.   Interestingly enough, the authors proposed that cooperativity might depend upon the state of ∩ 55,000 D glycoprotein which can polymerize and which would be the "specific Con A-receptor"; glutaraldehyde treatment
would polymerize this glycoprotein and, therefore, a simple michaelian binding of
Con A to treated membrane would be obtained, with high affinity. Actually they
did not show polymerization of the 55,000 D protein by glutaraldehyde, but only its
disappearance from SDS-gels, due to cross-linking by glutaraldehyde. Also, it is
not clear why glutaraldehyde-induced polymerization of the monomeric receptor
would lead to many more binding sites; this would require unmasking of 55,000 D
proteins in the isolated membranes through glutaraldehyde action; it is difficult to
imagine. We obtained on whole cells a completely different result (34-35): either
with Con A or with Succinyl-Con A, there were fewer receptors available on the
cells when they were fixed with glutaraldehyde (1.0 x 10^6 versus 1.6 x 10^6 for Succinyl-Con A; 1.7 x 10^6 versus 3.8 x 10^6 for Concanavalin A); in both cases, the
equilibrium constant was higher than the constant for living cells. It is likely that
the observed glutaraldehyde-dependent increase of Con A binding sites in isolated
membranes (Figure 4 in reference 44) was due to the fact that glutaraldehyde was
not neutralized and could therefore react with Con A.

The binding of Soybean agglutinin to human and rabbit erythrocytes has recently
been shown to present positive cooperativity (45); this is another example of a conformational change in the membrane induced at low lectin concentration, this time
in an unnucleated cell.   Interestingly enough, trypsin-treated cells bind more lectin
molecules than untreated cells at low concentration.   Positive cooperativity has been

recently reported for the binding of Con A, Wheat Germ Agglutinin and Wax Bean Agglutinin to teratoma cells (46).

The coupling between binding and activation should involve, in the case of lectins, a long list of biological effects, depending on the cellular systems used. As far as membrane enzymes are concerned, there are now two examples clearly showing cooperativity in the action or inhibition of enzymes upon Con A binding (47); it is the case for Acyl-ConA-L-Lec acyl transferase (47) and for 5'-nucleotidase (48). We can also, in this section, mention the recent work from Barnett et al (49) showing evidence that mitogenic lectins induce changes in lymphocyte membrane fluidity following a dose-response curve which superimposes the dose-response curve of mitogenic effect.

II. The Effects of Con A on Receptor Mobility

We will recall here only the well-established effects of Con A on patching and capping observed at 37°C on lymphoid cells. We will then turn to the question of the existence of post-binding events in the cell membrane which could be detected at the molecular level; such events have been suggested to us in the preceding section by the comparison of the binding of Con A and Succinyl-Con A at 4°C to normal and glutaraldehyde-treated cells.

A. Effect on Receptor Mobility at 37°C

A striking effect of Con A on the mobility of the receptors has been shown by Yahara and Edelman (50); first, it fails to induce redistribution of Con A receptors at 37°C except for very low concentrations of Con A. Secondly, if Con A is added to lymphocytes at room temperature or 37°C prior to treatment with anti-immunoglobulin, then both patch formation and cap formation are inhibited.

In contrast to these effects, if Con A is added to the cells at 4°C and the cells are then brought to 37°C, Con A forms patches and caps with its own receptors (51-52). The same occurs if Con A is added after exposure of the cell to colchicine. Unlike the native molecule, dimeric Succinyl-Con A has neither the capacity to induce patches and caps on splenic lymphocytes nor the capacity to inhibit patch and cap formation by its own and other receptors (52). The immobilization of cell surface receptors by Con A could be due to simple external cross-linking of a surface glycoprotein, since capping of Con A receptors leads in some conditions to co-capping of surface immunoglobulin or H_2 or Θ antigen (53). But experiments with Con A-derived nylon fibers showed, on mouse B lymphocytes, that local interaction of Con A with its receptors strongly modifies in a cooperative fashion the behavior of cellular structures that are involved in the movement of cellular receptors and also in cell movement (54). This effect was antagonized by colchicine. From this type of result, models were proposed representing the microtubular system as exerting trans-membrane control over surface receptors (55-56); the more elaborate forms were proposed by Edelman and his group (57-59).

B. Effects of Con A on Cell Surface at 4°C

As described above (section I, C), we had observed cooperative binding of Con A at 4°C and interpreted this result as a membrane modification leading to the recruitment of new receptors. To understand this effect of Con A binding on cell membranes, we developed a method to follow quantitatively the accessible active sites on cell-bound lectin molecules by subsequent fixation of horse-radish peroxidase,[1] a glycoprotein which interacts with Con A (60-61). Cells were incubated with ^{125}I-labeled lectins, washed, and then incubated with a constant amount of HRP; specific binding of HRP to cell-bound lectins was obtained. The molar ratio HRP/cell bound lectins

was calculated for living or pre-fixed cells when incubated with ^{125}I-labeled Con A
or ^{125}I-labeled Succinyl-Con A (Figure 2). It was observed that HRP binding by
cell-bound lectins was higher when cells were pre-fixed with glutaraldehyde before
incubation with lectins; this was true either with Con A or Succinyl-Con A; HRP binding appeared to be related to the immobilization of membrane receptors: the more
they were immobilized, the more receptor-associated lectin could bind HRP. The molar ratio HRP/cell-bound lectin was constant in all cases except for living cells incubated with Con A; in that case the ratio HRP/Con A varied along a sigmoidal curve
(Figure 2) from 0. 04 (1 HRP/25 Con A) at low concentration of Con A to 0. 17 (1 HRP
/6 Con A) at 100 µg/ml of Con A. A constant and high value of 0. 34 corresponding
to 1 HRP/3 Con A was obtained with pre-fixed cells incubated with Con A. For Succinyl-Con A, which is divalent, the values were lower than for native Con A. When
cells were fixed, a constant ratio corresponding to 1 HRP/20 Succinyl-Con A was observed. For living cells, the ratio was also constant for any concentration of Succinyl-Con A, corresponding to 1 HRP/100 Succinyl-Con A. The decreased values of
the ratio for living cells compared to pre-fixed cells was interpreted as the result of
a post-binding event which could occur only when lectins were incubated with living
cells and which led to masking of free sugar binding sites on the cell-bound lectin
molecules. This post-binding event might be a conformational change, a site-site
interaction of receptors, micropatches such as those shown for anti-immunoglobulin
receptors or any other molecular event at the surface; whatever it is, we call it micro-
redistribution.

From this we interpreted the sigmoidal curve for the molar ratio HRP/lectin obtained living cells incubated with Con A as showing that cell receptor micro-redistri-

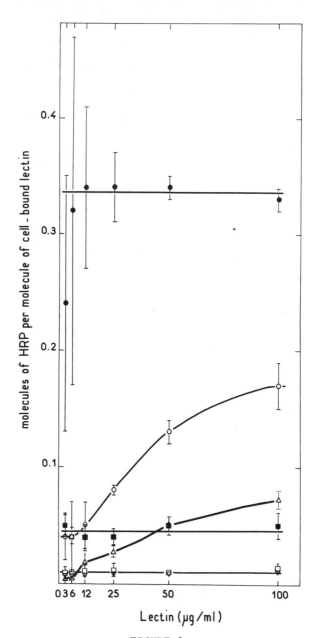

FIGURE 2.

Relative binding of horseradish peroxidase (HRP) to cell-
bound Con A or Succinyl-Con A, at 4°C.
O Molar ratio HRP/ Con A for normal cells.
● Molar ratio HRP/ Con A for glutaraldehyde-treated cells.
□ Molar ratio HRP/ Succinyl-Con A for normal cells.
■ Molar ratio HRP/ Succinyl-Con A for glutaraldehyde-treated
  cells.
△ Molar ratio HRP/ Con A for colchicine plus cytochalasin B
  treated cells.
▽ Molar ratio HRP/ Succinyl-Con A for colchicine plus
  cytochalasin B treated cells.

bution was progressively restricted as Con A concentration was increasing, just in parallel to what is observed for the dose effect of Con A on redistribution at 37°C. That HRP binding is related to the immobilization of cell receptors has already been suggested by a report from Collard and Temminck (62).

III. Receptors Recruitment and Cooperative Restriction of Receptor Micro-redistribution

The sigmoidal curve observed for peroxidase binding to cell-bound Con A (Figure 2) indicated that immobilization of the cell surface receptors by increasing concentrations of Con A occurred in a cooperative fashion. We had seen that the binding of Con A to cells produced a cooperative modification of the membrane leading to the appearance of new receptors (see section I, C). The cooperative immobilization of the receptor would thus be the concomitant effect of this membrane modification. The actual mechanism for such an effect at 4°C is unknown. Succinyl-Con A did not restrict micro-redistribution at all; this indicated that the valence of the lectins was likely to be important. Again, it could be some external cross-linking which could be formed by native Con A, but not by Succinyl-Con A; we have recently shown that the effect of Con A on membrane at 4°C was modulated by drugs like colchicine and cytochalasin B (63) and, therefore, that membranes of cells maintained at 4°C are still a complex system in which external receptors relate to internal constituents sensitive to drugs of the cytoskeleton.

Results can be summarized as follows (see Figure 1): cooperative binding of Con A to cells was slightly modified by each drug alone; binding at low concentrations was significantly reduced when the two drugs were used simultaneously. For higher doses, Con A progressively antagonized the synergistic action of the two drugs; a normal binding was recovered: the same number of Con A molecules could be bound per treated cell as compared to normal cells. The binding of Succinyl-Con A was lowered by the two drugs at all doses of lectin. Quite interestingly, the cooperative restriction of receptor micro-redistribution induced by increasing doses of Con A was inhibited in a synergistic way by colchicine and cytochalasin B together (Figure 2).

A synergism in the action of colchicine and cytochalasin B on cell receptors redistribution at $37°C$ has already been reported (53, 64-65). Our present findings showing a comparable synergistic effect on receptor recruitment and immobilization by high doses of Con A at $4°C$ suggest the existence, at the membrane level, of a colchicine and cytochalasin B-sensitive system related to surface receptors. A formal model, which is simply a modification of the current models (59), can be drawn as follows: Con A receptors are assumed to be either exposed or masked, this transition being correlated with the anchorage modulation of the receptors with the submembrane system. This is shown in Figure 3. The model supposes that: (i) the exposed state of Con A receptors is obtained when receptors are anchored to a fully associated submembrane system (state A or An); (ii) the non-exposed state of Con A receptors is obtained when Con A receptors are non-anchored or anchored to an incompletely associated submembrane system (state B); (iii) the various components of the whole system interact reversibly; (iv) long-range interactions of the components of the submembrane system occur only if stable nucleation centers are present. This is a general assumption for protein polymerization (66).

Tetravalent Con A would be able to induce and stabilize bidirectional nucleation centers of the drug-sensitive components of the cortical system. The polymerization of submembrane components would lead to an increased concentration of them in the

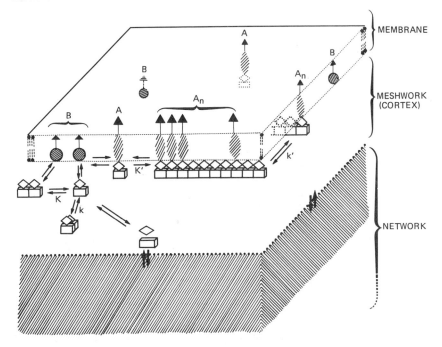

FIGURE 3.

Schematic representation of Con A receptor interaction with submembrane components. This simple model does not exclude that Con A receptor modulation is realized by more than two components in the submembrane meshwork.

 Con A receptor which is not accessible to Con A; it is not anchored to the submembrane system.

Con A receptor in   exposed state; it is anchored to the submembrane system.

Submembrane components which bind colchicine or cytochalasin B.

Various equilibrium constants exist between the components of the whole system; the relative values of K and k will favor one type of polymerisation of submembrane components; such a state is stabilized by Con A (An state) and this constitute a nucleating center allowing the other type of association of the submembrane components (see text); this leads to the cooperative recruitment and immobilisation of Con A receptors. Succinyl-Con A stabilizes only the state A.

The model supposes that the two intracellular compartments, i.e. the cortical meshwork and the network, interact in a dynamic way; a change in one compartment would lead to a change in the steady state of the other.

cortical zone of the cell; association of Con A receptors in B state of these components would be favoured, leading to their conversion to A state; recruitment of receptors would occur.   A minimum number of nucleating centers would be critical to induce polymerization, explaining the sigmoidal curve for recruitment.

The sigmoidal increase of peroxidase binding to cell-bound Con A would be easily explained in this model: at low Con A concentration, lectin binding only stabilizes nucleating centers (transition A to An state): low-peroxidase binding occurs.   At higher concentration, rapid polymerization of the submembranous system occurs and new receptors are exposed at defined distances from each other: Con A molecules cannot bind more than one receptor, free binding sites on these Con A molecules are now available for peroxidase binding.

Divalent Succinyl-Con A will not allow the formation of nucleating centers in the submembranous system; cooperative binding cannot occur and increase in perodidase binding to cell-bound lectin is not obtained.

Drugs like colchicine and cytochalasin B bind to molecules of the submembrane complex, probably tubulin and actin, and impair the association of these molecules to each other; state B for Con A receptors is favoured and the binding of Succinyl-Con A to cells is lowered.   In these conditions, long-range polymerization cannot take place; recruitment of receptors occurs only through stabilization of dispersed nucleating centers and need higher concentrations of free Con A.   Low peroxidase binding to cell-bound Con A should be obtained in such a situation.   This is what is observed (Figure 2).

The model supposes that the two intracellular compartments, i.e., the cortical meshwork and the network (67) interact in a dynamic way; a change in one compartment would lead to a change in the steady state of the other.

The important point is that the whole membrane complex, i.e., the membrane plus the submembrane meshwork, is still functional at 4°C, as judged by drug effects. The question of the relationship between the micro-redistribution, which probably can be detected only at 4°C and by non-visual techniques, and the mobility at 37°C can be asked. One could imagine very small clusters at 4°C which do not further coalesce, because of the effect of cold on colchicine binding protein (53), that is to say because of the fact that the meshwork is no longer connected with a functional network.

This study at 4°C has shown the subtle modulation of the membrane complex that Con A is able to achieve; this could help to explain the great variety of biological effects that Con A is able to affect, such as increase in adhesive properties (68), mimic of insulin effect on fat cells (69) suppression of the negative cooperativity of the binding of insulin (70), inhibition of the transport of metabolites (71), mitogenic signal for thymus-derived cells (72), etc. For all these effects, the valence of Con A is probably important; Succinyl-Con A, when tested, does not show the same activity. Along this line, the recent work of Goldman et al (73) on the effect of polymerized lectins on macrophages is particularly interesting; it shows that tetravalent lectins such as Con A and Wheat Germ Agglutinin induce vacuolation while the divalent Succinyl-Con A or Soybean Agglutinin do not; polymerization of Soybean Agglutinin transforms it into a vacuolating lectin.

IV. Receptor Immobilization and Cell Agglutination

The result that micro-redistribution leads to masking of free sugar binding sites on the cell-bound lectin raises a question about the current models used to explain agglutination, which imply clustering of receptors (74-76). In our study, cell agglutination was observed with Con A at 4°C, although less than what was obtained at

37°C. Agglutination at 4°C was maximum at 25 μg/ml of Con A and for higher con-

centrations. There was almost no agglutination with Succinyl-Con A. Our results

would mean that the effect of receptor mobility is to quickly mask free binding sites

and to avoid agglutination; it is also possible that preventing agglutination is the

biological function of receptor mobility. And it would be because receptor cluster-

ing is progressively inhibited by higher concentrations of Con A that free binding sites

are available for agglutination. However, we obtained no agglutination at all with

fixed cells, which were even protected from agglutination. Therefore, agglutination

would need restriction of receptor mobility but not a total freezing of the membrane.

Results along the same line have been reported by Rutishauser and Sachs (77) who used

nylon fibers coated with lectin molecules in different conditions to follow the binding

of normal and fixed cells. They concluded by affirming the necessity of short-range

lateral mobility.

V. Conclusion

Although far from clarified, the role of membrane, fluidity and receptor mobil-

ity for achieving cortical control on cell activity is obviously important. A direct

relationship between receptor mobility and a particular biological function however

cannot be drawn; a good illustration of such a statement is seen in the mitogenic

properties of Con A: both Con A and its Succinyl-derivative are potent mitogens of

thymus-derived cells at low concentrations. But a striking difference exists in the

dose-response curves of the two lectins (58). Whereas the dose-response curve for

Con A falls off very rapidly, the response to Succinyl-Con A shows near-maximal

stimulation over a ten-fold concentration range. Yahara and Edelman (60) were not

able to give a function to receptor patching in this phenomenon. However, our re-

sults (36) show that there is a correspondence between inhibition of the blastogenic stimulation at 37°C and restriction of the micro-redistribution observed at 4°C. In other words, there will be mitogenesis only for lectin occupancy on the cell surface which does not restrict the very subtle micro-redistribution that one can observe.

The coupling between the external milieu and the cell activity seems to be realized by the cytoskeleton; the structure of this cytoskeleton, its real functions and the control of its activity are unknown in great part. The experiments of enucleation in the past few years have shown the fascinating autonomy of the cytoskeleton (78-80). It is likely that the most significant steps to come in cell biology will be brought about by the understanding of the cytoskeleton role.

VI. Abbreviations

Con A = Concanavalin A; Succinyl-Con A = Succinyl-Concanavalin A;

HRP = Horse Radish Peroxidase.

VII. References

1.   Singer, S. J. and Nicholson, G. L., Science, 175:720, 1972.
2.   Nicolson, G. L., Int. Rev. Cytol., 39:89, 1974.
3.   Singer, S. J., Ann. Rev. Biochem., 43:805, 1974.
4.   Cuatrecasas, P., Ann. Rev. Biochem., 43:169, 1974.
5.   Vitetta, E. S. and Uhr, J. W., Science, 189:964, 1975.
6.   Unanue, E. R. and Karnovsky, M. J., Transplant. Rev. 14:184, 1973.
7.   de Meyts, P., Bianco, A. R. and Roth, J., J. Biol. Chem., 251:1877, 1976.
8.   Jarett, L. and Smith, R. M., Proc. Natl. Acad. Sci. U. S. 72:3526, 1975.
9.   Jacobs, S. and Cuatrecasas, P., Biochim. Biophys. Acta, 433:482, 1976.
10. Pochet, R., Boeynaems, J. M. and Dumont, J. E., Biochem. Biophys. Res. Commun., 58:446, 1974.
11. Roy, C., Barth, T. and Jard, S., J. Biol. Chem., 250:3149, 1975.
12. Rodbard, D., Endocrinology, 94:1427, 1974.
13. Levitzki, A., Segel, L. A. and Steer, M. L., J. Mol. Biol. 91:125, 1975.
14. Loor, F., in press in B and T cells in immune recognition, Edited by F. Loor and G. E. Roelants, J. Wiley and Sons, Ltd., Chichester, England.
15. Avrameas, S. and Guilbert, B., Eur. J. Immunol., 1:394, 1971.
16. Rabellino, E., Colon, S., Grey, H. W. and Unanue, E. R., J. Exp. Med. 133:156, 1971.
17. Engers, H. D. and Unanue, E. R., J. Immunol., 110:465, 1973.
18. Antoine, J. C. and Avrameas, S., Eur. J. Immunol., 4:468, 1974.

19. Taylor, R. B., Duffus, W. P. H., Raff, M. C. and de Petris, S., Nature New Biology, 233:225, 1971.
20. Hammerling, U. and Rajewsky, K., Eur. J. Immunol., 1:447, 1971.
21. Loor, F., Forni, L. and Pernis, B., Eur. J. Immunol., 2:203, 1972.
22. Unanue, E. R., Perkins, W. D. and Karnovsky, M. J., J. Exp. Med., 136: 885, 1972.
23. Antoine, J. C., Avrameas, S., Gonatas, N. K., Stieber, A. and Gonatas, J. O., J. Cell Biol., 63:12, 1974.
24. de Petris, S. and Raff, M. C., Eur. J. Immunol., 2:523, 1972.
25. Ault, K. A., Karnovsky, M. J. and Unanue, E. R., J. Clin. Invest., 52:2507, 1973.
26. Stackpole, C. W., Jacobson, J. B. and Lardis, M. P., Nature, 248:232, 1974.
27. Sell, S., J. Exptl. Med., 125:289, 1967.
28. Ling, N. R. and Kay, J. E., Lymphocyte Stimulation, p. 171, North-Holland /American Elsevier, 1975.
29. Karsenti, E. and Levy, F., In preparation.
30. Majerus, P. W. and Brodie, G. N., J. Biol. Chem., 247:4253, 1973.
31. Stein, M. D., Sage, H. J. and Leon, M. A., Arch. Biophys. Biochem., 150: 412, 1972.
32. Hammarström, S., Scand. J. Immunol., 2:53, 1973.
33. Cuatrecasas, P. C., Biochemistry, 12:1313, 1973.
34. Bornens, M., Karsenti, E. and Avrameas, S., in: Membrane Receptors of Lymphocytes, Edited by M. Seligmann, J. L. Preud'Homme and F. M. Kourilsky, p. 377, 1975. North-Holland/American Elsevier.
35. Bornens, M., Karsenti, E. and Avrameas, S., Eur. J. Biochem., 65:61, 1976.
36. Philips, P. G., Furmanski, P. and Lubin, M., Expt. Cell Res., 86:301, 1974.
37. Stobo, J. D., Rosenthal, A. L. and W. E. Paul, J. Immunol., 108:1, 1972.
38. Betel, I. and van den Berg, K. J., Eur. J. Biochem., 30:571, 1972.
39. Krug, U., Hollenberg, M. D. and Cuatrecasas, P., Biochem. Biophys. Res. Commun., 52:305, 1973.
40. Cuatrecasas, P. C., Biochemistry, 12:1312, 1973.
41. Gunther, G. R., Wang, J. L., Yahara, I., Cunningham, B. A. and Edelman, G. M., Proc. Natl. Acad. Sci. U. S., 70:1012, 1973.
42. Karsenti, E. and Avrameas, S., FEBS Letters, 32:238, 1973.
43. Painter, R. G. and White, A., Proc. Nat. Acad. Sci. U. S., 73:837, 1976.
44. Schmidt-Ullrich, R. and Wallach, D. F. H., Biochem. Biophys. Res. Commun., 69:1011, 1976.
45. Reisner, Y., Lis, H. and Sharon, N., Expt. Cell Res., 97:445, 1976.
46. Gachelin, G., Buc-Caron, M. H., Lis, H. and Sharon, N., in Press, 1976.
47. Resch, K. and Ferber, E., Proceedings of the Ninth Leucocyte Culture Conference; Immune Recognition, Edited by Rosenthal, A. S., p. 529, Academic Press, Inc., New York/London/San Francisco, 1975.
48. Carothers Carraway, C. A., Jett, G. and Carraway, K. L., Biochem. Biophys. Res. Commun., 67:1301, 1975.
49. Barnett, R. E., Scott, R. E., Furcht, L. T. and Kersey, J. H., Nature, 249: 465, 1974
50. Yahara, I. and Edelman, G. M., Proc. Natl. Acad. Sci. U. S., 69:608, 1972.
51. Unanue, E. R., Perkins, W. D. and Karnovsky, M. J., J. Exp. Med., 136:885, 1972.
52. Yahara, I. and Edelman, G. M., Exp. Cell Res., 81:143, 1973.

53. de Petris, S., J. Cell Biol., 65:123, 1975.
54. Rutishauser, U., Yahara, I. and Edelman, G.M., Proc. Natl. Acad. Sci. U.S., 71:1149, 1974.
55. Berlin, R.D., Oliver, J.M., Ukena, T.E. and Yin, H.H., Nature, 247:45, 1974.
56. Yahara, I. and Edelman, G.M., Nature, 236:152, 1973.
57. Edelman, G.M., Yahara, I. and Wang, J.L., Proc. Natl. Acad. Sci. U.S., 70:1442, 1973.
58. Yahara, I. and Edelman, G.M., Proc. Natl. Acad. Sci. U.S., 72:1579, 1975.
59. Edelman, G.M., Science, 192:218, 1976.
60. Avrameas, S., C.R. Hebd., Sceances Acad. Sci. Paris, 270:2205, 1970.
61. Bernhard, W. and Avrameas, S., Exp. Cell Res., 64:232, 1971.
62. Collard, J.G. and Temminck, J.H.M., Exp. Cell Res., 86:81, 1974.
63. Karsenti, E., Bornens, M. and Avrameas, S., in preparation.
64. de Petris, S., Nature, 250:54, 1974.
65. Ryan, G.B., Borysenko, J.Z. and Karnovsky, M.J., J. Cell Biol., 62:351, 1976.
66. Oosawa, F., J. Theor. Biol., 27:69, 1970.
67. Goldman, R.D., The J. Histochem. Cytochem., 23:529, 1975.
68. Mori, Y., Akedo, H. and Tanigaki, Y., Exptl. Cell Res., 78:360, 1973.
69. Cuatrecasas, P. and Tell, G.P.E., Proc. Nat. Acad. Sci. U.S., 70:485, 1973.
70. de Meyts, P., Gavin, J.R., III, Roth, J. and Neville, D.M., Jr., Diabetes, 23:355, 1974.
71. Inbar, M., Ben-Bassat, H. and Sachs, L., J. Memb. Biol., 6:195, 1971.
72. Powell, A.E. and Leon, M.A., Exptl. Cell Res., 62:315, 1970.
73. Goldman, R., Sharon, N. and Lotan, R., Exptl. Cell Res., 99:408, 1976.
74. Nicolson, G.L., Nature New Biology, 239:193, 1972.
75. Noonan, K.D. and Burger, M.M., J. Cell Biol., 59:134, 1973.
76. Andersson, J., Sjöberg, O. and Möller, G., Transplant. Rev., 11:131, 1972.
77. Rutishauser, U. and Sachs, L., Proc. Natl. Acad. Sci. U.S., 71:2456, 1974.
78. Prescott, D.M., Kates, J. and Kirkpatrick, J.B., J. Mol. Biol., 59:505, 1971.
79. Wise, G.E. and Prescott, D.M., Expt. Cell Res., 81:63, 1973.
80. Shay, J.W., Porter, K.R. and Prescott, D.M., Proc. Nat. Acad. Sci. U.S., 71:3059, 1974.

Chapter 4

IMMUNOLOGICAL PROBES INTO THE MECHANISM
OF CHOLERA TOXIN ACTION

S.W. Craig[*] and P. Cuatrecasas[+]
*Department of Physiological Chemistry
Johns Hopkins University, School of Medicine
Baltimore, Maryland 21205

and

[+]Wellcome Research Laboratories
3030 Cornwallis Road
Research Triangle Park, North Carolina 27709

Abstract

The use of antibodies to specific cell surface proteins or to ligands
which interact with cell surface receptors is a powerful tool for analyzing
the properties of membrane proteins and the consequences of specific cell
surface ligand-receptor interactions. Two central observations concerning
membrane structure and function, - the diffusibility of membrane proteins
(1) and ligand-triggered modulation of specific receptors (2), have derived
from the use of antibodies to analyze the properties of membrane proteins.
In our study of the mechanism of action of cholera toxin, a protein which
binds to a specific cell surface receptor and results in the activation of
adenyl cyclase, considerable information has been gained through the use of
immunological techniques. This review will briefly summarize the data
underlying our current concept of cholera toxin action at the cell membrane
and will emphasize those observations made through the use of immunological
approaches.

Cholera, a disease characterized by massive loss of body fluid through

the intestinal epithelium, is caused by a protein (cholera toxin or

choleragen) produced by Vibrio cholerae bacteria (3). Substantial evidence

indicates that choleragen exerts its physiological effects by activating the

plasma membrane adenyl cyclase of the intestinal epithelial cell (4,5,6,7).

Moreover, it has been found that choleragen activates adenyl cyclase and mimics the biological effects of cyclic AMP (cAMP) and of hormones which increase intracellular levels of cAMP in every cell type studied (3,8).

Like hormones which activate plasma membrane adenyl cyclase, choleragen initiates its interaction with a cell by binding to a specific cell surface receptor. That the monosialoganglioside, GMI, might be the natural receptor for choleragen was first suggested by the observation that preincubation of choleragen with GMI before exposure to cells neutralized the toxic activity (9). Studies with ^{125}I labelled choleragen demonstrated that choleragen binds to intact cells and purified plasma membranes with high avidity and specificity and that this binding can be inhibited by preincubating labelled choleragen with GMI and, to a much lesser extent by structurally related gangliosides, glycolipids, and glycoproteins (10). Similar results concerning specificity of toxin binding to GMI have also been reported by other laboratories (11,12). Evidence that GMI serves as the functional membrane receptor for choleragen comes from the observation that cells preincubated in GMI spontaneously incorporate GMI (as evidenced by an increased number of binding sites for choleragen) and become more sensitive to the biologic action of a fixed amount of choleragen. That is, the increased number of choleragen binding sites function as true receptors since the dose-response curve of the cells to choleragen is shifted to the left (10).

Studies on the subunit structure of choleragen (10), which in the essential features are in harmony with other studies (13-15), indicate the existence of two major components, the "binding" (MW = 66,000) and "active" (MW = 36,000) subunits. The former is identical to the toxin derivative, choleragenoid (16), which possesses membrane and ganglioside binding properties identical to those of choleragen, yet is biologically inert; thus it is an antogonist of toxin action. This subunit dissociates into 8,000 molecular weight components in acid or upon heating in sodium

dodecyl sulfate. The "active" subunit, which is highly hydrophobic,does

not compete for toxin binding or activity; it consists of very tightly but

not covalently bound 27,000 and 8,000 molecular weight components.

Antisera against choleragen crossreact with the "active" and "binding"

subunits but not with the 27,000 molecular weight component. Choleragen

contains some antigenic determinants which are not present in choleragenoid

(15), presumably, these determinants are present on the 27,000 molecular

weight component. Recent studies have shown that the active subunit alone

is capable of activating adenyl cyclase in intact cells (17,18) and that

this activation does not require the binding subunit since the process

cannot be inhibited by gangliosides (17,18) or by choleragenoid (17).

Thus the two subunits of choleragen have distinct functions; the 8,000

molecular weight components apparently enable choleragen to bind to the

cell while the 36,000 component activates the adenyl cyclase.

Unlike hormones, which activate adenyl cyclase almost immediately

upon occupancy of their cell surface receptor sites, choleragen exhibits

a lag of 20-60 min., after binding has occurred, before activation of

adenyl cyclase or change in cAMP stimulated cell function occurs (3,8).

Furthermore, while hormone activation of adenyl cyclase is a transitory

event, declining rapidly once the serum level of the hormone drops,

choleragen-stimulated adenyl cyclase appears to be irreversible and can last

for days (19,20).

Kinetic analysis of choleragen activation of adenylate cyclase provides

several important facts and ideas concerning the mode of toxin action (20,

21). The duration of the lag (the time between binding of choleragen and

activation of adenyl cyclase) is not markedly affected by the concentration

of choleragen. This is in marked contrast to the lag phase observed

with diptheria toxin which varies from 30 min. to 6 hrs., depending on
the toxin concentration, and argues against a catalytic mode of action for
choleragen. Choleragen activation of adenyl cyclase follows an exponential
time course and the rate of cyclase activation, once the lag is over,
depends on the number of toxin molecules bound. In toad erythrocytes, the
half-maximal increase in the rate of exponential activation occurs with
about 2,200 toxin molecules bound per cell. The extent of enzyme activation at near equilibrium also depends on the amount of bound choleragen
with an apparent K_A of about 1,500 toxin molecules per cell. These
results suggest that one toxin molecule does not lead to progressively
greater stimulation of many enzyme molecules such as would be expected if
toxin acted catalytically. If toxin acted catalytically, half-maximal
activation would require fewer toxin molecules with increasing length of
incubation.

The inactive nature of the initial toxin-ganglioside complex
implies that the biological effects of toxin are due to interaction with
other secondary receptor sites. The simplest possibility, in view of
the lack of evidence for a catalytic mechanism, is that the "receptor"
is the adenyl cyclase itself (20). The possibility that choleragen
interacts directly with adenyl cyclase was tested by assessing the ability
of anti-choleragen, anti-36,000 subunit, and anti-27,000 piece to precipitate adenyl cyclase activity from solubilized membranes prepared from
cells incubated for various periods of time with choleragen. The results
(Fig. 1, 17,22) suggest that following initial binding to the cell
surface, choleragen or some portion of the molecule forms a detergent-
stable complex with adenyl cyclase. From the absolute nature of the lag
phase, it would be predicted that choleragen would not become directly
associated with the adenyl cyclase until after the lag. In this respect,

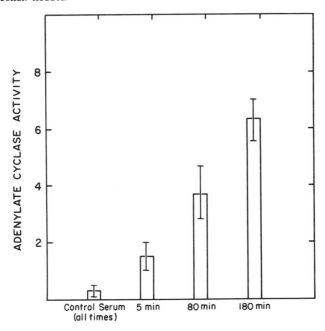

FIGURE 1

Patterns of fluorescence on rat lymphocytes incubated in F·CT. (a) Lymphocytes (10^7) were incubated ($0°$, 30 min) in 1 ml of Hanks' solution containing 1 μg of F·CT. The cells were washed in cold Hanks' solution before fixation and mounting. (b and c) Lymphocytes stained at $0°$ were washed and incubated at $37°$ for 20-30 min before fixing. All $37°$ incubations were carried out in a humidified incubator equilibrated with 95% air - 5% CO_2. Data from ref. 25.

it is important to realize that immunoprecipitation of adenyl cyclase activity with anti-choleragen is evidence, in the strictest sense, only for the close association of toxin and cyclase molecules. Thus immunoprecipitation of cyclase activity after only 5 min. of incubation with toxin (Fig. 1) is not contrary to the hypothesis that toxin eventually directly interacts with cylase. In fact, the observation that more cyclase activity is immunoprecipitated with increasing times of choleragen preincubation is most consistent with the idea that with time, toxin comes into progressively closer association with the adenyl cyclase molecules.

    The events occurring during the absolute lag between toxin binding
and activation of the adenyl cyclase have remained enigmatic. However,
the recent demonstration (17) that choleragen activates adenylate
cyclase in purified toad erythrocyte plasma membranes and maintains the
characteristic lag offers direct evidence that the lag reflects some
process occurring at the cell membrane which is essential in converting
the toxin-ganglioside complex to an active form. The fact that purified
"active" subunit activates adenyl cyclase of intact toad erythrocytes
without a lag phase suggests that the necessary event in toxin action is
its dissociation at the membrane into "active" and "binding" subunits.
Once dissociated from its "binding" subunit, the "active" subunit, which
is quite hydrophobic, might readily partition into hydrophobic regions of
the cell membrane. Evidence that toxin does dissociate at the cell
surface is provided by the finding that an increasing proportion of the
binding subunit will exchange between separate cells with increasing
length of incubation, while the active subunit is retained (17).

    It is pertinent to note that van Heyningen and King have recently
reported activation of adenylate cyclase by choleragen in lysates of pigeon
erythrocytes, and by active subunit in intact cells as well as cell lysates
(18). In these experiments, both the active subunit and whole toxin
activated adenyl cyclase of intact cells with a lag and the adenyl
cyclase of lysed cells without a lag. The reason for the discrepancy
between these results and ours (17,22) concerning the existence of a lag
during activation of adenylate cyclase in intact cells by the "active"
subunit is not clear at this time. Gill has recently reported (23) that
although choleragen can activate adenyl cyclase in purified pigeon
erythrocyte ghosts, it requires cell cytoplasm and exogenous NAD in
order to do so. Since we have not been able to demonstrate an NAD
requirement for choleragen activation of adenyl cyclase in purified

toad erythrocyte membranes (17), we are reluctant to accept the

hypothesis (23) that NAD is an obligatory component of choleragen

action. However, the discrepancy between our results and those of Gill

and colleagues must be explained before a clear picture of choleragen

action can emerge.

If our hypothesis that choleragen must release its active subunit

at the cell surface in order to activate adenyl cyclase is correct, then

any chemical treatment which facilitates toxin dissociation might

artifactually be considered obligatory to toxin action. It is not unlikely

that different cell types might contain diverse intracellular enzymes

which when added as a lysate to an in vitro system of choleragen

activation might facilitate toxin dissociation. However, such effects

would not necessarily represent the physiological pathway of toxin

action. In this respect it is significant that Bitensky et al (24)

have recently described that in the presense of cell cytosol

dithiothreitol can convert the toxin into a form that can activate

adenyl cyclase directly (ie. without a lag phase), even in the absence

of NAD. The activated form of the toxin appeared to be the active

subunit, which was generated from the parent molecule by exposure to

cytosol proteins and dithiothreitol (24).

In order to obtain further information concerning the events which

occur at the cell membrane during the lag phase, the interaction of

biologically active, fluorescein-labelled choleragen (F·CT) with viable

rat lymphocytes was studied (25). It was found that cells stained at

0° C with F·CT show a diffuse localization of F·CT (Fig. 2a); however,

if these cells are warmed at 37° C for 20 min, the F·CT forms a cap at one

pole of the cell (Fig. 2b and 2c). Binding and redistribution of fluore-

scent choleragen can be observed using concentrations of fluorescent toxin

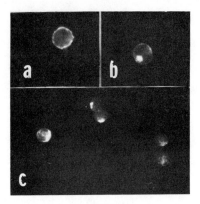

FIGURE 2

Immunoprecipitation of adenylate cyclase with antiserum to the
36,000 molecular weight subunit of choleragen. Fat cells were incubated
with choleragen at 37° for various periods, the particulate cell fractions
were solubilized, and 200-300 μl were incubated with 50μl of immune or
normal rabbit serum for 12 min at 30° followed by 2 hr in ice. Goat
antiserum against rabbit gamma globulin (200 μl) was added and the
mixture incubated at 30° for 5 min and in ice for 1 hr. The precipitate
was washed twice with 1 ml of 25 m M Tris-HCl, pH 7.7, suspended in
200 μl, and assayed (50 μl). 15-20% of the ^{125}I label was precipitated
when ^{125}I-labeled choleragen was used. Activity is in pmol of cAMP
produced per 50 μl/12 min at 37° $\pm$ SEM. Data from ref. 17.

as low as 10^{-10} M, which is well within the physiological range for choler-

agen activation of adenyl cyclase in rat mesentric lymph node lymphocytes.

Thus, the events described probably reflect the properties of the bio-

logically relevant toxin receptors. By triple layer immunofluorescence

techniques, capping of toxin ganglioside complexes has also been demon-

strated in human tonsil lymphocytes (26), in GMI-deficient human leukemic

cells into which exogenous GMI receptors were incorporated (26), and in a

small percentage of mouse thymocytes (27).

Ligand-triggered cell surface receptor mobility has been demonstrated

with antibodies to cell surface proteins and with lectins in lymphocytes

and other cell types. In most cases it has been found that the ligand must

be bivalent in order to induce capping (2,28). Choleragen is also at least

bivalent since choleragen treated lymphocytes bind specifically to GMI

agarose beads (Table I), indicating that choleragen can bind to at least
two receptors at the same time (25). Anti-choleragen has the effect of
reducing the effective valence of cell-bound toxin as evidenced by the fact
that choleragen treated cells incubated in anti-choleragen lose their
ability to bind to GMI-agarose beads. In addition, anti-choleragen markedly
suppresses, in a concentration-dependent fashion, patching and capping of
cell-bound F·CT. This inhibitory effect of anti-choleragen can be partially

TABLE I

Binding of cholera toxin-treated rat lymphocytes
to ganglioside-Sepharose beads

Exp.	Addition to G-beads	Cells/bead		Beads with > 20 cells*
		< =5	> 20	
		no. of beads		%
1	RMLN	526	5	0.8
	CT-RMLN	342	150	28
	G_{M1} blocked CT-RMLN	430	31	6
2	RMLN	474	1	0.2
	CT-RMLN	679	135	13
	G_{M1} blocked CT-RMLN	429	32	6
	Anti-toxin CT-RMLN	578	12	2

One x 10^7 lymphocytes in 1 ml of Hanks' solution
were incubated (15 min. 37°) with (CT-RMLN) or without
(RMLN) 1μg of toxin and washed in cold Hanks' solution.
One milliliter of packed, washed ganglioside-Sepharose
beads (G-beads) were suspended in 2 ml of Hanks' solution
and 200 μl was added to 1 x 10^7 cells in 100 μl of Hanks'
solution. After incubating at 37° for 40 to 60 min followed by chilling to 0°, the suspensions were diluted (0°)
and the beads were allowed to settle out several times.
Toxin-treated cells were blocked with G_{M1} (G_{M1}-blocked
CT-RMLN) by incubating (0°,40-60 min) with 10μg/ml of
G_{M1} before adding to the beads. Toxin-treated cells were
blocked with anti-toxin IgG (anti-toxin blocked CT-RMLN)
by incubating (0°,40-60 min) in 50μl of anti-toxin IgG
(14mg/ml) before exposure to the beads. Data from ref.
25.
* 500-600 beads were counted.

(50%) reversed by incubating the inhibited cells in anti-Fc (IgG) which apparently can cross-link the anti-choleragen IgG molecules bound to F·CT and thus cause the multi-layered complex to cap (25).

Since capping of F·CT occurs during the lag phase, the question of interest is whether multivalent binding of toxin at the cell surface and subsequent redistribution of toxin ganglioside-complexes are part of the mechanisms of toxin action. Studies with fluorescein-labelled choleragenoid show that choleragenoid is also capable of triggering capping of ganglioside receptors (25,26). Since choleragenoid is biologically inactive (3,8), capping of ganglioside toxin receptors cannot be sufficient for choleragen activation of adenyl cyclase. However, the fact that purified "active" subunit can activate cyclase (17,18) whereas choleragenoid, which is composed only of binding subunits, cannot activate cyclase show that the active subunit is essential for activation of cyclase. Therefore, the fact that choleragenoid caps but does not activate cyclase should not be taken as evidence against an essential role of capping in choleragen action. In fact, it appears that at least multivalent binding of choleragen at the cell surface (of which capping is a consequence) may be a necessary event in toxin action. If choleragen patching and capping are suppressed with anti-choleragen, choleragen activation of adenyl cyclase is markedly suppressed (Table II).

However, if anti-toxin is added after considerable toxin receptor reorganization has occurred, the inhibitory effect is diminshed. The effect of anti-choleragen IgG is abolished by preabsorption with choleragenoid and non-immune rabbit IgG has no effect on cyclase activity. It is possible that multivalent binding of choleragen to cell surface gangliosides places a strain on the molecule and thereby facilitates release of the "active" subunit and, consequently, transduction of the signal across the membrane to the adenyl cyclase.

TABLE II

Effect of anti-cholera toxin on activation
of adenylate cyclase

Exp.	Addition	Adenylate cyclase activity*	
		Toxin stimulated	Basal
1	None	11.1	2.2
	Anti-CT IgG (30 min, 0°)[†]	1.8	1.8
	Anti-CT (after 30 min, 37°C)	6.8	3.1
	N-R IgG (0°)	11.3	N.D.
	N-RIgG (after 30 min, 37°)	9.6	N.D.
2	None	99.1	13.2
	Anti-CT IgG (30 min, 0°)[†]	62.6	22.8
	Anti-CT (after 30 min, 37°)[‡]	71.1	21.4
	Toxoid absorbed anti-CT (0°)	96.9	N.D.
3	None	33.8	18.7
	Anti-CT IgG (30 min, 0°)[†]	18.6	24.9
	Anti-CT (after 30 min, 37°)[‡]	28.1	20.3

* pmol/15 min per mg of protein; mean of duplicate determinations, 600-2000 cpm/pmol; data from ref. 25.
[†] Cells (2×10^8) previously incubated with toxin (1.5 μg/ml, 30 min, 0°, in 1.8 ml) were washed and incubated (30 min, 0°) with 200μl of anti-CT IgG (28 mg/ml) and then transferred to 37° for 1.5 hr.
[‡] Cells previously incubated with toxin (1.5 μg/ml, 30 min, 0°) and washed were transferred to 37°. After 30 min, 200μl of anti-toxin IgG was added, the incubation continued at 37° for 1 hr, and the cells were washed.

From the above experiments, a scheme of the probable events in

choleragen action has been developed (Fig. 3). Choleragen interacts

with a cell by binding (via the 8,000 molecular weight subunits) to

cell surface ganglioside GMI. After binding, there is a period during

which each toxin becomes attached to more than one ganglioside. During

this period (the lag phase) the "active" subunit slowly dissociates

from the "binding" subunit, a process which may be facilitated by multi-

point attachment of choleragen. The free "active" subunit rapidly

inserts into the hydrophobic region of the membrane and, by lateral

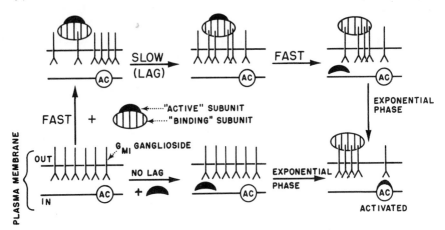

FIGURE 3

Diagrammatic summary of postulated sequence of events in the activation of adenylate cyclase (AC) by native choleragen and by the "active"
subunit   The specific mechanism by which the "active" subunit modifies
the cyclase complex is unknown and is therefore unspecified. Although
the "binding" subunit is fundamentally not an essential component, it
serves to give specificity and orientation, to enhance the affinity for
the toxin, as well as to provide a specialized water-soluble vehicle
for direct delivery of the active molecular species. From ref. 17.

diffusion, eventually collides with, binds to, and activates adenyl

cyclase. The mechanism by which the "active" subunit activates adenyl

cyclase is unknown and the ability of the "active" subunit to diffuse

laterally in the membrane has not yet been demonstrated. It also

remains to be determined if multivalent binding of toxin to gangliosides

can dissociate or facilitate dissociation of the "active" subunit from

choleragen.

References

1. Frye, L.D., and Edidin, M., J. Cell Sci., 7 : 319, 1970.

2. Taylor, R.B., Duffus, W.P.H., Raff, M.C., and dePetris, S., Nature
        New Biol., 233 : 225, 1971.

3. Finkelstein, R.A., CRC Crit. Rev. Microbiol., 2 : 533, 1973.

4. Field, M., Plotkin, G.R., and Silen, W., Nature, 217 : 469,1968.

5. Schafer, D.E., Lust, W.D., Sircar, B., and Goldberg,N.D., Proc. Nat.
    Acad. Sci., USA, 67 : 851, 1970.

6. Sharp, G.W.G., and Hynie, S., Nature, 229 : 266, 1971.

7. Pierce, N.F., Greenough, W.B., III, and Carpenter, C.C.J., Bacteriol.
    Rev., 35 : 1, 1971.

8. Bennett, V., and Cuatrecasas, P., in Specificity and Action of Animal,
    Bacterial, and Plant Toxins, Ed. P. Cuatrecasas, Chapman and
    Hall, in press, 1976.

9. van Heyningen, W.E., Carpenter, W.B., Pierce, N.F., and Greenough,
    W.B., III, J. Infect. Dis., 124 : 415, 1971.

10. Cuatrecasas, P., Biochemistry, 12 : 3547, 3558, 3567, 3577, 1973.

11. Holmgren, J., Lonnroth, I., and Svennerholm, L., Infect. and
    Immunity, 8 : 208, 1973.

12. King, C.A., and van Heyningen, W.E., J. Infect. Dis., 127 : 639, 1973.

13. Lonnroth, I., and Holmgren, J., J. Gen. Microbiol., 76 : 417, 1973.

14. Finkelstein, R.A., Boesman, M., Neoh, S.H., La Rue, M.K., and
    Delaney, R., J. Immunol., 113 : 145, 1974.

15. Holmgren, J., and Lonnroth, I., J. Gen. Microbiol., 86 : 49, 1975.

16. Finkelstein, R.A., Peterson, J.W., and LoSpalluto, J.J., J. Immunol.
    106 : 868, 1971.

17. Sayhoun, N., and Cuatrecasas, P., Proc. Nat. Acad. Sci., 72 :
    3438, 1975.

18. van Heyningen, S., and King, C.A., Biochem. J., 146 : 269, 1975.

19. O'Keefe, E., and Cuatrecasas, P., Proc. Nat. Acad. Sci., USA,
    71 : 2500, 1974.

20. Bennett, V., and Cuatrecasas, P., J. Memb.Biol., 22 : 29, 1975.

21. Bennett, V., and Cuatrecasas, P., J. Memb.Biol., 22 : 1, 1975.

22. Bennett, V., O'Keefe, E., and Cuatrecasas, P., Proc. Nat. Acad.
    Sci., USA, 72 : 33, 1975.

23. Gill, M., Proc. Nat. Acad. Sci., USA, 72 : 2064, 1975.

24. Bitensky, M.W., Wheeler, M.A., Mehta, H., and Miki, N., Proc.
    Nat. Acad. Sci., USA, 72 : 2572, 1975.

25. Craig, S.W., and Cuatrecasas, P., Proc. Nat. Acad. Sci., USA, 72 :
    3844, 1975.

26.   Revész, T., and Greaves, M., Nature, <u>257</u> : 103, 1975.

27.   Holmgren, J., Lindholm, L., and Lonnroth, I., J. Exp. Med.,
        <u>139</u> : 801, 1974.

28.   Gunther, G.R., Wang, J.L., Yahara, I., Cunningham, B.A., and
        Edelman, G.M., Proc. Nat. Acad. Sci., USA, <u>70</u> : 1012, 1973.

Chapter 5

CHOLERA TOXIN, GANGLIOSIDE RECEPTORS AND
THE IMMUNE RESPONSE

J. Holmgren and L. Lindholm
Institute of Medical Microbiology
University of Göteborg
Sweden

Abstract

Cholera toxin activates plasma membrane adenylate cyclase in all
mammalian cell types. The structure-function relationship of the toxin
has recently been clarified, and the cell membrane receptor identified.
This information has made cholera toxin the "agent of choice" for studies
in many biological systems of the possible regulatory role of adenylate
cyclase/cyclic AMP.

This article describes briefly our current knowledge about the toxin
and its receptor. It then reviews recent research which has revealed that
cholera toxin has strong modulating influences on the proliferation of
normal and malignant lymphocytes as well as on the initiation and expression
of immune responses. The toxin has been found to inhibit DNA synthesis of B
and T lymphocytes in vitro without inducing cell death and also to inhibit
the in vivo proliferation of a virus transformed lymphoma cell line. It
seems to decrease antibody secretion from plasma cells in vitro, and might
also interfere with the release of other soluble immunological mediator
substances. In vivo cholera toxin induces a transient involution of the
spleen and a more prolonged lymphocyte depletion of the thymus in mice;
these effects appear to be mediated through the adrenal glands. The toxin
inhibits allograft rejection, and either stimulates or suppresses antibody
formation depending on the timing of the toxin in relation to the antigen
administration. It increases the capacity of the spleen cells to induce
graft-vs-host reactions and the "allogeneic effect" on antibody production.
An inhibitory effect on a normal suppressor population among the spleen
cells has been identified. The findings illustrate the complex effects
induced on the immune system by the probably most discriminative investigative tool available for stimulation of the adenylate cyclase/cyclic AMP
system.

The role of cyclic nuecleotides in regulation of various cell activities

has recently attracted much interest. It has been implied that the intra-

cellular levels of cyclic 3'5'-adenosine monophosphate (cyclic AMP) may be

of direct consequence for cell division and other cellular activities, and

77

substances which affect intracellular cyclic AMP levels have become
important tools in cell biology (1, 2).

Cholera toxin (CT) is a potent specific activator of cellular adenylate
cyclase. This protein, produced and secreted by Vibrio cholerae bacteria,
is responsible for the diarrhea in cholera by increasing the cyclic AMP
levels in the small intestine mucosa (3-5). In natural disease little, if
any, active toxin reaches extraintestinal tissues. However, applied
parenterally or to tissues in vitro, CT can affect many cell types,
apparently due to the ubiquitous presence of the appropriate membrane
receptor as well as of a responsive adenylate cyclase. This has made CT a
useful investigative tool in many biological systems for clarification of
a possible regulatory role of adenylate cyclase/cyclic AMP.

In recent years the molecular properties and mode of action of CT have
been clarified in considerable detail. The toxin has been purified, its
subunit structure defined, the cell membrane receptor identified, and the
structure-function relationship of the toxin explained. This information
has given firm knowledge of the structure of a mammalian cell membrane
receptor, and shed light on the general problem of how external molecules
via receptors can convey signals to the cell interior.

This article will attempt to illustrate the potentials this bacterial
toxin might have as a probe in cellular immunology, in which field the
receptor problem has been central for two decades, and where recent research
suggests that cyclic nucleotides might have an important controlling influence on the functions of the participating cells. We will briefly
summarize current knowledge on the CT molecule and its mode of action,
describe the evidence identifying a ganglioside as the cell receptor for
this toxin, and on this basis review and discuss in greater detail recent
work from this laboratory dealing with the effects of CT on lymphocyte proliferation and on the immune response. The specific antitoxic cholera immunity that can be induced by immunization with CT or toxoid is outside the
scope of the present article but has recently been discussed elsewhere (5).

The toxin molecule and its cellular action

The purified CT has a molecular weight of 84 000 . It consists of two
types of noncovalently linked subunits, H (heavy) and L (light), having
molecular weights of about 28 000 and 9 000 respectively; alternative de-
signations for these subunits are A and B (6-9). The two types of subunit
have been found to have different roles in the intoxication process (10;

Fig 1). Subunit L, present in about six copies in the toxin molecule, mediates the tight, probably multivalent binding of the toxin to the cell membrane receptors. These subunits are biologically inactive, however, in the sense that they do not directly partake in the toxin activation of adenylate cyclase; in fact the isolated L subunit is an effective competitive inhibitor for CT. The H subunit, which consists of two disulfide linked peptides, effectuates activation of the adenylate cyclase. On its own, this "effector subunit" is incapable of effective membrane binding, however, which makes it virtually nontoxic for intact cells unless it is assisted by the binding L subunits (6-14).

Characteristically, there is a distinct delay, 10-60 min varying with cell type, between the almost instantaneous binding of the toxin to the cell

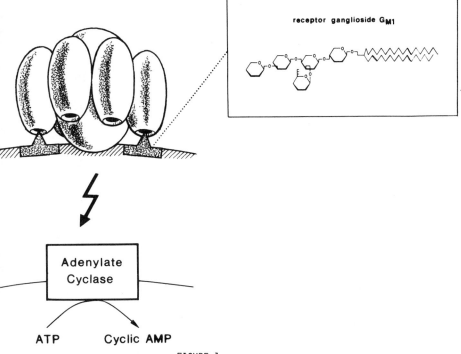

receptor ganglioside G_{M1}

Adenylate
Cyclase

ATP   Cyclic AMP

FIGURE 1

Cellular action of cholera toxin. The smaller subunits bind the toxin to GM1 ganglioside receptors on the cell surface. After a "lag" period the larger subunit activates membrane adenylate cyclase, resulting in increased cell production of cyclic AMP.

membrane receptors and the first noticeable effect on adenylate cyclase activity. Probably this "lag" phase involves lateral as well as vertical reorientation of the initial, inactive toxin-receptor complex in the membrane with
translocation of the effector subunit to a "productive" position. The details
in this process have just begun to be clarified, however (15-17). The stimulation of adenylate cyclase by toxin lasts for several hours and is not
reversed by cell washing, membrane isolation, or even solubilization of the
enzyme by means of detergents.

Identification of ganglioside GM1 as the cell receptor

    Strong evidence indicates that a ganglioside, GM1 (galactosyl-N-acetyl-
galactosaminyl-(sialyl)-galactosyl-glucosyl-ceramide) is the receptor for CT
(Fig 1). This is probably the first instance in which the chemical structure
of a biological membrane receptor in an animal cell has been fully clarified,
and has stimulated interest in the possible roles of other gangliosides and
related glycolipids as receptors (16).

    Ganglioside is the generic term for glycosphingolipids containing sialic
acid. Although present in highest concentrations in the brain, gangliosides
are probably normal components of the plasma membrane of all natural mammalian
cells. They contain a nonpolar, lipophilic ceramide moiety and a polar, hydro-
philic oligosaccharide portion, the latter differing in length, composition
and number of sialic acid residues between the many individual gangliosides
(18). The amphipathic nature of gangliosides suggests that the orientation in
the plasma membrane is such that the ceramide portion is buried within the
outer half of the lipid bilayer and the oligosaccharide portion protrudes
from the cell surface. It is possible that the ceramide might form hydrophobic
bonding to membrane protein. The described properties of gangliosides are
obviously quite consistent with a potential role as membrane receptors.

    The understanding that the ganglioside GM1 most likely is the membrane
receptor for cholera toxin emerged as a result of independent work in our
laboratory and in those of Cuatrecasas and van Heyningen. Several lines of
evidence now constitute firm support for the concept.

1. Specific binding and inactivation of toxin by GM1: The proposal that GM1
is the tissue receptor for CT was based on observations of the specific ability
of this ganglioside to fix and inactivate the toxin (19-21). We showed that
concentrations of GM1 down to equimolar with cholera toxin inhibited the toxic
activity. Of other gangliosides and glycolipids only GD1a and GA1 reacted with

CT; their affinity was, however, 500 times less than that of GM1. GM1 in contrast to all the other substances also gave rise to a specific precipitation band with CT in Ouchterlony type double diffusion-in-gel tests, thus providing direct evidence for specific binding between toxin and this ganglioside in a system independent of biological assays (19). Such binding has also been demonstrated in a "ganglioside receptor sorbent" system, where toxin binding to polystyrene-adsorbed ganglioside was measured immunologically or by means of radiolabelling (22) as well as in column affinity gel systems with ganglioside coupled to agarose beads (7).

2. Correlation between cellular GM1 content and number of toxin receptors: Studies of small bowel mucosal cells of various species established a linear relationship between the number of membrane receptors for CT and the cell content of GM1 but not of other glycolipids (23). Subsequent analyses of various other tissues have further documented the close relationship between the number of binding sites and the chemically measured content of GM1 ganglioside (24).

3. Chemical modification of toxin concomitantly affects binding to cells and to GM1: Artificial toxin analogs containing the H and L subunits in different proportions have demonstrated very similar properties in binding to isolated GM1 and to intact cells (9, 10). Furthermore, it was found that modification of the toxin structure by means of amino acid residue-specific reagents affects binding to isolated GM1 to the same extent as it affects binding to the receptor on intact cells. Derivatives without GM1-binding capacity were always nontoxic (14, 15).

4. Specific blocking of membrane GM1 with toxin: Recent studies by Mullins et al. (25) are important in excluding the faint possibility that a membrane glycopeptide with the same carbohydrate structure as GM1 would be the true receptor for CT. They noticed that gangliosides in thyroid plasma membranes were tritiated by treatment of the membranes with galactose oxidase followed by reduction with 3H-labelled sodium borohydride. The strongest labelling occurred with GM1, which contains an unblocked terminal galactose residue (Fig 1). In contrast, prior binding of CT to the membranes completely prevented the labelling of GM1, and instead increased labelling of other gangliosides took place.

5. Incorporation of GM1 into the cell membrane as functional receptor: Cuatrecasas (26) made the important observation that incubation of fat cells

with GM1, followed by removal of ganglioside in the medium by washing, enhanced the capacity of these cells both to bind CT and to respond to the
lipolytic action. This suggested that the exogenous ganglioside could be
inserted into the membrane and there interact with the toxin in a manner
identical to that of the natural receptors. We found (23) that soaking of
intestinal cells or intact intestine mucosa with 3H-labelled GM1 resulted in
an increase in number of binding sites for CT on the membrane which closely
correlated to the number of inserted GM1 molecules. Furthermore, such incorporation of GM1 into the small bowel mucosa of rabbits enhanced the
susceptibility of the gut to toxin-induced diarrhea. This proves that exogenous GM1 can be incorporated into the natural target tissue for toxin and
thereby enhance the physiologic response of that tissue to toxin in vivo.

Binding of cholera toxin lymphocytes and effects on intracellular cyclic AMP

Although antisera to GM1, probably of relatively low affinity, were
reported to fail to detect GM1 on mouse thymus cells (27) our studies provide
compelling evidence that this ganglioside is the receptor for CT also on
these cells: (i) GM1 has been isolated from such cells (24), (ii) the K_A in
binding of CT to thymus cells and to isolated GM1 is the same ($\sim 10^9$ liters/
/mole), (iii) subunit and other structural requirements of toxin derivatives
are similar in biologic action on this and other cell types; derivatives with
decreased GM1-binding capacity have correspondingly reduced toxic activity;
isolated GM1-binding L subunits can block the biologic receptors for the
toxin and can also dissociate already receptor-bound toxin (15), (iv) inser-
tion of exogenous GM1 into the cell membrane, or sialidase hydrolysis of more
complex membrane gangliosides to GM1 increases the cellular binding of CT
and, concomitantly, the cyclic AMP response (Table I).

Immunofluorescence studies of thymus lymphocytes which had been incubated
with various concentrations of CT revealed that all cells could specifically
bind the toxin although with different efficiency (10, 28). This indicates
that receptor GM1 ganglioside is present on all cells but differs in concentration between individual cells. On an average, a single thymus cell can
maximally bind about 5×10^4 toxin molecules as quantitated using ^{125}I-labelled
CT (10). Binding is rapid, being completed in less than 5 min (80% at 1 min)
at room temperature (10). Provided that the incubation temperature exceeds
$24^\circ C$ the cell-bound toxin, after a "lag" of about 15 min, gives rise to an
increase in the intracellular level of cyclic AMP which reaches a peak after

TABLE I

Effects of pretreatment of mouse thymocytes with GM1 ganglioside or
with V. cholerae neuraminidase on CT binding and
stimulation of cyclic AMP formation

(J. Holmgren, I. Lönnroth and L. Svennerholm, unpublished data)

Cell pretreatment[a]	% Increase in number of CT binding sites	% Increase in cyclic AMP response to CT conc.	
		10^{-11}M	10^{-10}M
Ganglioside GM1, 0.3 μM	30	50	30
Neuraminidase, 25 U/ml	100	150	45

[a] 37°C, 30 min. followed by washing

about 45 min (29). A detectable response requires only a few toxin molecules
per cell, and the response becomes maximal at about 25 per cent receptor
occupancy (Fig 2).

Effects of cholera toxin on proliferation of lymphoid cells in vitro and
in vivo

The effects of CT on proliferation of lymphoid cells in vitro have been
studied using murine spleen and thymus lymphocytes stimulated with concanavalin A (ConA) or phytohemagglutinin (PHA) as model systems (10). Lectin
induced DNA synthesis was found to be drastically inhibited by CT. This
observation was independant from the similar one by Sulzer and Craig (30),
who further found that lipopolysaccharide stimulation of spleen cells was
also prevented by CT. We found no difference in cell viability between lectin-
-stimulated lymphocytes cultivated in the presence or absence of CT as
evaluated with trypan blue exclusion. The results indicate that CT can inhibit
lectin-induced proliferation of T as well as B lymphocytes by an action retaining cell viability for at least many hours.

Observations indicated that the inhibitory effect on DNA synthesis by CT
in vitro was mediated by cyclic AMP. With a large number of CT derivatives
it was found that the capacity to inhibit lectin-induced (ConA) DNA synthesis

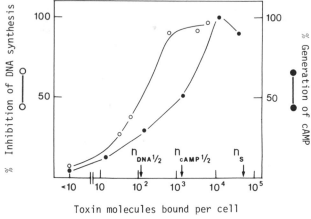

Toxin molecules bound per cell

FIGURE 2

Inhibition of lectin-induced (Con A) DNA synthesis and stimulation of cyclic AMP in mouse thymus cells in relation to the number of CT molecules bound per cell. n-s denotes the maximum number of CT molecules bound per cell. n-DNA 1/2 and n-cAMP 1/2 indicate the numbers of bound molecules needed for half-maximal inhibition of DNA synthesis and stimulation of cyclic AMP formation, respectively.

in thymus cells corresponded closely to the capability to stimulate cyclic AMP formation in the cells as well as to the potency to elicit toxic mani-festations in a rabbit skin assay. The extent of inhibition of DNA synthesis was dependant on the toxin concentration in a manner roughly parallelling the dose-response curve for CT induced cyclic AMP formation in resting cells. From measurements of the amount of CT bound to mouse thymocytes at different incubation concentrations it can be calculated that about 100 toxin molecules bound per cell inhibit the DNA synthesis by 50%, whereas it requires about 1000 bound molecules per cell to increase cyclic AMP formation to 50% of the maximally attainable stimulation (Fig 2). Thus, provided that the relation is of a causal nature, it is obvious that submaximal cyclic AMP formation in lymphocytes can mediate full inhibition of DNA synthesis.

The evidence that GM1 is the receptor for CT on mouse thymocytes has been described in the previous section. Consistent with this notion the inhibition of lectin induced DNA synthesis by CT was abolished by free GM1, but not by other gangliosides. Mere binding to the cellular receptor was not enough to trigger the events leading to inhibition of DNA synthesis. Choleragenoid,

which consists of the GM1 binding L subunits only, was unable to inhibit
Con A induced DNA synthesis although it had identical cell binding properties
as the active toxin.

The possibility that CT and Con A exert their effects through the same
cellular receptor was shown to be very unlikely since there was no competition
between the two proteins for binding to mouse thymocytes as tested with radiolabelled substances. It therefore seems clear that CT acting via the GM1
receptor can affect the response of lymphocytes to potent stimulators binding
to other cellular receptors. Proably physiological substances like hormones
could act similarly to CT in this respect.

The influence of CT on lymphoid cell proliferation in vivo has been
studied using a Moloney virus induced mouse lymphoma, YAC, which has been
adapted to growth as dispersed cells in the peritoneal cavity of syngeneic
mice (31). Following inoculation of 1×10^6 viable cells these will have increased to 100×10^6 after 8 days. During this phase of rapid growth about 50%
of the cells synthesize DNA at a given moment as studied by autoradiography.
After 8 days only little net growth occurs and the percentage of cells being
in the S phase decreases drastically.

Immunofluorescence studies showed that all tumour cells carried toxin on
their surface a few minutes after an intraperitoneal (i.p.) injection of 10 µg
of CT, although the subjective impression was that individual cells varied in
their content of toxin. In vitro binding studies confirmed that CT binds to
specific receptors on the YAC cells and further showed that rapidly and slowly
growing cells have a similar number of receptors for toxin (about 7×10^5 on the
average). After i.p. injection of a nonsaturating amount of CT the number of
cells carrying toxin identifiable by immunofluorescence decreased with 30%
from 1 to 30 min. At the same time there was about a doubling (from 4 to 8%)
in the number of cells showing laterally redistributed toxin ("capping").
Studies using ^{125}I-labelled toxin showed that, despite the apparent loss of
CT as judged from immunofluorescence, most of the initially bound toxin remained associated with the cell membrane.

I.p. injection of 1 µg of CT caused a marked increase in cyclic AMP in
the YAC cells irrespective of the growth phase. No certain difference in
"pre-toxin" cyclic AMP content could be demonstrated between rapidly and
slowly growing YAC cells but the "lag" phase in the cyclic AMP response to CT
was significantly shorter in the rapidly growing cells. This different kinetics in the response between rapidly and slowly growing cells was verified in

in vitro experiments where the cells could be compared at the same cell and
toxin concentrations.

Administration of 1 µg of CT, either i.p. or intravenously (i.v.) to YAC
bearing mice during the rapid cell growth period (day 4-6) inhibited the
multiplication of the tumour cells profoundly (Table II). Choleragenoid,
binding to the cells without activating adenylate cyclase, had no effect on
the cell multiplication. Preincubation of the cells with CT did not inhibit
cell proliferation but rather caused a slight increase in the number of
tumour cells present 7 days after inoculation.

In addition to inhibiting the cell multiplication CT was found to
decrease the incorporation of ^{3}H-thymidine into the YAC cells as judged by
autoradiography. There was a lag between injection of toxin and the inhibition
of DNA synthesis, the latter being manifest six but not three hours after
injection of toxin. Since the viability of the YAC cells as tested with trypan
blue exclusion was not affected by CT either in the mice in vivo or on incuba-

TABLE II

Effect of 1 µg of CT or choleragenoid on the growth of
YAC cells in A/Sn mice
(From experiments in ref. 31)

Treatment[a]	Cells counted day	Per cent inhibition of cell density[b]
CT		
Preincubation of cells	7	-29±10
I.p. day 4	7	42±2
I.v. day 4	5	76±6
I.v. day 4	7	68±5
Choleragenoid		
I.v. day 4	5	-3±13

[a] In relation to the inoculation of YAC cells

[b] Mean of 5 mice ± SEM

tion in vitro it seems likely that decreased DNA synthesis of live cells was at least partly responsible for the observed inhibition of YAC cell multiplication.

In most of our in vivo studies the effects of CT on YAC cells after i.p. injection of the toxin were obtained under conditions such that the total number of GM1 receptors on the YAC cells exceeded the number of injected CT molecules by a factor of 3-4. In view of the rapid and strong binding of CT to its receptor one might therefore suppose that very little toxin escaped the peritoneal cavity and that the effects obtained thus were due to direct effects of the toxin on the YAC tumour cells.

Effects of cholera toxin on antibody formation and lymphoid tissue morphology

The initiation and expression of immune responses are complex phenomena involving interactions of different cells with synergistic as well as antagonistic effects. The administration of a potent, specific adenylate cyclase activator like CT at different stages of the immune response development might aid in elucidating these cellular mechanisms. Our studies (28) and those of others (32-35) on the influence of CT on the antibody response have shown that the toxin can cause marked stimulation or suppression of antibody formation. The major determining factor for the effect is the timing of the CT administration in relation to that of the antigen.

We found that 1 μg of CT (i.v. injection) stimulated 19S and 7S antibody formation (plaque forming cells, PFC) to sheep erythrocytes (SRBC) when the toxin was given simultaneously with the antigen. Increased numbers of 19S antibody producing cells were found 4 but not 7 or 11 days after the primary antigen injection. The increase in 7S antibody producing cells was most marked on the 7th day after antigen administration, was still present on the 11th day but had not become evident on the 4th day after antigen. Injection of 1 μg of choleragenoid along with the antigen did not affect either 19S or 7S antibody production. Given 2 days after the antigen, CT caused a pronounced suppression of the primary antibody response involving 19S as well as 7S antibodies.

The effect of CT on secondary response 7S antibody production was found to be similar to that on primary 7S antibody formation in that the toxin, given simultaneously with the secondary antigen dose, increased the number of antibody producing cells (Table III). However, in contrast to the effect on the primary response CT also stimulated secondary 7S antibody formation when administered after the secondary antigen dose. In addition, it was found that injection of CT two or three days before the secondary antigen injection

TABLE III

Effect of i.v. injection of CT on the secondary antibody
response to SRBC in DBA/1xC57B1/6 mice
(From data published in ref. 28)

CT	7S PFC per spleen[a]
-	43,560±4,476
1 μg, day -3[b]	13,640±967
1 μg, day -2[b]	17,160±546
1 μg, day 0[b]	262,240±1,042
1 μg, day +2[b]	66,440±4,310

[a] Mean ± S.E.M. determined 21 days after a primary dose
of 10^8 SRBC and 7 days after a booster dose of 10^8 SRBC.

[b] In relation to the secondary SRBC dose.

decreased anamnestic 7S antibody formation (Table III). CT given in relation
to the primary antigen dose had rather small effects on the antibody response
to a booster injection of the antigen two weeks later. When the toxin was
given simultaneously with or two days after the first antigen dose secondary
antibody formation was increased. Toxin given one day before the primary
antigen injection, however, slightly inhibited the anamnestic response.

The effects of CT on antibody formation in mice are thus complex and
critically dependant on the timing of toxin in relation to antigen. It seems
likely that different types of cells vary in their susceptibility to the
toxin at various stages of stimulation. Inspection of the lymphoid tissues
following in vivo administration of CT revealed dramatic toxin influences
on the lymphoid system. Thus, although CT is not directly lethal to lymphoid
cells it was found that i.v. injection of 1 μg of CT into 4 weeks old CBA
mice caused a marked decrease in spleen and thymus weight but not in body or
kidney weight (28). The spleen weight was reduced by 50% within three days
and thereafter increased to reach the normal size after six days. The weight
of the thymus decreased to about 20% three days after injection of CT and had

not started to recover 8 days after toxin injection. The decrease in organ weight was accompanied on the histologic scale by death of cells, more pronounced in the thymus than in the spleen. Since CT is known to increase secretion of hormones from adrenal cells in vitro (36), the influence of the adrenals on the CT induced involution of lymphoid organs was studied. No decrease in spleen or thymus weight was observed 24 hours after injection of CT into adrenalectomized mice making it probable that the effect is dependant upon substances secreted by the adrenals following administration of CT. Since 1 µg of CT was lethal for adrenalectomized animals within 36 hours, the study of the effects of CT on lymphoid organs in these animals could not be extended further. Similar findings have been reported by Morse et al. (37).

Effects of cholera toxin on graft-versus-host reactions and the "allogeneic effect"

CT has also a strong modulating influence on the capacity of lymphoid cells to effectuate cell-mediated immune reactions. In vitro the toxin has been found to decrease the cytotoxic activity of immune lymphocytes on allogeneic target cells (38). Also lymphocytes obtained from immunized mice given CT in vivo have shown reduced in vitro cytotoxicity (39). Furthermore, CT administration has been found to prolong allograft survival in mice (40, 41) as well as to suppress various cell-mediated immunological reactions related to experimental schistosomiasis (40).

In the described in vivo systems it is difficult to distinguish between possible influences of CT on the immunocompetent cells themselves and effects on other components, e.g. inflammatory, integrated in the cellular immune reactions. To avoid this problem we have studied the influence of CT on graft-versus-host reactions (GVHR) and the "allogeneic effect", in which systems lymphoid cells from CT treated animals are transferred to nontreated recipients (28,42). It was found that i.v. injection of 1 µg of CT into adult DBA/1 mice increased the ability of spleen cells to induce GVHR in newborn DBA/1xC57Bl/6 F1 hybrid mice. CT was efficient in augmenting the GVHR inducing capacity of spleen cells, when given to the cell donor mice one or two days, but not three, before cell transplantation. Peculiarly, the spleen cells from the CT treated mice showed an unusual dose-response relationship resulting in a marked deviation from the linear relation obtained with normal cells when $\log_{10}$ cell number is plotted against $\log_{10}$ spleen index. Thus, the

stimulation of the GVHR inducing ability was marked when low numbers of cells
were tested but not demonstrable at high numbers of cells (Table IV).

The increased potency of spleen cells from CT treated mice to cause GVHR
could be explained in a number of ways. One possibility was that CT preferentially acted by inhibiting some type of suppressor cell thereby releasing
the effector cells in GVHR from inhibition. To test this hypothesis $4-8 \times 10^6$
spleen cells from CT treated mice were admixed with small numbers ($0.2-4 \times 10^6$)
of cells from normal animals immediately before the transfer to recipients.
It was found that 1×10^6 or more normal spleen cells normalized, i.e.
suppressed, the capacity of spleen cells from CT treated mice to induce GVHR.
10^6 normal thymocytes were also suppressive but to a lesser extent than
spleenocytes. Spleen and thymus cells from normal neonatal (<7 days of age)
animals were less active than adult cells in suppressing spleen cells from
CT treated animals. In order to be effective suppressors, it seems that the
added spleen cells have to be allogeneic with the recipients, i.e. DBA/1x
C57B1/6 cells could not replace DBA/1 cells in the strain combination used.
In conclusion, pretreatment of mice with CT appears to reduce a suppressive
influence but leave the effector function of the spleen cells intact. This
results in much increased GVHR when low numbers of cells are transferred.

TABLE IV

Effect of 1 μg of CT, given i.v. to DBA/1 cell donor mice 24 h
before sacrifice, on the GVHR inducing capacity of their
spleen cells in DBA/1xC57B1/6 F1 hybrid mice
(From data in ref. 42)

Number of cells transferred	$\log_{10}$ spleen index ± SEM obtained with cells from	
	Untreated mice	CT treated mice
3×10^6	0.192±0.013	0.274±0.030
8×10^6	0.289±0.039	0.451±0.034
2×10^7	0.389±0.027	0.425±0.032
4×10^7	0.464±0.017	0.420±0.045

The "allogeneic effect" (43) is manifested by the ability of parental lymphocytes to increase antibody formation in Fl hybrid animals. In Fl hybrid mice that have been primed with sheep erythrocytes injection of parental spleen or thymus cells can elicit a secondary antibody response (44). Using this model, it was found that pretreatment of animals with CT increased the ability of their spleen cells to mediate the "allogeneic effect". Preliminary experiments have also indicated a suppressive influence of spleen cells from untreated animals on the ability of cells from CT treated mice to induce the "allogeneic effect".

Discussion of possible sites of action of cholera toxin on immune responses in vivo

As described CT has but one known primary influence on mammalian cells, i.e. to stimulate adenylate cyclase following binding to a specific membrane receptor, the GMl ganglioside. Several variables might, however, affect the outcome of this action: (i) the receptor density on different cells creating competition for toxin, (ii) the ratio between the stimulated adenylate cyclase and the inherent phosphodiesterase activities, determining the net effect on intracellular cyclic AMP levels, (iii) the phase in the cell cycle when the raise in cyclic AMP occurs. It is important to remember that when CT is administered systemically the toxin will reach many tissues and affect a variety of cell systems. Obviously CT could therefore modulate immune responses not only directly by an action on the lymphoid cells but also indirectly by effects on the milieu in which these cells perform.

In figure 3 the cells participating in immune responses are schematically represented. It is clear that CT could interfere with the immune system by effects on the cellular milieu (e.g. mediated by an action of the toxin on the endocrine system), or by effects on effector, helper, suppressor or accessory cells. Since CT is in itself an antigen, it might also cause "antigenic competition" (45) for the immune response under study. This seems to be of minor importance, however, in view of the lack of effect on unrelated immune responses of choleragenoid which carries the strong antigenic determinants of the toxin. There exists experimental evidence about effects of CT on the cellular milieu, on suppressor cells and on effector cells. We are not aware of any studies so far where the influence of CT on accessory cells has been studied.

The rapid involution of lymphoid organs with cell death following i.v. injection of CT and the subsequent cellular repopulation to the normal

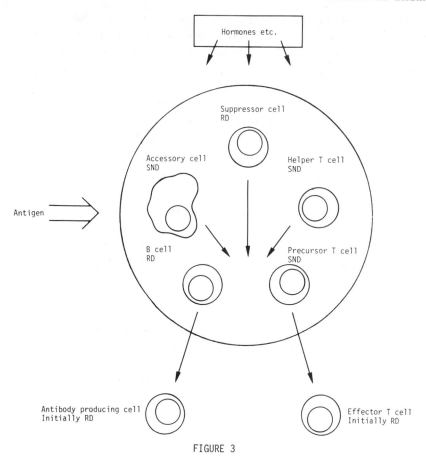

FIGURE 3

   Schematic representation of the cells of the immune response. RD =
Rapidly dividing, SND = Slowly or not dividing.

tissue size suggest obvious possibilities for changes in the class composition of lymphocytes which could contribute to the modulation of immune
responses by CT. These changes largely appear to be mediated through a toxin
influence on the adrenal glands, for in adrenalectomized mice the lymphoid
organs seem to retain normal size and histology following injection of CT.
Steroid hormones could be the mediator substances, since CT has been observed
to increase secretion of such hormones from the adrenals both _in vitro_ (36)
and _in vivo_ (37).

It is evident, however, that CT modulates immune responses also by its direct action on the immunocompetent cells. As described the toxin binds to lymphoid cells in vitro as well as in vivo and raises the intracellular cyclic AMP levels. Very similar toxin effects to those described on in vivo antibody formation have been reported for secondary response antibody formation in vitro (35). In the latter system dibutyryl cyclic AMP mimicked the effects of CT.

CT has been described to decrease in vitro secretion of antibodies from immunized spleen cells (46) as well as the production of migration inhibitory factor by cultured granuloma tissue (40). This raises the possibility that some effects of CT on immune responses in vivo might be due to inhibited secretion of signal substances or antibody molecules. Pretreatment of spleen cells with CT in vitro reduces the number of PFC (28, 46). However, secretion of antibody molecules from cells taken from mice given CT 4-8 days earlier seems to be unaffected as judged from the finding that the developing times amd the size distribution of the plaques did not differ between spleen cells from CT-treated and untreated mice (28). Therefore, the observed effects of CT on in vivo antibody formation are probably not due to changed antibody secretion from the cells but reflect true differences in the number of antibody synthesizing cells.

The possibility that CT modulates immune responses by inhibiting lymphoid cell proliferation is attractive in view of the documented inhibitory effects of the toxin on DNA synthesis of B and T lymphocytes, YAC lymphoma cells and other, nonlymphoid, cell types. Inhibition of the proliferation of effector cells, or helper cells, could result in many of the observed suppressive effects of CT on the immune response in vivo. The stimulatory effects of CT on antibody formation and on the capacity of spleen cells to cause GVHR and the "allogeneic effect" could be explained by an inhibitory effect of the toxin on suppressor cell proliferation. As described there exists strong evidence that spleen cells from CT treated mice have a diminished suppressor activity for immune responses to histocompatibility antigens. Shearer et al. (47) described that fractionation of spleen cells on columns derivatized with histamine, epinephrine or isoproterenol removed a hormone-binding cell population which had a suppressive regulatory influence on 19S and 7S antibody formation. Transfer of the fractionated cells along with SRBC antigen to irradiated syngeneic recipients resulted in enhanced PFC responses. The differences in response between fractionated and unfractionated cells were

marked at low cell inocula, but disappeared at inocula above $2x10^7$ cells.
Addition of $1x10^6$ cells eluted from the columns to $5x10^6$ fractionated, non-
-binding spleen cells normalized the response (48). These observations
resemble our findings on the capacity of spleen cells from CT treated mice
to cause GVHR and the "allogeneic effect", suggesting that the suppressor
activity removed by the toxin in vivo might be similar to that eliminated in
the amine-derivatized columns.

The stimulatory effect of CT on antibody formation has been interpreted
in terms other than diminished suppressor cell activity. From the results of
experiments involving transfer of lymphoid cells from CT treated donors it
has been suggested that CT directly stimulates helper T cells (34). We think,
however, that inhibition of suppressor cell activity cannot be ruled out in
these experiments.

In conclusion, although CT has only one primary effect, common to all
cell types, i.e. to activate plasma membrane adenylate cyclase, the modulating
influence of this toxin on the immune response in vivo might nevertheless be
multifactorial due to the many cells contributing to the response. It is our
belief that a fuller clarification of the effects of CT on immune responses
will yield a greater insight into the physiological role of cyclic AMP in
regulating the functions of immunocompetent cells. It might also provide useful
information for the design of efficient immunomodulating agents.

Acknowledgement

We are grateful to Drs. S. Lange and I. Lönnroth for collaboration in
much of the work presented here.

References

1.   Braun, W.B., Lichtenstein, L.M. and Parker, C.W. (eds.) Cyclic AMP,
      Cell Growth, and the Immune Response, Springer-Verlag New York, 1974.

2.   Pastan, I.H., Johnson, G.S. and Anderson, E.B., Ann. Rev. Biochem.,
      44:491, 1975.

3.   Field, M., N. Engl. J. Med., 284:1137, 1971.

4.   Pierce, N.F., Greenough, W.B., III and Carpenter, C.C., Bacteriol.
      Rev., 35:1, 1971.

5.   Holmgren, J. and Svennerholm, A.-M., J. Inf. Dis. Suppl., in press.

6.   Lönnroth, I. and Holmgren, J., J. Gen. Microbiol., 76:417, 1973.

7. Cuatrecasas, P., Parikh, I. and Hollenberg, D., Biochemistry, 12: 4753, 1973.

8. van Heyningen, S., Science, 183:656, 1974.

9. Holmgren, J. and Lönnroth, I., J. Gen. Microbiol., 86:49, 1975.

10. Holmgren, J., Lindholm, L. and Lönnroth, I., J. Exp. Med., 139:801, 1974.

11. van Heyningen, S. and King, C.A., Biochem. J., 146:269, 1975.

12. Gill, D.M. and King, C.A., J. Biol. Chem., 250:6424, 1975.

13. Sahyoun, N. and Cuatrecasas, P., Proc. Natl. Acad. Sci. USA, 72: 3438, 1975.

14. Lönnroth, I. and Holmgren, J., J. Gen. Microbiol. 91:263, 1975.

15. Holmgren, J. and Lönnroth, I., J. Infect. Dis. 133:(suppl.), S 64, 1976.

16. Holmgren, J., in Proc. 5th Int. Con. Endocrin., edited by V.H.T. James, Excerpta Medica, in press.

17. Craig, S. and Cuatrecasas, P., This series.

18. Svennerholm, L. In Methods in Carbohydrate Chemistry, edited by Whistler, R.L. and BeMiller, J.N., p. 464, Academic Press, New York, Vol. 6.

19. Holmgren, J., Lönnroth, I. and Svennerholm, L., Infect. Immun., 8: 208, 1973.

20. King, C.A. and van Heyningen, W.E., J. Infect. Dis. 127:639, 1973.

21. Cuatrecasas, P., Biochemistry, 12:3547, 1973.

22. Holmgren, J., Infect. Immun., 8:851, 1973.

23. Holmgren, J., Lönnroth, I., Månsson, J.E. and Svennerholm, L., Proc. Natl. Acad. Sci. USA, 72:2520, 1975.

24. Hansson, H.-A., Holmgren, J. and Svennerholm, L., Submitted for publication.

25. Mullin, B.R., Aloj, S.M., Fishman, P.H., Lee, G., Kohn, L.D. and Brady, R.O., Proc. Natl. Acad. Sci. USA, 73:1679, 1976.

26. Cuatrecasas, P., Biochemistry, 12:3558, 1973.

27. Stein-Douglas, K.E., Schwarting, G.A., Naiki, M. and Marcus, D.M., J. Exp. Med. 143:822, 1976.

28. Lindholm, L., Holmgren, J., Lange, S. and Lönnroth, I., Int. Arch. Allergy, 50:555, 1976.

29. Lönnroth, I. and Lönnroth, C., Exp. Cell. Res., in press.

30. Sulzer, B.N. and Craig, J.P., Nature New Biol., 244:178, 1973.

31. Holmgren, J., Lange, S., Lindholm, L., Lönnroth, C. and Lönnroth, I., Manuscript in preparation.

32. Chisari, F.V., Northrup, R.S. and Chen, L.C., J. Immunol. 113:729, 1974.

33. Kateley, J.R., Kasarov, L. and Friedman, H., J. Immunol. 114:8, 1975.

34. Kateley, J.R. and Friedman, H., Cell. Immunol. 18:239, 1975.

35. Cook, R.G., Stavitsky, A.B. and Schoenberg, M.D., J. Immunol. 114:426, 1975.

36. Donta, S.T., King, M. and Sloper, K., Nature New Biol., 243:246, 1973.

37. Morse, S.I., Stearns, C.D. and Goldsmith, S.R., J. Immunol. 114:665, 1975.

38. Strom, T.B., Carpenter, C.B., Garovoy, M.R., Austen, K.F., Merril, J.P. and Kaliner, M., J. Exp. Med. 138:381, 1973.

39. Henney, C.S., Lichtenstein, L.R., Gillespie, E. and Rolley, R.T., J. Clin. Invest. 52:2853, 1973.

40. Warren, K.S., Mahmoud, A.A.F., Boros, D.L., Rall, T.W., Mandel, M.A. and Carpenter, C.C., J. Immunol. 112:996, 1974.

41. Ljungström, I., personal communication.

42. Lange, S., Lindholm, L. and Holmgren, J., Submitted for publication.

43. Katz, D.H., Transplant. Rev. 12:141, 1972.

44. Lindholm, L. and Strannegård, Ö., Int. Arch. Allergy, 47:508, 1974.

45. Pross, H.F. and Eidinger, D., Adv. Immunol. 18:133, 1974.

46. Bourne, H.R., Melmon, K.L., Weinstein, Y. and Shearer, G.M., in Cyclic AMP, Cell Growth, and the Immune Response, edited by W. Braun, L.M. Lichtenstein and C.W. Parker, p. 99, Springer Verlag, New York 1974.

47. Shearer, G.M., Weinstein, Y. and Melmon, K.L., J. Immunol. 113:597, 1974.

48. Shearer, G.M., Weinstein, Y., Melmon, K.L. and Bourne, H.R., in Cyclic AMP, Cell Growth, and the Immune Response, edited by W. Braun, L.M. Lichtenstein and C.W. Parker, p. 135, Springer Verlag, New York, 1974.

Chapter 6

POST RECOGNITION ION DEPENDENT EVENTS IN MITOGEN INDUCED
LYMPHOCYTE PROLIFERATION AND IN CYTOTOXIC EFFECTOR CELL RESPONSES

M.H. Freedman[*] and E.W. Gelfand[+]
*Faculty of Pharmacy and Institute of Immunology, University of Toronto,
Toronto, Ontario M5S 1A1; and [+]Research Institute, Department of
Immunology, Hospital for Sick Children, Toronto, Ontario M5G 1X8

Abstract

This review discusses the increased permeability and enhanced uptake/
transport of monovalent and divalent cations following mitogen-induced
lymphocyte proliferation. The observed Ca^{2+} uptake is discussed in terms
of gated Ca^{2+} channels. The importance of divalent cations, particularly
Ca^{2+}, is discussed in terms of triggering of cytotoxic effector cell responses for three model systems (antibody dependent cytotoxicity, mitogen-
induced cytotoxicity and cell mediated cytotoxicity); (85).

The activation of lymphocytes, characterized by a very complex series
of changes in the cell membrane, cytoplasm and nucleus, has proven to be
an important model cell system for studying mechanisms by which repressed
mammalian cells are induced to proliferate. One of the key problems in
cellular biology is to determine how antigens interact with cell surface
receptors, and how this interaction is linked to the activation of cellular
metabolic processes which result in cell proliferation and specific immunological function. The thymus-dependent T lymphocytes (T cells) may become
killer cells or secrete a variety of non-antigen specific factors which play
an important role in various cell-mediated immune responses. The thymus-
independent B lymphocytes (B cells) mature into antibody secreting cells.
The parameters used in many of these studies are cell division (or DNA
synthesis) or the development of effector function (antibody synthesis,
lymphokine secretion or cytotoxicity) after 4-72 hr.

Lymphocyte activation has been defined as the initiation of new DNA
synthesis in a resting population of lymphocytes, and the progression of
these cells into and through mitosis. The process begins with the surface
binding of a variety of specific and nonspecific activating agents (ligands)

to the surface receptors of the lymphocyte plasma membrane (the "first message") (1). The "second message" is responsible for changes in the cytoplasmic metabolism and culminates in changes in the intranuclear enzyme
activity resulting in synthesis of new DNA and eventually in cell division
(1). One of the main goals in the study of the sequence of events leading
to lymphocyte activation has been the attempts to define an early event
which may initiate all the subsequent biochemical changes seen in activated
lymphocytes.

Lymphocytes may be induced to divide or differentiate into effector
cells by a variety of different ligands (e.g. specific antigens, antibodies
directed against some lymphocyte cell surface determinants and some plant
lectins) which bind to their surface. The response of the cell is characteristic of the class of lymphocyte and not of the ligand. One of the major
problems in the elucidation of the biochemical basis of lymphocyte activation by antigens is the small part of the lymphocyte population ($\sim$0.1%)
which participates in a specific immunological response (2). On the other
hand, a variety of nonspecific lymphocytic mitogens stimulate many clones,
independent of their antigen specificity. It is generally accepted that
the study of nonspecific or polyclonal activation of lymphocytes by mitogens is relevant to the problem of antigen activation, since apart from
the specificity of the antigenic response, lymphocytes,activated by mitogens,
exhibit very much the same properties as antigenically stimulated cells.
Polyclonal mitogens often show class specificity (e.g. T cell=soluble Con A
and PHA; B cell=LPS) (3). Studies with nonspecific mitogens have provided
a uniform system in which sufficient numbers of cells are activated to
allow a detailed analysis of the sequence of events involved in the induction of activation. It is not necessary for the mitogen to enter the cell
for it to cause activation (4). Since the mitogen does not need to enter
the cell, then it must act directly at the membrane, or its attachment to
the membrane must result in the formation of a "second messenger" which is
released into the cytoplasm to bring about the alteration in cellular
metabolism.

The first message for lymphocyte activation is antigen binding to a
receptor, but often this alone is insufficient to stimulate. The macrophage,
and not the lymphocyte, appears to be the initial antigen-binding cell in
T cell antigen stimulation (5). Also, for most antigens to activate B
cells, the co-operation of macrophages and specific T cells is required
(6). It seems likely however that at least some of the nonspecific mito-

gens activate lymphocytes directly without the requirement for cell-cell
interactions. However, since most of them bind to more than one type of
membrane molecule, it is difficult to determine which one is the important
activating receptor (1).

A number of membrane, cytoplasmic and nuclear events have been
observed within minutes to hours, following the addition of nonspecific
mitogens to a suspension of purified lymphocytes. These events include:
lectin binding with removal and resynthesis of receptors; receptor redistribution ("capping") (7,8); altered surface charge of the membrane (9);
increased membrane fluidity (10,11); increased permeability and enhanced
uptake/transport of small molecules and metabolites (e.g. K^+ (ouabain
sensitive Na^+,K^+ pump activity and Mg^{2+} and Na^+,K^+ dependent ATPase activity) (12-17); Na^+ (18); Ca^{2+} (19-25); inorganic phosphate (26); choline
(27); sugars (non-utilized 3-0-methylglucose, glucose) (13,28); nucleosides
(uridine) (27); amino acids (non-utilized 2-aminoisobutyric acid, leucine)
(29,30)); increased metabolism/synthesis of membrane phospholipids (particularly phosphatidylinositol) and increased incorporation of long chain fatty
acids in plasma membrane lecithin, and saturated fatty acids are replaced
by polyunsaturated fatty acids (predominantly arachidonic acid) in membrane
phospholipids (31-33); activation of membrane bound enzymes (e.g. microsomal
Na^+,K^+ independent ATPase) (34); adenylate cyclase-resulting in elevated
cAMP (35-37); guanylate cyclase-resulting in elevated cGMP (38-41)); histone
acetylation (42,43) and non-histone protein phosphorylation (44,45); increases in RNA polymerase I activity and initial decreases in RNA polymerase
II activity resulting in increased RNA synthesis (46); increases in protein
synthesis and ultimately increased DNA synthesis. Each of the events described, although related to the induction of DNA synthesis, remains to be
established as an obligate requirement for DNA synthesis and therefore an
essential feature of mitogenesis. It has been difficult to establish which
of these rapid cellular changes are necessary, which are sufficient, which
are non-contributory and which are secondary to the process of activation.
Almost all of the experiments are complicated by the fact that what is good
for a cell to be initially triggered may be antagonistic to subsequent
proliferation.

This review will confine its discussion on cell activation only to
those events resulting in increased permeability and enhanced uptake/trans-
port of cations following mitogen stimulation of quiescent lymphocytes into
a derepressed state of active growth.

CATION REQUIREMENTS IN LYMPHOCYTE PROLIFERATION

The addition of PHA or Con A to human peripheral lymphocytes results in a rapid increase in the uptake of $^{42}K^+$ which is discernable within 30 sec (13,15). The K^+ transport influx, in the presence of mitogen, is an active transport mechanism and is probably not due to a decreased K^+ efflux (47,48).

Ouabain, a specific inhibitor of the plasma membrane-bound Na^+,K^+, ATPase and of monovalent cation (Na^+,K^+) membrane transport, reproducibly and reversibly inhibited PHA-induced lymphocyte transformation (47,49). The inhibition of the lymphocyte response by ^{3}H-ouabain could be overcome by the addition of excess K^+. Na^+,K^+,ATPase is intimately involved in the uphill transport of the monovalent cations, and ouabain acts as a competitive inhibitor of K^+ at a site located at the outside surface of the cell membrane (16). The parameters of lymphocyte transformation which are inhibited by ouabain include: morphological (blast transformation and mitosis), physiological (cell respiration and the increased cell respiration which follows stimulation (12)) and biochemical (synthesis of DNA, RNA and protein) changes (16). Recently, the inhibitory effects of ouabain have been extended to include B cell activities (16,50).

The kinetics of the increased K^+ uptake under the influence of PHA-stimulated cells show that the maximum rate of uptake (V_{max}) is affected but not the apparent affinity constant (K_m) (12,13,47), whereas the increased uptake of Ca^{2+} following PHA stimulation was, on the contrary, due to an effect on the K_m and not on the V_{max} (22). The increased V_{max} of transport was not caused by PHA-stimulated synthesis of new Na^+,K^+,ATPase (12,16). The K^+ kinetic results have been interpreted as indicating that the same type of transport sites are increased rather than a new type being involved. PHA causes expression of active K^+ sites on preexisting cryptic membrane Na^+,K^+,ATPase transport sites by a conformational change involving membrane glycoproteins or by aggregation of inactive subunits of active oligomers in the membrane. The increase in the number of functional sites would result in increasing the internal K^+ level above a threshold concentration required for almost all of the events which link the membrane stimulus to the synthesis of DNA and cell division that ensues (16,17). The possibility that mitogens also increase the rate of cation transport by each pump site cannot be excluded (15); however, the data (16,17) strongly indicate an increase in total number of sites at the cell surface, which are active as monovalent cation pumps, occurring within minutes of addition of mitogen to the lymphocyte cultures.

It has been suggested that the K^+ level or perhaps the K^+/Na^+ ratio is an essential aspect of the signal to the nuclear apparatus. The possible regulatory role of Na^+ cannot be ruled out, since its intracellular concentration will vary inversely with that of K^+.

Activation of the Na^+,K^+,ATPase and K^+ transport may also be associated with the cyclic nucleotide levels (51) and with the fluidity of the membrane lipids (52). Observed changes in phospholipid metabolism in activated lymphocytes may be associated with Na^+,K^+,ATPase (32,52,53).

One of the earliest events following lymphocyte stimulation by mitogens is an increase in the rate of the K^+ influx via the Na^+,K^+,ATPase system of the membrane. With the exception of a few early events (membrane clumping; early synthesis of RNA; interaction of PHA with its receptor) virtually every event, morphological, physiological and biochemical, which characterizes the transformed activated lymphocyte, has an absolute requirement for the increased level of internal K^+ which follows stimulation. The threshold level of internal K^+ required for some events (e.g. blastogenesis) are lower than those for others (e.g. DNA synthesis) (16). The exact intracellular function of K^+ at the molecular level is still a matter of conjecture.

The possibility that certain divalent ions might also play a role in the mitogenic activation of lymphocytes was noted by Alford (19) and Allwood et al. (20) when they studied the addition of PHA to human lymphocytes. Experiments using Ca^{2+}- and Mg^{2+}-free medium confirmed that Ca^{2+} rather than Mg^{2+} was the cation of principal importance. The apparent synergistic effect of Ca^{2+} and Mg^{2+} implies that each has independent functions which can only be partially fulfilled by the other. Earlier, Whitfield and Youdale (55) demonstrated that Ca^{2+} stimulates the early phases of mitosis of rat thymocytes. The observation of an early Ca^{2+} uptake is analogous to the role of Ca^{2+} in muscle and secretory cells. Also Zn^{2+} and Fe^{2+} appear to be essential for in vitro PHA-induced lymphocyte transformation (19).

Whitney and Sutherland (21,22,56) have considerably extended these early initial observations. Addition of PHA to human peripheral lymphocytes causes an increased uptake of Ca^{2+} by an energy-independent mechanism within a minute after addition of the mitogen (22,57). The rate of uptake progressively increases to a maximum value 60 hr after PHA treatment (near time for maximum DNA synthesis). This suggests that Ca^{2+} may be required for the initiation of the response and for maintenance of later cellular synthetic processes (21). PHA somehow disturbs membrane structure resulting in an increased passive permeability to Ca^{2+}. Calcium efflux seems to be an active process requiring expenditure of metabolic energy to expell Ca^{2+} against a high electrochemical gradient, and in some cell types a Ca^{2+}-activated ATPase, which is independent of the Na^+,K^+,ATPase, has been found.

Cyclic AMP, dibutyryl cAMP and theophylline also altered Ca^{2+} transport providing evidence for an effect of cAMP on an early event in the process of transformation (21). Little is known about the specific role of intracellular Ca^{2+} in the lymphocyte activation process. The role of cyclic nucleotides in transformation is also not clear (35-41).

The Ca^{2+} uptake of PHA-stimulated lymphocytes follows saturation kinetics implying participation of a membrane carrier mechanism in lymphocyte Ca^{2+} transport. Mitogens affect the Ca^{2+} transport system by increasing the Ca^{2+} influx initially by decreasing the K_m for the binding of Ca^{2+} to the carrier while leaving the V_{max} unaltered. The kinetic changes of Ca^{2+} uptake in stimulated cells differ from those for K^+ (see above) and metabolite transport (e.g. 2-aminoisobutyric acid (29)). Also, both K^+ and amino acid uptake are active transport processes in contrast to the energy-independent accumulation of Ca^{2+} (56). Mn^{2+} altered K_m but not V_{max} which indicated that Mn^{2+} is a competitive inhibitor of Ca^{2+} uptake. Mg^{2+} had no effect on Ca^{2+} uptake (56). The Ca^{2+} influx kinetic data has been interpreted to mean that PHA increases the affinity of membrane sites for Ca^{2+} rather than increasing the rate of transport by these sites or by increasing the number of sites. The kinetics of Ca^{2+} uptake by PHA seem to be independent of de novo protein synthesis or new RNA synthesis. These results may indicate that the binding of Ca^{2+} to the carrier is initially altered, for instance, through a conformational change in the membrane, rather than affecting the activity of the sites.

ROLE OF GATED CALCIUM CHANNELS IN LYMPHOCYTE ACTIVATION

There is increasing evidence for the importance of Ca^{2+} in lymphocyte activation: (a) the great majority of early and late events following activation of lymphocytes by antigens or nonspecific mitogens require extracellular Ca^{2+} (19,23,33,57,58); (b) increased Ca^{2+} uptake is one of the earliest events that has been measured following mitogen activation (24,25); and (c) the Ca^{2+} ionophore, A23187, has been shown to induce DNA synthesis and blast transformation in pig (59) and human (60) lymphocytes, and more recently to induce some of the early events in lymphocyte activation, such as a rise in intracellular cGMP (61) and increased phosphatidylinositol

turnover (62). Although these observations do not establish that Ca^{2+} is
the "second message" in lymphocyte activation, they are compatible with
this view, and encouraged us to try to directly demonstrate the importance
(or lack of importance) of gated Ca^{2+} membrane channels in lymphocyte
activation.

We studied the $^{45}Ca^{2+}$ uptake in mouse spleen T lymphocytes stimulated
with Con A, PHA and PWM (24,25). The Con A-induced $^{45}Ca^{2+}$ uptake was
measureable by 30-45 sec after addition of the mitogen and was complete by
1 min suggesting that the response was of brief duration. In the absence of
T cells(anti-θ^{+}+complement treated cells or nu/nu mouse spleen cells) these
T cell mitogens had no effect. The lower concentration end of the dose-
response curve for Con A-induced $^{45}Ca^{2+}$ uptake and DNA synthesis was similar.
Compounds which would be expected to raise the intracellular levels of cAMP
(dibutyryl cAMP,cholera toxin and theophylline) inhibited the Con A-induced
$^{45}Ca^{2+}$ uptake, while dibutyryl or 8-bromo cGMP enhanced it. Treating cells
to deplete ATP had no effect on the $^{45}Ca^{2+}$ uptake. The simplest interpretation of these findings is that Con A binding to (and probably cross-linking
of) the relevant membrane glycoprotein molecules, directly or indirectly
open Ca^{2+} channels in the membrane, allowing Ca^{2+} to enter down a steep
concentration gradient ($\geq 10^{-3}$M outside, $\leq 10^{-5}$M inside (63)). One can calculate that large amounts of Ca^{2+} are taken up by Con A-stimulated T cells
($\sim 0.23 \times 10^{-9}$M/10^{6} cells), sufficient to raise the intracellular Ca^{2+} concentration by ~ 1.0 mM if all of the cell water were available to the incoming
Ca^{2+}. Although we have not definitely proven that Con A-induced $^{45}Ca^{2+}$
uptake is related to accumulation within the cell as opposed to adsorption
at the cell surface, it seems likely since (a) only T cells respond; (b)
the effect is not seen if cells are pre-incubated with Con A for 5 min before
adding $^{45}Ca^{2+}$; (c) the effect is modulated by cyclic nucleotides; (d) dis-
placing surface bound Ca^{2+} with ruthenium red or EGTA does not inhibit the
Con A-induced $^{45}Ca^{2+}$ uptake and (e) the amount of Ca^{2+} involved is extremely
large.

In the early experiments, using uncorrected $^{45}Ca^{2+}$ uptake data, we
obtained a 1.2-1.5 fold $^{45}Ca^{2+}$ increase in T cell mitogen-induced mouse
spleen lymphocytes compared to non-induced control cells (24,25). With
modifications to the assay procedure, a 1.5-3.0 fold $^{45}Ca^{2+}$ uptake (Table
I) for T cell mitogen-induced mouse spleen cells and cortisone-resistant
mouse thymocytes has been observed (64). In the early experiments, the B
cell mitogen LPS did not show any measurable $^{45}Ca^{2+}$ uptake in mouse spleen

Table I

Effect of mitogens on $^{45}Ca^{2+}$ uptake in mouse and rabbit
spleen lymphocytes and RBC*

Species	Cells	Mitogen	Concentration (μg or μl ml^{-1})	No. of experiments	$^{45}Ca^{2+}$ uptake**
Mouse & Rabbit	RBC	None	–	–	1.0
		Con A	0.5 – 1.0	5	0.96±0.01
		LPS	5.0 –10.0	5	0.97±0.01
Mouse	Spleen	None	–	–	1.0
		Wheat germ agglutinin	15.0 –50.0	4	0.99±0.01
		Con A	0.5 – 1.0	40	2.16±0.41
		LPS	5.0 –10.0	20	2.00±0.53
Rabbit	Spleen	None	–	–	1.0
		Wheat germ agglutinin	15.0 –50.0	2	1.01±0.01
		Con A	0.5 – 1.0	20	2.05±0.48
		LPS		2	1.94±0.35
		anti-allotype antiserum	0.1 – 0.15	16	1.99±0.44

* Cells, isotopes and mitogens were incubated at $37°$ for 30 min; each
experiment was done in triplicate.

** $^{45}Ca^{2+}$ uptake is calculated by dividing the $^{45}Ca^{2+}/^{3}H_2O$ value of the
sample by that of the control cells incubated with medium alone, so that
control ratio is normalized to 1.0. The increase above 1.0 is a measure
of the mitogen-induced $^{45}Ca^{2+}$ uptake. The uptake is expressed as mean
± s.d. of the means of each experimental triplicate.

cells. Using a purer preparation of LPS (containing more lipid A and less
carbohydrate), we obtained a 1.5-2.5 fold $^{45}Ca^{2+}$ uptake (Table I) in mouse
spleen cells and in nu/nu mouse spleen cells (64). Similar results can be
seen when rabbit spleen lymphocytes were stimulated with various T and B
cell mitogens including specific anti-allotype serum (Table I) (65).

The presence of Ca^{2+} may be required for certain processes occurring
early in the sequence of events which leads to DNA synthesis and mitosis.
It should be noted that EDTA and ouabain, which alters Na^+ and K^+ transport, both could rapidly turn off nucleic acid synthesis implying that other
divalent and monovalent cations were important in maintaining the trans-

formation process once it was underway. EGTA had a marked inhibitory effect
on the uptake of the amino acid analogue, 2-aminoisobutyric acid in stimulated
lymphocytes. The inhibition can be reversed by the addition of excess Ca^{2+}
but not Mg^{2+} (57). These results suggest that Ca^{2+} may not be essential to
the initial interaction of PHA with the cell, but does influence amino acid
transport which may be a critical preparatory event for later increased
protein synthesis. Diamantstein and Ulmer (23) studied the Ca^{2+} requirement
for stimulation of DNA synthesis in vitro by PHA and LPS using mouse spleen
cells. Their results suggest that Ca^{2+} is required for a step preceding DNA
synthesis but not for the early initial phase of transformation. The Ca^{2+}
requirement for mitogenic stimulation seems to be a general phenomenon since
both B and T cell mitogenic responses can be inhibited by removal of Ca^{2+} by
EGTA. Their results argue against the possibility that Ca^{2+} may be involved
in generation or regulation (21,22) of the early signal mechanism at the
level of cyclic nucleotides (35-41).

Whitfield et al. (66), using different cells including rat thymic
lymphoblasts and mouse 3T3 cells, suggested that in cellular "prolifero-
genesis" a brief post-mitotic burst of cGMP might serve to prevent certain
activated cells from being diverted into a non-proliferative (but still
activated) G_0 state. This is followed by two waves of cAMP accumulation
(the second of which crests immediately before the beginning of deoxyribo-
nucleotide synthesizing enzymes in the latter part of the G_1 phase). The
cAMP drives Ca^{2+} from the mitochondria into the cytosol to activate newly
synthesized thymidylate synthetase (or other primed enzymic assemblies)
either directly or indirectly by inactivating a repressor. Alternatively,
these ions might cause a timely release of such enzyme(s) from inactive
complexes in the same way they release neurotransmitters from secretory
granules and synaptic vesicles.

CATION REQUIREMENTS IN CYTOTOXICITY

The demonstration of a role for divalent cations, particularly Ca^{2+}
in a number of cytotoxic effector cell reactions, may also be consistent with
a role for this cation as a messenger in the triggering of cellular cytotoxic activity. The in vitro destruction of isotope-labelled target cells
has been most extensively studied in three systems: antibody dependent
cytotoxicity (ADC), cell mediated cytotoxicity (CMC), and mitogen-induced
cytotoxicity (MIC) (67). In each system, specificity of the reaction is
imparted by the nature of the triggering ligand. Thus, in ADC, triggering

of the effector cell is through binding to the IgG-Fc receptor on the effector cell and specificity is provided by the Fab portion of the antibody
molecule. In CMC, sensitized effector cells carry a receptor for the sensitizing antigen on the target cell and binding through this receptor
presumably initiates a sequence of events resulting in target cell lysis.
For MIC, the lectin serves as a bridge between the target cell and the
effector cell and activates the effector cell cytolytic mechanism. In each
of these assays the nature of the effector cell remains somewhat controversial
with both non-T and non-B cells mediating ADC and non-T and T-cells thought
to mediate CMC and MIC (68-72). Current understanding of the molecular events
by which cytolysis is effected is also limited.

On the basis of data provided by several laboratories (73-75), cytotoxic reactions have been dissected and analysed in three discrete stages.
Divalent cation-dependence and -independence of these stages of lysis have
also been documented. In the first or recognition stage, effector cells and
target cells establish contact through receptors at the surface of the effector cell (specific antigen receptors, Fc receptors or lectin receptors). In
the second post-recognition lytic (75) or programme for lysis (73) step, an
irreversible lesion of the target cell is created. Data from a number of
experiments in which effector cells are inactivated by heat, antisera, or
the addition of chelating agents, have supported the presence of a third
effector cell-independent phase. This third stage appears to be divalent
cation-independent. In the first stage the binding of lectin-coated (MIC)
or antibody-coated (ADC) target cells to effector cells also appears to be
cation-independent. However, the adsorption of specifically sensitized
lymphocytes onto target cells (i.e. the recognition stage in CMC) has proven
to be Mg^{2+}-dependent with Ca^{2+} serving as a poor substitute (74,76).

Lectin binding to lymphocyte membranes induces a series of changes as
discussed above which include increased membrane movement and fluidity, Ca^{2+}
uptake, and uropod formation. These changes may also be prerequisite for
triggering of the effector cells to become cytotoxic effector cells. In
order to study divalent cation-dependency of the second or lethal hit phase
of the reaction, two main approaches have been used: (a) divalent cation-
free media supplemented with varying concentrations of Ca^{2+} or Mg^{2+}, and
(b) the addition of EDTA to incubation mixtures. Discordant results from
laboratory to laboratory may be accounted for by differences in tissue culture medium, serum source,   form (disodium vs. tetrasodium) and concentration of EDTA used.

Antibody Dependent Cytotoxicity. We have studied ADC in man using two
target cells, antibody-coated chicken red blood cells (CRBC) and antibody-
coated Chang liver cells (71,72). As shown in Table II, ADC for CRBC is
divalent cation-independent to a large extent, and the addition of either
Ca^{2+} or Mg^{2+} can restore normal cytolytic activity. Chang cell lysis is
Ca^{2+}-dependent confirming the results of others (77). Using sheep red blood
cells (SRBC) as target cells, Golstein (77,78) has shown that ADC for SRBC
in man is both cation-independent and Mg^{2+}-dependent to varying degrees.
It is unclear as yet why the requirements for SRBC lysis in humans is
somewhat different from that for CRBC lysis but could reflect different
mechanisms of lysis or different populations of cells responsible for lysis.
We have previously suggested that the predominant effector cell for CRBC
lysis in both man and rabbit is a monocyte-macrophage (71,72,79). Using
mouse spleen effector cells, SRBC lysis was Mg^{2+}-dependent (78) whereas
mouse peritoneal exudate cell-induced lysis of SRBC was divalent cation-
independent (80). Furthermore, the degree of Ca^{2+}-dependence of mouse spleen
effector cell ADC may be corollated with the concentration of antibody

Table II

Divalent cation requirements for ADC

Addition	% specific ^{51}Cr-release	
	CRBC	CHANG
Normal Medium	49	29
0	24	0
Ca^{2+} 2mM	50	26
5mM	44	27
Mg^{2+} 1mM	48	0
2mM	48	0
5mM	46	0
Ca^{2+} 2mM + Mg^{2+} 1mM	49	30
Ca^{2+} 5mM + Mg^{2+} 2mM	42	30
10mM EDTA	22	0

Mononuclear cells from human peripheral blood were incubated with
antibody-coated target cells for 4 hr in Ca^{2+}- and Mg^{2+}-free Dulbecco's
medium plus 1% fetal calf serum. Divalent cations or EDTA were added at
time 0. Effector cell:target cell ratio was 2:1 for CRBC and 5:1 for
Chang cells.

present in the incubation mixture (75). The apparent cation-independent nature of CRBC lysis is not likely due to contaminating Ca^{2+} or Mg^{2+} released from cell surfaces or exchanged from intracellular pools since the addition of EDTA at time 0 did not alter the cation-independent portion of lysis (Table II).

Mitogen-Induced Cytotoxicity. MIC for CRBC is predominantly cation-independent whereas MIC of nucleated (non-erythrocyte) target cells is cation-(Ca^{2+}) dependent (81). We have previously shown that the cells responsible for ADC and MIC of CRBC are likely the same and are non-T-cell in origin (72).

The Ca^{2+}-independent lysis of erythrocyte targets in both ADC and MIC may suggest that (a) effector cells for erythrocyte targets represent a different population of effector cells from those leading to lysis of non-erythrocyte targets and therefore lysis of erythrocytes is not mediated through a conventional extracellular Ca^{2+}-dependent stimulus secretion mechanism (82),(b) these different effector cells may not be triggered in the same way, that is, the erythrocyte killer cells are not dependent on a Ca^{2+} gate or,(c) similar cells which are not dependent on an increased uptake of Ca^{2+} for triggering may effect the lysis of both forms of target cells, but the requirements for Ca^{2+} in the cytolytic reaction (e.g. a secreted molecule or involving membrane enzyme activation) are different. Thus, Ca^{2+} is not required for erythrocyte target cell lysis, but is for nucleated, non-erythrocyte target cell lysis. In this way Ca^{2+} may have a role as an enzyme cofactor in the destruction of certain target cells and not others.

Cell Mediated Cytotoxicity. In virtually all systems using allosensitized effector cells, a clear cut and absolute requirement for Ca^{2+} has been demonstrated in the post-recognition phase (73-75,83). Earlier studies suggested that addition of Mg^{2+} could restore cytotoxic activity to EDTA treated cultures (84). However, since these studies were not carried out in cation-free media, the addition of Mg^{2+} may merely have caused dissociation of chelated Ca^{2+}, thus accounting for the improved effector cell activity. In several studies the addition of Mg^{2+} to Ca^{2+}-supplemented incubation mixtures further enhances the degree of cytotoxicity.

ROLE OF CALCIUM UPTAKE IN CYTOTOXICITY

A requirement for Ca^{2+} is a major feature of many induction systems including lymphocyte transformation and both immunological and non-immunological secretion mechanisms. Calcium influx may also influence cytotoxic systems as discussed above. Beyond the recognition phase in which adhesions

between target cells and effector cells are established, an absolute requirement for Ca^{2+} has been documented in all systems where non-erythrocyte target
cells have been assayed. Calcium uptake studies are now required to determine if the cytolytic event is dependent on a Ca^{2+} "gating" phenomenon as
has been suggested for lectin-induced proliferation (24,25). We have
attempted to study this possibility in an indirect fashion using the Ca^{2+}
ionophore A23187. Addition of this drug prior to the addition of target
cells (Table III) or to an ongoing reaction failed to enhance or induce ADC
for CRBC or Chang using human peripheral blood mononuclear cells (or macrophages) as effector cells. Indeed, a reduction in cytotoxicity was seen at
higher concentrations of the drug.

There is an obligate requirement for the presence of extracellular
Ca^{2+} for both lymphocyte activation by mitogens and for all three model
systems for cytotoxicity with non-erythrocyte target cells. The increase
in Ca^{2+} uptake may be incidental to rather than critical for cell activation and cytotoxic killing. It is possible that the mitogen-induced Ca^{2+}
uptake and Ca^{2+} gating that we and others have demonstrated cannot be a
sufficient or irreversible signal leading to DNA synthesis. We have

Table III

Effect of calcium ionophore A23187 on ADC

Ionophore	% specific ^{51}Cr-release CRBC	CHANG
–	26.8	16.5
10^{-7}	0	0
5×10^{-8}	5.7	0
10^{-8}	10.1	2.5
5×10^{-9}	13.0	6.6
10^{-9}	19.3	8.0
5×10^{-10}	29.6	14.1
10^{-10}	29.0	13.9

Mononuclear cells from human peripheral blood were incubated with
antibody-coated target cells for 4 hr in 1 ml Dulbecco's medium plus 10%
fetal calf serum. Dilutions of ionophore were prepared in DMSO and added
to effector cells 1-15 min before the addition of target cells. Control
tubes contained DMSO alone.

attempted to demonstrate the relevance of gated Ca^{2+} channels for mitogenesis.
To date, most attempts to demonstrate such a requirement have failed. Calcium gating has not been looked at in cytotoxic effector cell responses, and
further studies are necessary to determine if Ca^{2+} uptake, similar to that
observed for mitogen-induced lymphocyte activation, is also observed
following activation of the cells in the various cytotoxic model systems.
At the moment, the early sequence of events following mitogen binding to
lymphocyte surface receptors, which leads to DNA synthesis and the sequence
of events involved in cytotoxic effector cell responses following binding
of Fc receptors, lectin or specific antigen are still a mystery, and the
role(s) of Ca^{2+} and other cations in these events remain to be determined.

Acknowledgements

The authors would like to acknowledge the financial support
provided by the Medical Research Council of Canada.

References

1.   Raff, M.C., in Cell Surfaces and Malignancy, Fogarty International
     Proceedings No. 24, edited by P.T. Mora, p.159, DHEW Publications, 1974.

2.   Ada, G.L., Transplant. Rev., 5:125, 1970.

3.   Greaves, M. and Janossy, G., Transplant. Rev., 11:87, 1972.

4.   Greaves, M.F. and Bauminger, S., Nature New Biol., 235:67, 1972.

5.   Rosenthal, A.S., Blake, J.T. and Lipsky, P.E., J. Immunol., 115:1135,
     1975.

6.   Feldmann, M., Basten, A., Boylston, A., Erb, P., Gorczynski, R., Greaves,
     M., Hogg, N., Kilburn, D., Kontiainen, S., Parker, D., Pepys, M. and
     Schrader, J., in Progress in Immunology II, edited by L. Brent and J.
     Holborow, Vol.3, p.65, American Elsevier, New York, 1974.

7.   Taylor, R.B., Duffus, W.P.H., Raff, M.C. and De Petris, S., Nature New
     Biol., 233:225, 1971.

8.   Loor, F., Forni, L. and Pernis, B., Eur. J. Immunol., 2:203, 1972.

9.   Currie, G.A., Nature, 216:694, 1967.

10. Ferber, E., Reilly, C.E., Pasquale, G. de and Resch, K. in Lymphocyte
     Recognition and Effector Mechanisms, p.529, Academic Press, New York,
     1973.

11. Barnett, R.E., Scott, R.E., Furcht, L.T. and Kersey, J.H., Nature, 249:465, 1974.

12. Quastel, M.R., Dow, D.S. and Kaplan, J.G., in Proc. 5th Leucocyte Culture Conference, edited by J.E. Harris, p.97, Academic Press, New York, 1970.

13. Averdunk, R., Hoppe-Seyler's Z. Physiol. Chem., 353:79, 1972.

14. Wright, P., Quastel, M.R. and Kaplan, J.G. in Proc.7th Leucocyte Culture Conference, edited by F. Daguillard, p.87, Academic Press, New York, 1973.

15. Averdunk, R. and Lauf, P.K., Exptl. Cell Res., 93:331, 1975.

16. Kaplan, J.G. and Quastel, M.R., in Immune Recognition, edited by A.S. Rosenthal, p.391, Academic Press, New York, 1975.

17. Kaplan, J.G., Quastel, M.R. and Dornand, J., in Biogenesis and Turn-over of Membrane Macromolecules, edited by J.S. Cook, p.207, Raven Press, New York, 1976.

18. Dent, P.B., J. Natl. Cancer Inst., 46:763, 1971.

19. Alford, R.H., J. Immunol., 104:698, 1970.

20. Allwood, G., Asherson, G.L., Davey, M.J. and Goodford, P.J., Immunology, 21:509, 1971.

21. Whitney, R.B. and Sutherland, R.M., Cell. Immunol., 5:137, 1972.

22. Whitney, R.B. and Sutherland, R.M., J. Cell. Physiol., 82:9, 1973.

23. Diamanstein, T. and Ulmer, A., Immunology, 28:121, 1975.

24. Freedman, M.H., Raff, M.C. and Gomperts, B., Nature, 255:378, 1975.

25. Raff, M.C., Freedman, M. and Gomperts, B., in Membrane Receptors of Lymphocytes, edited by M. Seligman, J.L. Preuh'homme and F.M. Kourilsky, p.393, American Elsevier, New York, 1975.

26. Cross, M.E. and Ord, M.G., Biochem. J., 124:241, 1971.

27. Peters, J.M. and Hausen, P., Eur. J. Biochem., 19:502, 1971.

28. Peters, J.M. and Hausen, P., Eur. J. Biochem., 19:509, 1971.

29. Mendelsohn, J., Skinner, S.A. and Kornfeld, S., J. Clin. Invest., 50:818, 1971.

30. van den Berg, K.J. and Betel, I., Exptl. Cell Res., 66:257, 1971.

31. Fisher, D.B. and Mueller, G.C., Biochim. Biophys. Acta, 248:434, 1971.

32. Resch, K., Gelfand, E.W., Hansen, K. and Ferber, E., Eur. J. Immunol., 2:598, 1972.

33. Maino, V.C., Hayman, M.J. and Crumpton, M.J., Biochem. J., 146:247, 1975.

34. Novogrodsky, A., Biochim. Biophys. Acta, 266:343, 1972.

35. Smith, J.W., Steiner, A.L., Newberry, W.M., Jr. and Parker, C.W., J. Clin. Invest., 50:432, 1971.

36. Parker, C.W., Sullivan, T.J. and Wedner, H.F., in Advances in Cyclic Nucleotide Research, edited by P. Greengard and A.G. Robinson, p.1, Raven Press, New York, 1974.

37. Wedner, H.J., Bloom, F.E. and Parker, C.W., in Immune Recognition, edited by A.S. Rosenthal, p.337, Academic Press, New York, 1975.

38. Hadden, J.W., Hadden, E.M., Haddox, M.K. and Goldberg, N.D., Proc. Nat. Acad. Sci. USA, 69:3024, 1972.

39. Hadden, J.W., van den Berg, K.J. and Resch, K., in Progress in Immunology II, edited by L. Brent and J. Holborow, Vol.2, p.320, American Elsevier, New York, 1974.

40. Hadden, J.W., Johnson, E.M., Hadden, E.M., Coffey, R.G. and Johnson, L.D. in Immune Recognition, edited by A.S. Rosenthal, p.359, Academic Press, New York, 1975.

41. Hadden, J.W., Hadden, E.M., Sadlik, J.R. and Coffey, R.G., Proc. Nat. Acad. Sci. USA, 73:1717, 1976.

42. Pogo, B.G.T, Allfrey, V.G. and Mirsky, A.E., Proc. Nat. Acad. Sci. USA, 55:805, 1966.

43. Desai, L.S. and Foley, G.E., Arch. Biochem. Biophys., 141:552, 1970.

44. Kleinsmith, L.J., Allfrey, V.G. and Mirsky, A.E., Science, 154:780, 1966.

45. Johnson, E.M., Karn, J. and Allfrey, V.G., J. Biol. Chem., 249:4990, 1974.

46. Johnson, L.D. and Hadden, J.W., Biochem. Biophys. Res. Comm., 66:1498, 1975.

47. Quastel, M.R. and Kaplan, J.G., Exptl. Cell Res., 63:230, 1970.

48. Quastel, M.R. and Kaplan, J.G., Exptl. Cell Res., 94:351, 1975.

49. Quastel, M.R. and Kaplan, J.G., Nature, 219:198, 1968.

50. Milthorp, P., Quastel, M.R. and Kaplan, J.G., Cell. Immunol., 14:128, 1974.

51. Borasio, P.G. and Vassale, M., Amer. J. Physiol., 226:1232, 1974.

52. Grimsham, C.M. and Barnett, R.E., Biochemistry, 12:2635, 1973.

53. Resch, K. and Ferber, E., Eur. J. Biochem., 27:153, 1972.

54. Stahl, W.L., Arch. Biochem. Biophys., 154:56, 1973.

55. Whitfield, J.F. and Youdale, T., Exptl. Cell Res., 43:602, 1968.

56. Whitney, R.B. and Sutherland, R.M. in Proc. 7th Leucocyte Culture Conference, edited by F. Daguillard, p.63, Academic Press, New York, 1973.

57. Whitney, R.B. and Sutherland, R.M., Biochim. Biophys. Acta, 298:790, 1973.

58. Maino, V.C., Green, N.M. and Crumpton, M.J. in Immune Recognition, edited by A.S. Rosenthal, p.417, Academic Press, New York, 1975.

59. Maino, V.C., Green, N.M. and Crumpton, M.J., Nature, 251:324, 1974.

60. Luckasen, J.R., White, J.G. and Kersey, J.H., Proc. Nat. Acad. Sci. USA, 71:5088, 1974.

61. Coffey, R.G., Hadden, E.M. and Hadden, J.W., Fed. Proc., 34:1029, 1975.

62. Crumpton, M.J., personal communication.

63. Rasmussen, H., Science, 170:404, 1970.

64. Freedman, M.H., unpublished results.

65. Freedman, M.H., Dubiski, S. and Cinader, B., unpublished results.

66. Whitfield, J.F., MacManus, J.P., Rixon, R.H., Boynton, A.L., Youdale, T. and Swierenga, S., In Vitro, 12:1, 1976.

67. Perlmann, P. and Holm, G., Adv. Immunol., 11:117, 1969.

68. MacLennan, I.C.M., Transplant. Rev., 13:67, 1972.

69. Perlmann, P., Perlmann, H. and Wigzell, H., Transplant. Rev., 13:91, 1972.

70. Cerottini, J.C. and Brunner, K.T., Adv. Immunol., 18:67, 1974.

71. Gelfand, E.W., Transplantation, 21:73, 1976.

72. Gelfand, E.W. in Immune Reactivity of Lymphocytes, edited by M. Feldman and A.Globerson, p.301, Plenum Press, New York, 1976.

73. Martz, E., J. Immunol., 115:261, 1975.

74. Plaut, M., Bubbers, J.E. and Henney, C.S., J. Immunol., 116:150, 1976.

75. Golstein, P. and Smith, E.T., Eur. J. Immunol., 6:31, 1976.

76. Stulting, P.D. and Berke, G., J. Exp. Med., 137:932, 1973.

77. Golstein, P. and Fewtrell, C., Nature, 255:491, 1975.

78. Golstein, P. and Gomperts, B.D., J. Immunol., 114:1264, 1975.

79. Gelfand, E.W., Resch, K. and Prester, M., Eur. J. Immunol., 2:419, 1972.

80. Scornik, J.C. and Cosenza, H., J. Immunol., 113:1527, 1974.

81. Gelfand, E.W., unpublished observations.

82. Ruben, R.P., Calcium and the Secretory Process, Plenum Press, New York, 1974.

83. MacDonald, H.R., Eur. J. Immunol. 5:251, 1975.

84. Mauel, J., Rudolf, H., Chapuis, B. and Brunner, K.T., Immunology, 18:517, 1970.

85. Abbreviations: T cells, thymus-dependent T lymphocytes; B cells, thymus-independent B lymphocytes; DNA, deoxyribonucleic acid; Con A, concanavalin A; PHA, phytohaemagglutinin; PWM, pokeweed mitogen; LPS, lipopolysaccharide cAMP, adenosine 3':5'-cyclic monophosphate; cGMP, guanosine 3':5'-cyclic monophosphate; EDTA, ethylenediaminetetraacetic acid; EGTA, ethyleneglycol-bis-(β-aminoethyl ether) N,N'-tetraacetic acid; ADC, antibody dependent cytotoxicity, CMC, cell mediated cytotoxicity; MIC, mitogen-induced cyto-toxicity; CRBC, chicken red blood cells; SRBC, sheep red blood cells; DMSO, dimethyl sulfoxide.

Chapter 7

CONTROL OF IMMUNE RESPONSES BY CYCLIC AMP AND
LYMPHOCYTES THAT ADHERE TO HISTAMINE COLUMNS

Y. Weinstein* and K.L. Melmon[‡]
*Department of Hormone Research, Weizmann
Institute of Science, Rehovot, Israel; ‡Division
of Clinical Pharmacology, Departments of Medicine and
Pharmacology, and the Cardiovascular Research Institute, University
of California, San Francisco, California 94143

Abstract

Mixed lymphocytes from human peripheral blood, murine spleens, lymph nodes or thymus glands have pharmacologically specific receptors for histamine, beta mimetic catecholamines and prostaglandins. When these cells are exposed to the panoply of drugs mentioned above, their intracellular cyclic AMP concentrations increase. The biologic consequences of such an increase were at first elusive. Now we know that the immune potential of some murine spleen cells may be modulated and the release of lysosomal enzymes and histamine from human leukocytes may be inhibited.

This paper concentrates on the effects that manipulation of cells with amine receptors has on their immune function. Recent studies have revealed that a subpopulation of splenic suppressor T cells responds to increases in its cyclic AMP content by reversing its suppressive effects on the humoral antibody response. When these T cells are removed from the murine cell population by their differential adherence to insolubilized conjugates of histamine with albumin, the remainder of the cells are more responsive to sheep cell antigen, as tested by transferring the spleen cells together with the antigen into lethally irradiated recipient animals. The suppressor T cells that adhere to the insolubilized conjugates of histamine-albumin (called histamine-rabbit serum albumin-Sepharose, or HRS) are Ia positive, they appear to have receptors for histamine, beta adrenergic amines and prostaglandins of the E series, and when stimulated by these agents their *in vivo* and *in vitro* suppressor actions are reversed. The reversal seems quantitatively dependent on cyclic AMP accumulation.

Receptors for the amines and prostaglandins are found on the T cell precursors of cell mediated immunity. They develop on some T effector cells in selected models of allogeneic target cell lysis. The receptors also appear to develop on selected B cells once these cells become committed to antibody production.

The distribution of receptors on all leukocytes has not been adequately studied nor has their full potential in the immune response been studied in detail.

Receptors for Low Molecular Weight Hormones on Leukocytes

Histamine, beta adrenergic agonists (e.g., epinephrine), and prostaglandins of the E series (PGE) elevate the cellular concentration of
cyclic AMP (cAMP) in human and murine leukocytes (1-3). Antihistamines
specifically and competitively block only the effect of histamine. Beta
blockers such as propranolol block only the effect of beta agonists (e.g.,
isoproterenol, norepinephrine, and epinephrine) (2). None of these
blockers affects the response of leukocytes to PGE (1). We have therefore
assumed that the response to histamine is independent of the response
of epinephrine or PGE, and that, from a pharmacologic point of view,
each class of drug mediates its effects via its specific membrane receptor.

A few years ago we insolubilized histamine that had been conjugated
to protein by covalently linking the conjugates to Sepharose beads. We
did this by first coupling the histamine to a carrier [rabbit serum albumin
(RSA)] or a synthetic peptide (poly Ala-Tyr) via water soluble carbodiimide.
We then coupled the hormone conjugate to cyanogen bromide activated Sepharose
beads (4,5).

The free amine of the histamine molecule is probably linked via an
amide bond to a carboxyl group of the protein, as shown in the following
scheme:

$$R-CH_2-NH_2 \quad + \quad HOOC-Protein \xrightarrow{ECDI} R-CH_2-NH-CO-Protein$$
$$\text{Histamine}$$

The insolubilized hormone conjugate bound certain human blood
leukocytes and mouse splenic leukocytes, but did not stimulate cyclic AMP
to accumulate in the cells.

The specificity of leukocyte binding by insolubilized hormone conjugates
is being explored. We demonstrated that the hormone (histamine or norepinephrine)
was critical for binding of the cells because beads alone, or beads linked to
the carrier protein or peptide, did not bind the cells. Binding of cells
to histamine-RSA-Sepharose (HRS) or to norepinephrine-carrier-Sepharose
was blocked or reversed by the appropriate antagonist (antihistamines for
the former and beta adrenergic blocking agents for the latter) (4,5).
Similar results were obtained when the rosette technique was used by
others, who coated sheep red blood cells (SRBC) with histamine conjugated
to RSA (HR). Incubation of lymphocytes with HR coated SRBC resulted in
the formation of rosettes. The formation of rosettes is inhibited by
antihistamines, as well as by high concentrations of histamine (6). Such
experiments are compatible with the idea that the binding phenomena observed

with the insoluble hormones were dependent at least in part on coaptation of the amine moiety of the conjugates with cell membrane receptors for the amines.

We filtered human leukocytes and mouse splenic lymphocytes over hormone conjugate columns and over control columns. We knew that cAMP synthesis is increased when mixed leukocyte populations are stimulated by free amines; we reasoned that the responsiveness of the cells that passed through the column (i.e., nonadherent cells) would be less than normal to the corresponding free amines if cell separation were at all dependent upon the receptor complement of the cells. We found that human peripheral or mouse splenic leukocytes that had passed through HRS columns did not respond well to histamine, epinephrine, or PGE_1, but did respond normally to cholera toxin (6-9; Fig. 1). The responses of cells that did not adhere to control columns such as albumin Sepharose (RS), polylysine Sepharose, or histidinemethylester-RSA-Sepharose were no different from the responses of unchromatographed control cells (8,9; Fig. 1). Thus, leukocytes apparently can be separated on the basis of their complement

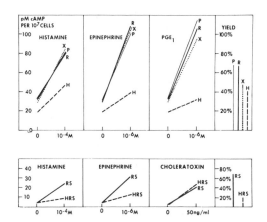

FIGURE 1. Two separate experiments are illustrated (upper and lower horizontal panels) and show baseline (0) and drug stimulated increases in cellular cAMP. Cells stimulated were those that did not adhere to RS (R), the o-methyl ester of histidine (x), or HRS (H). P designates cells that were not chromatographed. Note that the cells nonadherent to HRS responded significantly less to all agonists except cholera toxin. The column of o-methyl ester of histidine retained as many cells as HRS but did not concentrate cells that were responsive to vasoactive amines, prostaglandin E_1 or cholera toxin. The o-methyl ester of histidine is pharmacologically inactive.

of amine receptors. If insolubilized conjugates are made from inactive
amines (such as histidinemethylester) the binding of cells may depend on
the charge properties of the conjugates but the binding was not discriminating,
i.e., binding did not depend on cell receptors for amines and cells that
could not bind to such substances responded to free amines as if they had
not been chromatographed.

The mechanism by which the cells adhere to the columns of insolubilized
hormones is not known. Although binding of cells seems to correlate with
the membrane receptors for the amines that are used to construct the
conjugate, it is not known whether the cell actually is bound through its
receptor or through other sites on the membrane that are linked to, or
correlated with, the receptor. We can at least conclude from these
experiments that: 1) most of the cells separated by our procedure will
respond to all the agonists tested if they respond to one. Thus, the
histamine, beta catecholamines, and PGE receptors are likely to be
located on the same cell membrane. If different subpopulations of lymphocytes
possess different complements of receptors, they cannot be detected by the
methods that we are using; 2) cells that do not adhere to insolubilized
histamine or norepinephrine respond much less to PGE, histamine, and
beta catecholamines than do the nonchromatographed cells.

Immunologic Implications of Cell Binding by Hormone-conjugate Columns - The Humoral Antibody Response

HRS can effectively separate subpopulations of lymphocytes that have
different immunologic properties. Spleen cells from mice that had been
immunized with SRBC seven days before the experiment were chromatographed
over HRS columns and the nonadherent cells were injected, together with
SRBC, into lethally irradiated syngeneic recipients. These animals produced
a consistently higher number (2 to 12 fold) of direct and indirect plaque-
forming cells (PFC) than did animals that received unchromatographed cells
or cells that did not adhere to RS columns (10, 11). Similar results
obtained if the spleen cells were taken from unimmunized mice (10, 11).

To study the mechanism of the phenomena of enhancement of humoral
antibody responses, we first explored the possibility that any Sepharose
column could nonspecifically concentrate antigen-sensitive cells or remove
inhibitory cells from a mixed population. Spleen cells were incubated
and passed through RS or highly positively charged polylysine-Sepharose.
The nonadherent cells from such columns were transferred to irradiated

recipients. As shown in Table 1, chromatography of spleen cells over
RS or polylysine-Sepharose did not result in an enhanced immune response
to SRBC. Only when cells were passed through HRS did enhancement of the
humoral antibody response occur.

The patterns of humoral antibody response over time were similar for
direct and indirect PFC to SRBC. In these transfer experiments, cells that
were nonadherent to HRS or unfiltered control spleen cells were compared.
The results verified that the enhancement of humoral antibody response
could be detected for several days (11). The enhancement was not due to
an artifact of the timing of the assay resulting in a shift in the
peak response time of PFC by either group of spleen cells.

The cell type(s) responsible for or associated with these elevated
PFC responses appear(s) to be of thymic origin (11) because: 1) when
thymus cells were filtered over HRS, the nonadherent cells generated a
3.2 fold greater increase in PFC than an equal number of unseparated
thymocytes mixed with unseparated marrow cells; and 2) when bone marrow

TABLE 1. Direct PFC responses in irradiated mice injected with unfiltered
spleen cells (no. col.), spleen cells passed over RSA-Sepharose columns
(RSA-S), spleen cells passed over HRSA-Sepharose columns (HRSA-S) or
passed over polylysine Sepharose columns (polylysine-S). Splenic
leukocytes (22×10^6) were mixed in a plastic test tube with 0.6 ml
25% (weight/volume) conjugated Sepharose. After 15 min incubation
at 37° C the mixture was transferred to a small column plugged with
sponge and washed with 2 ml medium or buffer. The nonadherent (passed)
cells were collected for injection. (BALB/c X C57BL/6)F_1 mice, exposed
to 750 rads of ^{60}Co gamma irradiation, were injected with 5 x 10^6 spleen
cells (mixed with 3×10^8 SRBC) from unimmunized syngeneic donors (Exp. 2),
or donors immunized 7 days before with SRBC (Exp. 1; data was taken from
refs. 10 and 11).

Exp. No.	1	2
No Col[1]	1060 + 204[2]	544 + 143
RSA-S	1310 + 224	N.T.[3]
HRSA-S	4830 + 873	1260 + 248
Polylysine-S	N.T.	555 + 117

[1] No Col - Not passed over column S

[2] Standard errors

[3] N.T. - Not tested

cells that were nonadherent to HRS were mixed with unchromatographed
thymocytes, no elevation of PFC response was seen.

Using a different approach, we were able to mimic the enhancement
of the humoral immune response seen by removing lymphocytes that adhere
to HRS. For these experiments, we used fluoresceinated-HR (HRF) (3 moles
histamine per 1 mole RSA) or fluoresceinated RSA (control) - RF). The
soluble fluoresceinated chemicals were incubated in the presence of 10%
agammaglobulin horse serum with splenic leukocytes. Then the cells were
subject to a cell sorter that segregated them on the basis of the intensity
of the fluorescence that they emitted (8). Although we were not able to
distinguish by fluorescence microscopy which cells has been labeled with
the compounds, the sorter segregated those cells that accounted for the
20% of the total number of cells that had the most intense fluorescein
labeling from the 80% remaining. The latter nonlabeled cells were injected
with SRBC into lethally radiated syngeneic mice. Seven days later a
2 to 4 fold enhancement of the humoral antibody response was produced
by the cells that did not bind HRF. Other spleen cells were uniformly
labeled by RF and the enhancement was not seen when compared to transfer
experiments that use unsorted control cells or cells incubated with HRF or
RF, but that were not sorted. Further investigation is needed to
establish whether cells that adhere to HRS (columns) are identical to
the cells that adhered to HRF.

When antibodies are produced in an *in vitro* system (e.g., Mishel-
Dutton) (12), cells that do not adhere to HRS produce more PFC to
SRBC than do control cells that passed through RS columns or unchromatographed
spleen cells. However, the same number of PFC were found in all three
groups of cells when lipopolysaccharide (a T independent antigen) was
used instead of SRBC (a T dependent antigen) (8, 13). The most likely
explanation for these contrasting phenomena is that the HRS columns attract
and remove a subpopulation of T cells that can suppress proliferation of
antibody-forming cells.

The preliminary data implicating adherence of T suppressor cells to
HRS were extended. We decided to set up a model that would test the effects
of cell chromatography on immunotolerance. We reasoned that if tolerance
or immunosuppression results from active suppression of B cells by T cells,
rather than elimination of clones (14), then immune responses from
tolerant animals might be restored if HRS could be used to remove
suppressor T cells from the environment of the antigen. To test this
hypothesis, mice were made tolerant to SRBC by repeated injection of

SRBC lysates (15). When spleen cells from the mice that had been made
tolerant to the lysates were transferred to lethally irradiated syngeneic
recipients, their IgG response to SRBC remained suppressed (16, 17). When
the spleen cells from the mice that we had made tolerant to SRBC were
chromatographed over HRS columns and the nonadherent cells were transferred,
the IgG response was restored, i.e., the tolerance was broken (Table 2)
(16, 17). Some of the cells that had been bound to the HRS column were
recovered by mechanically shaking the beads. When these previously
adherent cells were added to the nonadherent cells from an HRS column
(in the ratio of 1:5), the IgG response was again suppressed. The suppressor
activity of the cells adherent to HRS was abolished after they had been
treated with specific anti θ anti serum plus complement (8, 17).

Once "suppressor" T cell activity was found in mice that responded
to SRBC, the response to other antigens was investigated. After being
immunized with S III pneumococcal polysaccharide, BALB/c mice produce
only direct (IgM) plaque forming cells. However, when spleen cells from
the immunized mice were passed through HRS columns, and the nonadherent cells

TABLE 2. The effect of cell chromatography with control RS or HRS columns
and elimination of θ sensitive cells on suppressor capacity of spleen cells
from mice tolerant to SRBC.

Group	Manipulation of spleen cells from tolerant mice before transfer to radiated recipient	Reciprocal hemagglutinin Titre at day 7[1]	
		$-2ME^2$ (IgM)	$+2ME$ (IgG)
1	Non-chromatographed	64	4
2	Depleted of cells adherent to albumin Sepharose	64	16
3	Depleted of cells adherent to HRS	512	128
4	Depleted of cells adherent to HRS plus 2.5 x 10^6 cells eluted from HRS	512	1
5	Group 4, but eluted cells treated with normal mouse serum plus complement	256	8
6	Group 4, but eluted cells treated with anti θ serum plus complement	256	128

[1] Sera pooled from 5 mice in each group; 10^7 splenic cells were transferred
unless otherwise indicated.

[2] ME - mercaptoethanol

were transferred into irradiated recipients that were then challanged with
S III pneumococcal polysaccharide, both direct (IgM) and indirect (IgG) PFC
were found (18). Again, some cells that had adhered to the HRS columns
were recovered. When a mixture of the previously adherent cells plus the
cells that passed through an HRS column was transferred, the formation of
indirect PFC (IgG) was abolished. The suppressive activity of the cells
eluted from HRS columns could be abolished by incubating them with: 1) specific
anti θ serum (anti-Thy-1.2) plus complement; or 2) specific anti-Ia (A.TH
anti-A.TL) plus complement. Normal mouse serum and complement had no effect.
These data further demonstrate that the cells that adhered to the HRS
column and that suppressed the IgG response to S III are Ia positive T
cells.

When B cells were studied, we found that HRS columns could bind
substantial percentages of splenic PFC to SRBC (7). This phenomena is
strain specific (7). No other specific subpopulation of B cells including
the precursor of the cells that produce antibody to SRBC bound to the
HRS columns.

Cell Mediated Immune Response

HRS columns have been used to separate precursors of cytotoxic cells
from other cells by using *in vitro* allogeneic systems (19, 20). Spleen cells
from BALB/c mice have been cultured with stimulatory or target cells from
irradiated C57BL/6 mice. Five days later the T cell mediated cytotoxic
activity of the effector cells was assessed by their ability to release
^{51}CR from target cells (20). Fractionation of the spleen cells over the
HRS columns before their initial exposure to target cells depleted or
reduced the cytotoxic potential of the spleen cell population. That
is, nonadherent cells eventually did not produce a substantive number
of effector cells. All the cytotoxic potential of the cell mixture was
recovered by eluting cells that had adhered to the HRS columns. In contrast,
the effector cell responsiveness was unaltered after cell chromatography
over RS columns.

HRS columns also can bind "initiator" T lymphocytes (21). In these
experiments, mouse lymphocytes were sensitized for a few hours *in vitro*
against immunogenic (allogeneic or syngeneic) fibroblasts. The lymphocytes
were then injected into the hind foot pad of a normal syngeneic animal.
The sensitized lymphocytes (that could be transferred after they had been
irradiated) actually recruited host lymphocytes that became killer cells

(21). The appearance of the effective or recruited phase is assayed by the accumulation of lymphocytes in the popliteal lymph node that drains the injected foot pad. The node enlarged as recruited lymphocytes accumulated in it. If the lymphocytes were chromatographed over HRS before the sensitization phase, the nonadherent cells failed to recruit host lymphocytes. The recruited, or killer, cells of the host did not adhere to HRS.

The Mechanism of Suppressor Cell Activity

We do not know how the suppressor cells involved in the various settings described above function. Several mechanisms have been postulated. They may secrete soluble factors such as lymphokines that contribute to their inhibitory action; they may inhibit the response to an antigen somehow by coming into physical contact with the antigen sensitive cells; or they may inhibit the response to an antigen either by competing for the antigen itself or by occupying the environment in which the antigen-sensitive cells develop.

We decided to explore the possibility that suppressor cells secrete an inhibitory lymphokine - interferon. Interferon (immunointerferon) is secreted from T lymphocytes that respond to antigens or mitogens (22-24). The interferon that inhibits viral replication also inhibits lymphocyte proliferation (25, 26). Spleen cells that are nonadherent to HRS responded better to PHA (as measured by [^{3}H]thymidine incorporation after 3 days in culture) than control cells that were not chromatographed or that were nonadherent to RS columns. Cells nonadherent to HRS produced the least amount of interferon of the 3 groups of cells tested (13). We conclude from these and other (8, 25, 26) experiments that: 1) cells that adhere to HRS columns secrete (or control the secretion of) immunointerferon; 2) immunointerferon inhibits lymphocyte proliferation; and 3) the type of regulation via this soluble mediator seems to be nonspecific, unless the interferon is secreted while the antigen-sensitive cells are responding to the antigen.

These experiments have in no way proven that the suppressor cells and interferon producing cells are the same. Even if they were identical, our data would not indicate that interferon is the only lymphokine they produce or that release of lymphokine is a method by which a portion of their suppressor action is achieved. The data indicate that the cells under consideration are likely to have in common at least their T cell origin and receptors for histamine and other amines that stimulate cellular adenylate cyclase.

The Role of cAMP in Modulating Immune Responses

If the amine receptors had a potential role in modulating the immune-related functions of the T cell, then it might be reasonable to assess such a role by stimulating unchromatographed murine spleen cells to determine whether the effects of such pharmacologic manipulation could mimic the results of cell separation via insolubilized amines. As was mentioned above, human leukocytes and mouse splenic lymphocytes that have passed through an HRS column (nonadherent cells) do not respond well to histamine, beta adrenergic catecholamines, or PGE (8, 9; Fig. 1). Logically, and in fact, cells adherent to HRS did respond to those drugs.

Recently, a number of investigators have found that accumulation of intracellular cAMP inhibits a number of immune-related functions of leukocytes (Table 3). An intracellular rise in cAMP levels or the external addition of DB-cyclic AMP blocks the response of lymphocytes to mitogens such as Con A and PHA (27-29). The same rises inhibit the cytolytic activity of sensitized lymphocytes (30, 31), as well as the production or release of the lymphokine, interferon (31). Elevated cAMP concentration in basophils inhibits the release of histamine from human basophils (31-33) and the synthesis or release of antibodies to SRBC from splenic B lymphocytes (3).

If the suppressor cells that adhere to the HRS columns are inhibited by drugs that cause accumulation of cAMP or that mimic or potentiate the effects of cAMP, then such drugs should enhance the immune responses controlled by these particular cells. The results from a number of

TABLE 3.   Inhibition of leukocyte functions by cAMP.

Function	Leukocyte type	Reference
1. Response to mitogens	Human and mouse lymphocytes	27-29
2. Immune cytolysis	Mouse and rat lymphocytes	30, 31
3. Interferon production	Human lymphocytes	31
4. Plaque formation to SRBC	Mouse lymphocytes	31, 33
5. Histamine release	Human basophil	3

models sustain the hypothesis as long as accumulation of cAMP is
brief. Long exposure of cells to cAMP blocks their DNA synthesis, as
well as cell division (27, 28). Prolonged exposure to high concentrations
of cAMP inhibits cell division and will block the cellular response to
an antigen.

However, in various *in vivo* and *in vitro* models of immunological
systems, short exposure of mixed cells to drugs that increase their
cAMP accumulation resulted either in an enhanced immune response, or
reversion of *in vivo* immune suppression. When cholera toxin (an agent that
elevates intracellular cAMP) or DB-cyclic AMP were injected into mice
that were simultaneously challanged with SRBC, the humoral antibody
response of both plaque forming cells (IgG and IgM) was enhanced over
control (8). The enhancement occurred only when the pharmacologic agent
was injected at the same time as the antigen. The cholera toxin had no
effect when injected before or after the SRBC were administered. Treatment
of a culture of spleen cells for the first 12 hours of culture with
DB-cyclic AMP or aminophylline (a phosphodiesterase inhibitor that causes
elevation of intracellular cAMP) results, after a period of 4 to 5 days,
in 7 fold more PFC to the SRBC antigen than if the drugs were omitted
(34).

When cultures of lymph node cells taken from Kehole limpet hemocyanin
primed mice were treated with DB-cyclic AMP or cholera toxin for the first
24 hours of culture, the antibody synthesis was enhanced 2 to 3 fold within
5 days (35).

In a different experimental setting, drugs that caused high intracellular
concentrations of cAMP restored immune response to a specific synthetic
antigen (36). The experiment was as follows: Mouse spleen cells were
exposed for the first time *in vitro* to the multichain synthetic polypeptide
antigen Poly(L-Tyr, L-Glu)-Poly L Pro-Poly Lys [(T,G)-Pro-L)]. The cells
were washed after 15 min and subsequently were injected into irradiated
syngeneic recipients. A limiting dilution number of spleen cells was
transferred so that only 65% of the irradiated mice responded. The cells
that had been exposed to soluble antigen before they were transferred
had suppressed immunological capacity and only 32% of the mice responded
to the antigen. Recipients that received cells that were not preincubated
with antigen developed normal humoral antibody responses to (T,G)-Pro-L.
However, if the lymphocytes were exposed *in vitro* simultaneously to (T,G)-Pro-L
and to either DB-cyclic AMP, histamine, cholera toxin, or PGE (in doses that
would increase their intracellular cAMP content) the antigen induced immune
suppression was reversed to normal (Table 4).

TABLE 4. Effect of histamine and DB-cyclic AMP on antigen-induced immune
suppression[1] (17).

Drug	Percent of positive responding sera[2]		Percent of control of cAMP accumulation[3]
Control	64 + 7[4]	(52)[5]	100
(T,G)Pro-L	32 + 7	(41)	104
Histamine 10^{-4} M	58 + 14	(12)	190
Histamine 10^{-4} M + (T,G)Pro-L	67 + 16	(9)	N.T.[6]
DB-cyclic AMP (10^{-4} M)	60 + 15	(10)	N.T.
DB-cyclic AMP (10^{-4} M) + (T,G)Pro-L	59 + 14	(12)	N.T.

[1] Spleen cells were injected into irradiated (750 rad) syngeneic recipients
immediately following *in vitro* incubation of cell suspensions. Recipients
were immunized intraperitoneally with 10 μg (T,G)Pro-L 24 hours after cell
transfer and bled 2 weeks later.

[2] Donor-derived responses showing anti-(T,G)Pro-L hemagglutination titers at
recipient sera dilutions greater than 1:4 (Mozes, E., Shearer, G.M., and
Sela, M., J. Exp. Med. 132:613, 1970).

[3] Percent of control (14 picomoles cAMP per 10^7 cells). On different days
basal cAMP varied from 10-15 picomoles per 10^7 cells.

[4] + Standard error of percent.

[5] ( ) indicates total number of recipients tested.

[6] N.T. - Not tested.

Some Limitations in the Use of HRS Columns

Binding of leukocytes to HRS columns appears to require the membrane
receptor for histamine but it is also likely to require as yet undefined
non-specific membrane sites. Use of the columns can result in separation
of cells that have receptors for the amine corresponding to the amine
used in the insolubilized conjugate but: a) the methods for quantitative
separation of cells with receptors from those without receptors has not
been determined; b) non-specific trapping of cells by the amine coated
Sepharose beads is always a factor that must enter into the interpretation
of experimental results; c) some species of mice have histamine responsive
B cells that are not adherent to HRS columns (7); and d) some T effector

cells have their cytolytic activity modulated by free histamine but are
not adherent to HRS (37). Thus, although the insolubilized conjugates of
amines may be useful in a variety of settings to separate subpopulations
of T or B lymphocytes with receptors for amines from those without amine
receptors, the extent of separation must always be measured for each system
by assessing the responsiveness of nonadherent cells to free amines.
Furthermore, the pertubation of studies on immunological models produced
by chromatography of participating cells over amine-Sepharose columns
cannot *a priori* be attributed to cell separation that is based on the
amine receptor complement of cells. Only after the effectiveness of the
separation via amine receptors is tested by pharmacologic stimulation
of nonadherent cells can such an assumption be considered.

The full potential of the usefulness of the HRS or analogous columns
in immunology has not been determined. Neither has the usefullness of
these columns in sequence with other types of cell chromatography been
assessed. Likewise, the mechanism by which the conjugates bind cells has
not been elucidated beyond the requirement of the membrane receptor for
the amine used to construct the conjugate.

Conclusions

In several models of immunological systems, the removal from mixed
murine spleen cells of those that are bound by HRS produces effects
similar to those seen when nonchromatographed cells are treated with
histamine and other drugs that elevate intracellular cAMP. Cell
populations from which HRS adherent cells were removed responded less,
or did not respond at all to amines that elevate cAMP (7-9). cAMP inhibits
a variety of immunologic functions of T lymphocytes (27-33) and it is
reasonable to speculate that the T suppressor cells that adhere to HRS
are sensitive to histamine, PGE, and other agents that activate adenylate
cyclase. Intracellular elevation of cAMP blocks the inhibitory effects
of T suppressor cells and enhances the immune response of other cells that
they ordinarily suppress. The mode of action of these suppressor cells
is not clear.

The data leading to the above conclusions also allows the prediction
that the administration of certain antigens, such as SRBC, will trigger the
humoral immune system by transiently elevating intracellular cAMP of T
suppressor cells. The cAMP likely inhibits the suppressor cell
function.

TABLE 5. Properties of leukocytes adhering to HRS

Leukocyte type	Function
Mouse thymocytes or splenic T lymphocyte	Suppress the response to SRBC
Mouse splenic T, Ia positive splenic	Suppress the IgG response to S III
Mouse splenic T lymphocyte	Precursor of cytotoxic killer cell
Mouse splenic T (or B) lymphocyte	Production of immune interferon
Mouse splenic T lymphocyte	Initiator T lymphocytes
Mouse splenic B lymphocyte	Antibody's secretion to SRBC (only in certain strains)
Human peripheral and mouse splenic	Responsiveness to histamine, epinephrine, and catecholamine

This prediction apparently is true. Indeed, when heterologous red blood cells were injected into mice, cAMP levels in splenic lymphocyte cells were raised (38). This elevation of cAMP in spleen cells probably results from the secretion of prostaglandins (39, 40).

Table 5 summarizes the properties of leukocytes that adhere to HRS columns. There is no reason as yet to assume that the T suppressor cells are identical to the precursors of the T killer cells or initiator T lymphocytes.

However, we do suggest that if the T suppressor cells are precursors of cellular immunity, then the mechanism of suppression of the humoral immune response might occur when the cell mediated response is enhanced. Such a reciprocal relationship between cell mediated vs. humorally mediated immune events is currently under investigation.

Acknowledgments

This work was supported in part by National Institutes of Health grants GM-16496, HL-06285, and GM-00001.

References

1. Bourne, H.R., Lehrer, R.I., Cline, M.J. and Melmon, K.L., J. Clin. Invest., 50:920, 1971.
2. Bourne, H.R. and Melmon, K.L., J. Pharmacol. Exp. Ther., 178:1, 1972.
3. Melmon, K.L., Bourne, H.R., Weinstein, Y., Shearer, G.M., Kram, J. and Bauminger, S., J. Clin. Invest., 53:13, 1974.
4. Melmon, K.L., Bourne, H.R., Weinstein, Y. and Sela, M., Science, 177:707, 1972.
5. Weinstein, Y., Melmon, K.L., Bourne, H.R. and Sela, M., J. Clin. Invest., 52:1349, 1973.
6. Kedar, E. and Bonavida, B., J. Immunol., 113:1544, 1974.
7. Melmon, K.L., Weinstein, Y., Shearer, G.M., Bourne, H.R. and Bauminger, S., J. Clin. Invest., 53:22, 1974.
8. Melmon, K.L., Weinstein, Y., Bourne, H.R., Shearer, G.M., Poon, T., Krasny, L. and Segal, S., in Proceedings of the Ninth Miles International Symposium, in press, 1976.
9. Poon, T., Weinstein, Y., Melmon, K.L. and Bourne, H.R., in preparation.
10. Shearer, G.M., Melmon, K.L., Weinstein, Y. and Sela, M., J. Exp. Med., 136:1302, 1972.
11. Shearer, G.M., Weinstein, Y. and Melmon, K.L., J. Immunol., 113:597, 1974.
12. Mishell, R.I. and Dutton, R.W., J. Exp. Med., 126:423, 1967.
13. Brodeur, B.R., Weinstein, Y., Melmon, K.L. and Merigan, T.C., submitted for publication, J. Immunol., 1976.
14. Dresser, D.W. and Mitchison, N.A., Immunology, 8:145, 1968.
15. Anderson, T.M., Roethle, J. and Auerbach, R., J. Exp. Med., 136:166, 1972.
16. Segal, S., Weinstein, Y., Melmon, K.L. and McDevitt, H.O., Fed. Proc., 33:723, 1974.
17. Segal, S., Weinstein, Y., Melmon, K.L. and McDevitt, H.O., in preparation, 1976.
18. Hammerling, G.J., Black, S.J., Segal, S. and Eichmann, K., in 10th Leukocyte Culture Conference, Amsterdam, 1975.
19. Simpson, E., Shearer, G.M., Weinstein, Y. and Melmon, K.L., Fed. Proc., 32:887, 1973.
20. Shearer, G.M., Simpson, E., Weinstein, Y. and Melmon, K.L., J. Immunol., in press, 1976.

21. Cohen, I.R. and Livnat, S., Transplantation Rev., in press, 1976.

22. Valle, M.J., Bobrove, A.M., Strober, S. and Merigan, T.C., J. Immunol.,
     114:435, 1975.

23. Stobo, J., Green, I., Jackson, L. and Baron, S., J. Immunol., 112:1589,
     1974.

24. Wheelock, E.F., Science, 149:310, 1965.

25. Lindahl-Magnusson, P., Leary, P. and Gresser, I., Nature (N.B.),
     237:120, 1972.

26. Gisler, R.H., Lindahl, P. and Gresser, I., J. Immunol., 113:438, 1974.

27. Smith, J.W., Steiner, A.L. and Parker, C.W., J. Clin. Invest., 50:442,
     1971.

28. Weinstein, Y., Segal, S. and Melmon, K.L., J. Immunol., 115:112, 1975.

29. Povogrodsky, A. and Katchalski, E., Biochem. Biophys. Acta, 215:291,
     1970.

30. Henney, C.S., Bourne, H.R. and Lichtenstein, L.M., J. Immunol., 108:1526,
     1972.

31. Bourne, H.R., Lichtenstein, L.M., Melmon, K.L., Henney, C.S., Weinstein, Y.
     and Shearer, G.M., Science, 184:19, 1974.

32. Bourne, H.R., Melmon, K.L. and Lichtenstein, L.M., Science, 173:743, 1971.

33. Lichtenstein, L.M., Bourne, H.R., Henney, C.S. and Greenough III, W.B.,
     J. Clin. Invest., 52:691, 1973.

34. Teh, H.S. and Paetkau, V., Nature, 250:505, 1974.

35. Cook, R.G., Stavitsky, A.B. and Shoenberg, M.D., J. Immunol., 114:426, 1975.

36. Mozes, E., Weinstein, Y., Bourne, H.R., Melmon, K.L. and Shearer, G.M.,
     Cell. Immunol., 11:57, 1974.

37. Plaut, M., Lichtenstein, L.M. and Henney, C.S., J. Clin. Invest., 55:856,
     1975.

38. Plescia, I.J., Yamamoto, I. and Shimamura, T., Proc. Natl. Acad. Sci. USA,
     72:888, 1975.

39. Yamamoto, I. and Webb, D.R., Proc. Natl. Acad. Sci. USA, 72:2320, 1975.

40. Osheroff, P.L., Webb, D.R. and Paulsrad, J., Biochem. Biophys. Res.
     Commun., 66:425, 1975.

Chapter 8

MEMBRANE RECEPTORS ON NEUTROPHILS

Peter M. Henson
Department of Immunopathology
Scripps Clinic and Research Foundation
La Jolla, California 92037

ABSTRACT

Neutrophils recognise humoral immunologic reactants through 'receptors' on their surface membrane. Most widely studied and probably of greatest biologic significance are the immunoglobulin (Fc), and complement (primarily C3b and C5a) receptors which enable the cell to react with, and be stimulated by, antigen-antibody, or antigen-antibody-complement complexes and their products. This interaction with the putative receptors and the consequent cell activation occurs most optimally on surfaces and plays a critical role in the mammalian host defense system.

Polymorphonuclear neutrophil leukocytes, or neutrophils, represent a key component of the first line defense system of the body. The neutrophils, in this regard, can be considered as an effector arm of the immune system. Specific antibody serves as the recognition unit and binds to the bacterium or foreign material. The neutrophil then recognises the bound antibody and disposes of the complex by a combination of phagocytosis, bacteriocidal activity and digestion. The complement system acts as an intermediate amplifying system by further coating the complex with materials capable of being recognised by the neutrophils. The chemotactic properties of other complement fragments also serve to attract neutrophils to the sites of complement activation.

131

This recognition of the presence of an activated immune system thus represents a critical aspect of the function of the neutrophil. Recent concepts of cell biology have emphasized the role of cell surface 'receptors' in such recognition phenomena. In fact, in the case of neutrophils, the presence of such cell membrane recognition units can be inferred from the early experiments of Levaditi (1) and Laveran and Mesnil (2) who showed the adherence of bacteria to the leukocytes of immune animals. These and other similar observations may represent one of the earliest observations of the functioning of specific cell surface receptors. Wright and Douglas (3) coined the term 'opsonin' to indicate the coating of a particle by serum factors which make it more readily ingested by phagocytes. A vast array of experiments since that time have been directed toward identification of these opsonic factors. The purpose of this discussion however, is not to review the opsonin literature but to more specifically concentrate on our present knowledge of neutrophil receptors.

Neutrophils belong to a family of cells with somewhat similar properties. We have loosely termed these immunological 'mediator' cells (4), primarily because of their common ability to release inflammatory and pharmacologic mediators to the external environment. The cell types include neutrophils, eosinophils, basophils, macrophages, monocytes, mast cells, platelets and probably lymphocytes. One of their characteristics is that they are all activated by the Fc region of one or more classes of immunoglobulin and by one or more fragments of the complement system. Another is their containment of mediators in intracytoplasmic granules. One might speculate on the evolutionary origin of these cells as a development from the primordial phagocytic macrophage-like cell and on their close association with the developing immune system. Since the emphasis of this series is on immunological receptors, the discussion will concentrate on receptors associated with the inflammatory, immunological and host/resis-

immune complexes (8) or antibody coated erythrocytes (5,6,7,9   ) would adhere to

human and animal neutrophils in a serum free medium. Other studies have demonstrated

the stimulation of a variety of cell functions by aggregated or particle-bound immuno-

globulin (e.g. 10-22 and see 4). Direct visualization of immunoglobulin binding to

neutrophils has been achieved (18) and binding of monomeric immunoglobulins to washed

neutrophils has also been measured (23,24) but the degree of binding is very low.

Neutrophils do not seem to bind 'cytophilic' antibody in the way that macrophages do.

They primarily react with complexed or aggregated immunoglobulin.

Specificity of neutrophil immunoglobulin receptors.

Adherence (rosette) experiments early showed that neutrophils exhibited

receptors for IgG but not for IgM (5,6,7, and 9). By studying the ability of aggre-

gated myeloma and macroglobulinemia proteins to stimulate secretion of lysosomal en-

zymes from neutrophils (19), the lack of receptors on human neutrophils for IgM was

also shown and was extended to IgE and IgD. Direct binding of immunoglobulins to

neutrophils confirmed this observation. In this same study (19) aggregated IgG of all

four subclasses stimulated neutrophil secretion although IgG1 and IgG3 were, on average,

slightly more efficient. These observations were recently confirmed using latex particles

coated with myeloma proteins (25). Messner and Jelinek (9) had suggested a preference

for binding of IgG1 and IgG3 immunoglobulins on the basis of inhibition of rosettes of

antibody-coated erythrocytes by myeloma proteins. However, they themselves pointed

out the possible dangers (see above) of this technique. Direct binding studies of non-

aggregated immunoglobulin however revealed significant binding of only IgG1 and

IgG3 subclasses (23). This discrepancy cannot be accounted for by the presence of con-

taminants in the IgG2 and IgG4 myeloma protein used for the stimulation study since the

same proteins were used for the binding experiments and because the level of contamination could not have been more than 5% and yet the threshold amount of aggregates for stimulation of neutrophils was approximately the same for the IgG1 and IgG2 myeloma protein (19). A major problem with all these studies lies in the use of myeloma proteins, which may aggregate differently from protein to protein. In addition, inhibition of erythrocyte adherence will critically depend on the density of immunoglobulin on the erythrocytes, and on the degree of mutual repulsion between the erythrocyte and neutrophil surfaces. Inhibition of the binding of immune aggregates by monomeric Ig, might be quantitatively very different.

A recent observation in two different laboratories (26,27) that macrophages may have different receptors for monomeric immunoglobulins and for aggregated (or complexed)immunoglobulins, may provide the explanation for the abovementioned discrepancies. Thus, if neutrophils also have these two types of receptors, the ability of all four subclasses of IgG to stimulate secretion when aggregated can be reconciled with our difficulties in inhibiting binding of these aggregates with monomeric immunoglobulin (19) and the inability of monomeric IgG2 and IgG4 myeloma proteins to bind to the neutrophils (23).

Direct binding of IgA myeloma proteins to the surface of neutrophils was suggested (23) albeit to a lesser degree than was shown for IgG. When aggregated by bisdiazotized benzidine however, IgA1 and IgA2 proteins proved to be potent activators of neutrophil secretion, and the aggregates were shown to be phagocytosed by the cells (19). This observation provided one of the first examples of possible effector functions for IgA. However, it is open to criticism for the use of artifically aggregated myeloma proteins. As another approach a mouse myeloma IgA with anti-DNP activity was allowed to precipitate with BSA-DNP, and shown to stimulate neutrophils (Spiegelberg and Henson, unpublished observations). However, secretory IgA in complex with antigen has not been adequately studied. Moreover, IgA antibody was shown to be inactive with regard to opsonization of bacteria (28). Further work is required to answer the question raised by these differences although once again, some of them might be explained on the basis of special receptors for aggregated immunoglobulins, which might therefore have broader specificity. Preliminary but not highly satisfactory evidence for different receptors for the IgG and IgA molecules was presented by Lawrance, et al. (23) but again further experiments are needed to conclusively show this.

Structural requirements in the immunoglobulin for binding to neutrophil receptors.

a) Fc binding. The primary binding of immunoglobulin to neutrophil receptors appears to be by the Fc region. Thus aggregated $F(ab)_2$ myeloma proteins were unable to stimulate neutrophils (19). Aggregated Fc however readily induced secretion and Fc fragments inhibited erythrocyte–antibody rosettes (9). While the Fc region appears to have the predominant binding site, the possibility that other portions of the molecule, particularly if aggregated, might react with membrane constituents cannot be excluded. MacLennon (29) showed the Fc binding site for neutrophil receptors to be in the C3H domain and to be different from that required for binding of C1q. On the other hand, using a myeloma half molecule, Spiegelberg concluded that binding sites in both the Cγ2 and Cγ3 regions might be involved (30). However the possibility of special receptors for aggregated immunoglobulins requires a reopening of this question.

b) Requirements for aggregation or complexing of the Ig. Monomeric immunoglobulin molecules bind only weakly to neutrophils (23) when compared with immune complexes or aggregates. They are also incapable of stimulating neutrophil function (e.g. 12,15,19). The question may be asked however, does the enhanced binding of aggregates result from cooperative binding of the Fc receptors of a number of molecules which exponentially increases the binding of the whole complex or is there a configurational change in the Fc upon aggregation or upon binding to antigen which presents a new site, capable of being recognised by the neutrophil receptor? If the observation of specific receptors for aggregated and monomeric immunoglobulins can be extended to neutrophils, this question will be answered in favor of the second alternative. On the other hand data to support the cooperative binding hypothesis was presented by Phillips-Quagliata, et al. (8) using mono and multivalent haptens. Moreover Kazmierowski, et al. (31) showed that bacteria could not be opsonised with hybrid molecules composed of both antibody and normal immunoglobulin even though these bound to the particle. This suggests either the requirement for a configurational change or, as they postulated, the requirement for a structurally intact Fc region. The question remains unanswered.

Neutrophil requirements for binding of immunoglobulin: Nature of the Fc receptor.

        Since the Fc receptor(s) has not been isolated, its molecular nature is unknown. Neutrophil plasma membranes have been prepared which retain Ig binding activity (Henson unpublished observations, Hawkins, D., personal communication) but the further isolation of the receptor has not yet been accomplished. The binding of erythrocyte-antibody complexes to neutrophils is independent of cations (6) occurs at 4° (24) and is resistant to both trypsin and neuraminidase (6,7). In this last respect it is clearly differentiated from the C3b receptor which is trypsin sensitive. The immuno-globulin may react in part with lipid moieties within the membrane since aggregated IgG1 and IgG3 can bind to, and perturb, the lipids in artificial liposomes (32).

COMPLEMENT RECEPTORS

        The best studied of these is the receptor for the large fragment of C3, C3b. However interactions with C5, C4, C5a, C3a and C567 have all been suggested.

C3b receptor.

The complement derived opsonic activity in serum which had been studied by the early investigators in this field (33,34) was probably due to C3b. Nelson (35) first showed the adherence of what would now be designated PnAC1423, to guinea pig neutrophils, where PnA represents antibody coated pneumococci. Later studies by Henson (5,6) and Lay and Nussenzweig (7) extended these observations to different species, showed that EAC1423 was adherent but that EAC142 was not. At the same time; the elegant study of Gigli and Nelson (36) demonstrated that the binding of C3 to EAC142 was required for erythrophagocytosis by neutrophils (but see below).

a) <u>C3 structure involved in binding to neutrophil receptors.</u> Erythrocytes with bound C3 lose their ability to bind to neutrophils upon incubation with serum (36, 37). The C3b inactivator in serum presumably cleaves C3b to C3c and C3d. Since neutrophils were independently shown to have receptors for C3b but not for C3d (38,39), in contrast to B lymphocytes which have both, this observation is explained. The neutrophil receptor appears to be a true C3b receptor and consequently does not react readily with native C3 (40). This has obvious teleological advantages for a biological effector system which should only be triggered by an immune reaction. A series of studies by Stossel and his associates have questioned whether the opsonic fragment in some (or all) circumstances might not be some smaller fragment of C3 than C3b (41,43). Nevertheless, purified C3b was shown to prevent binding of EAC1423b to neutrophils (39). The implication is that C3b can bind to neutrophils but that other fragments of C3b, but not the smaller fragments C3d and C3c, may also have this property.

b) <u>Neutrophil requirements for C3b binding: The nature of the C3b receptor.</u> The C3b receptor on neutrophils has not been characterised. However, its sensitivity to treatment with trypsin (but not neuraminidase) suggests that it is part or in whole, proteinaceous in nature (6,7). Attempts to isolate the receptor have not yet been successful although neutrophil membrane fragments with receptor activity have been reported (44) and we have recently been able to solubilize these with Nonidet p40 and retain the ability to inhibit the induction of neutrophil secretion by particle bound C3b (Giclas and Henson, unpublished observations). C3b receptor activity from human erythrocytes was suggested to have a mucoid or mucopeptide portion (45), and Dierich, et al. (46) suggested C3b receptors from lymphoid cells to be complex lipoproteins. Neither of these suggestions are inconsistent with the observations on neutrophils. However, the functional or biochemical equivalence of C3b receptors on different cells has not been proven, and in fact there is evidence that an individual whose erythrocytes were immune adherence negative, had normal monocyte C3b receptors (47).

Binding of EAC1423 to rabbit but not human neutrophils required the presence of divalent cations (6,7). However it seems unlikely that cations are directly required for the interaction between C3b and receptor but may be involved in adequate exposure of the receptors on the cell membrane.

C4b receptors.

Cooper (48) demonstrated that human erythrocytes exhibited an immune adherence reaction with EAC14 if enough C4 was present and Ross and Polley (49,50) and Bokisch and Sobel (51) later extended this to neutrophils. However their studies concentrated for the most part on lymphocyte C4b receptors and indicated either a complete or partial identity of C4b and C3b receptors both by cocapping of C3b and C4b receptors (50) and by competitive inhibition with soluble C4b and C3b (50,51). However some lymphoblastoid lines reacted only with C3b not with C4b suggesting only partial identity (51). Such experiments have not been performed for neutrophils but until proven otherwise, the receptors might be assumed to be similar.

C5b receptors.

        No direct evidence of a receptor for C5b has been presented for neutrophils,

or for that matter for any cell, either in terms of adherence studies or direct activation.

Nevertheless, indirect evidence does exist for some participation of the C5 molecule in

stimulation of some forms of neutrophil phagocytosis and this may represent the presence

of receptors for some portion of the C5 molecule. Thus patients with abnormal C5 have

been described whose serum exhibits a defect in opsonization of yeast particles for phago-

cytosis by neutrophils (52,53). A similar C5-associated defect occurred in citrated

plasma stored for 36 hours. Since only certain bacteria or yeast exhibit this C5 depen-

dence (52) and erythrocytes apparently do not (36), some property of the particle sur-

face is also implicated.

        A possible explanation for these observations lies in the now clearly

demonstrated neutrophil stimulating properties of the small cleavage fragments of C5,

C5a (see below). Cleavage of C5 on the yeast particle would generate C5a which on

this surface may not be released into the supernate. This could then act as an enhancing

stimulus for phagocytosis. The abnormal C5 could therefore be defective in the small

C5a region of the molecule only, thus explaining the normality of the hemolytic func-

tion and of the immunochemical analyses (53). Storage in plasma could lead to

partial cleavage of the C5a end of the molecule with the same result. It could therefore

be predicted that the opsonically abnormal C5 would also demonstrate decreased chemo-

tactic factor generation (a property of the C5a fragment) and this was in fact, found to

be true (53).

Receptors for C5a, small C3 fragments and $\overline{C567}$ ("chemotactic factors").

        The activity of certain cleavage fragments of proteins from the complement

system in initiating neutrophil chemotaxis has been known for some time. This property has been attributed to low molecular weight fragments of C3 (54,55) and C5 (56,57) and to the trimolecular complex $\overline{C567}$ (58,59) and there is an extensive literature on this subject (see 4,60-64). Recent evidence, from a variety of sources, however, brings into question whether three separate mechanisms of activation are responsible. For example, direct chemotactic activity of the anaphylatoxic C3a fragment has now been clearly refuted (65,66) and while some other fragment of C3 may therefore supply the chemotactic activity, the question of minute contaminating amounts of C5 in the C3 preparations must be considered as a possible source of the chemotactic factor since C5a is active in very low concentrations. Chemotactic factors from C3 and C5 and $\overline{C567}$ all induce cross desensitization of neutrophils (4,67) indicating that they may either react with the same receptor (or pathway) in the neutrophil or may be due to the same factor. Thus the possibility that the $\overline{C567}$ retains some C5a which is itself the active fragment has not been completely eliminated. Evidence that while $\overline{C567}$ is active, $\overline{C56}$ is not (58,59,66) would imply, if this hypothesis is correct, that C7 somehow makes the C5a in the complex available to the neutrophil.

The alternative hypothesis is that there are three separate factors with similar properties and similar receptor sites on the neutrophil. This site could be a common 'chemotactic factor' receptor as suggested by Showell, et al. (68,69 and see below). The questions require structural studies with purified chemotactic factors in order to be fully answered.

Nevertheless, C5a is a potent activator of neutrophils (56,57,59,61,65, 70,71,72), suggesting the presence on their membrane surface of a receptor for this molecule. Recent evidence from our laboratory suggests that C5a from which the C-terminal arginine has been enzymatically removed and which no longer has anaphyla-toxic (muscle contracting and vascular permeability increasing) properties may also be chemotactic (65).

NONIMMUNOLOGIC NEUTROPHIL RECEPTORS

Receptors for the vast array of materials known to interact with and stimulate neutrophils have not been directly demonstrated. However some of these materials have possible relationships with the immunologic receptor and will be briefly mentioned.

A mixture of peptides and lipids in the supernatant from bacterial cultures (bacterial factors) are efficient initiators of neutrophil chemotaxis (see 73,74). Moreover, a variety of synthetic peptides, particularly those containing an N terminal formylated methionine, are also chemotactic for neutrophils (75). Most active to date was Form-Met-Leu-Phe (68). As proposed by Becker (69) similar peptides may represent the common active sequence in a number of denatured proteins which appear to attract neutrophils (see 64). However, since large concentrations are required the biological significance of this response to denatured proteins remains unknown. An involvement of hydrophobic residues has been suggested (64). Kallikrein (76,77) and plasminogen activator (78), both proteolytic enzymes of the coagulation-fibrinolytic-kinin forming system, have also been reported to have neutrophil activating properties. Inhibition studies indicate that the enzymatic property of the molecules is involved but recent evidence suggests that the two molecules may be identical (79).

Concanavalin A interacts with the surface of neutrophils and induces metabolic stimulation of the cells (80). A redistribution of the surface bound con A occurs (81,82) resulting in 'capping'. Of great potential interest however was the observation of Hawkins (83) that con A blocked the stimulation of neutrophil secretion by immune complexes. It also prevents phagocytosis (84).

FUNCTIONAL ASPECTS OF NEUTROPHIL RECEPTORS

A variety of neutrophil functions can be stimulated through the Fc and C3b receptors. As described above these include phagocytosis, chemotaxis and secretion of lysosomal enzymes. In addition however, these stimuli also enhance glucose oxidation and metabolism (85,21,70,77,80,86), initiate the release of prostaglandins (87) and H_2O_2 (88), the elaboration of singlet oxygen and superoxide anion (20,22) and the Fc receptor-dependent cytolysis of tumor cells (89) or complement-dependent killing of parasites (90) to which antibody is bound.

Surface requirements for neutrophil activation: Chemotaxis, phagocytosis and the stimulation of neutrophils by chemotactic factors on surfaces.

Neutrophils function optimally on surfaces. They crawl rather than swim, and phagocytosis, which is probably their key function, is a surface phenomenon. The requirement of relatively large immune aggregates or complexes to stimulate phagocytosis and neutrophil secretion (12,14,17,21) is in keeping with this concept. Neutrophils can react with, and be stimulated by, antibody and complement on membranes in vivo (91) and we, as well as others, have shown that immunoglobulin or complement on artificial surfaces are potent stimuli for neutrophil secretion in vitro (12,13,22,92). Soluble immunoglobulin aggregates were inactive with regard to stimulation of secretion but gained potent reactivity if bound to the surface of a Petri dish or micropore filter (21).

Recent observations on the effects of chemotactic factors on neutrophils have revealed that they can stimulate granule secretion and oxidative metabolism as well as chemotaxis (68,70,71,93). However, unless cytochalasin B is included, secretion requires the factors to be surface bound (68,70). From these experiments we have developed the hypothesis that if presented in the appropriate form, one stimulus e.g. C5a, a bacterial chemotactic factor can initiate most if not all neutrophil functions. Certainly the latter was shown to stimulate in one experiment chemotaxis, secretion, oxidative metabolism and if bound to a particle, phagocytosis (91) and C5a has separately been demonstrated to initiate most of the changes mentioned at the beginning of this section. It is not yet known if interaction of C5a with one type of receptor stimulates all these functions but this is certainly the simplest interpretation.

We are further hypothesising that chemotaxis is a surface phenomenon and our preliminary data would support such a concept. Gradients of C5a preparations can be bound to micropore filters, which after extensive washing, still induced neutrophil migration (Henson and Oades unpublished observations). If so, the mechanisms of chemotaxis and phagocytosis may be similar and may each involve the successive 'activation' of receptors on a pseudopod or membrane extension which gradually surrounds a particle or moves through the pores of a filter or over a surface. Such a process has been termed 'zippering' by Silverstein and his associates (94) who has independently suggested a similar mechanism for chemotaxis and phagocytosis.

Stimulation of Fc and C3b receptors in the initiation of phagocytosis: Does C3b stimulate phagocytosis?

Earlier data indicating that C3b on a particle can itself initiate its phagocytosis have recently come into question. Thus it has been suggested that C3b induces adherence of particles to the neutrophil (or macrophage) surface but it is immunoglobulin reacting with the Fc receptor which actually stimulates the engulfment. Thus erythrocytes coated with IgM antibody and complement (EAIgMC) bound to neutrophils (95) and macrophages (96,97) or monocytes (98) but were not engulfed. EAIgGC were bound and phagocytosed. The engulfment of the latter particle into neutrophils was prevented by Fc fragments (95) or into macrophages and neutrophils by Fab anti IgG (97,99).

On the other hand, mouse peritoneal exudate macrophages elicited by thioglycollate, unlike their unstimulated counterparts, are perfectly capable of engulfing EAIgMC (94,96). Moreover, from our studies with C3b stimulation of platelets and neutrophils, we know that we need to bind much more C3b to the particle to induce secretion than to induce adherence to the platelet or neutrophil. Very few (100s) of molecules are required for C3b induced adherence of erythrocytes to C3b receptor bearing cells (100). Consequently C3b-induced attachment, with IgG-induced engulfment, might easily result if C3b density was too low to induce phagocytosis by itself.

Despite these questions as to whether C3b can stimulate phagocytosis, it seems clear from the abovementioned experiments that IgG is an extremely potent stimulus for this function. C3b therefore probably serves most importantly as an inducer of adherence to neutrophils. Studies of binding affinity to isolated receptors and the ability of defined preparations of IgG and C3b on particles are necessary to finally answer this question.

REFERENCES

1. Levaditi, C., Annals Inst. Pasteur 15:894,1901.
2. Laveran, A. and Mesnil, F., Annals Inst. Pasteur 15:673, 1901.
3. Wright, A.E. and Douglas, S.R., Proc. R. Soc. Lond. 72:357, 1903.
4. Becker, E.L. and Henson, P.M., Advanc. Immunol. 17:93, 1973.
5. Henson, P.M., The Biological Activities of C3, Ph.D., Cambridge Univ., 1967.
6. Henson, P.M., Immunol. 16:107, 1969.
7. Lay, W.H. and Nussenzweig, V., J. Exp. Med. 128:991, 1968.
8. Phillips-Quagliata, J.M., Levine, B.B. and Uhr, J.W., Nature 222:1290, 1969.
9. Messner, R.P. and Jelinek, J., J. Clin. Invest. 49:2165, 1970.
10. Movat, H.Z., Macmorine, D.R.L. and Burke, J.S., Life Sci. 3:1025, 1964.
11. May, C.D., Levine, B.B. and Weissmann, G., Proc. Soc. Exp. Biol. Med. 133: 758, 1970.
12. Henson, P.M., J. Immunol. 107:1535, 1971.
13. Henson, P.M., J. Immunol. 107:1547, 1971.
14. Henson, P.M., J. Exp. Med. 134:114s, 1971.

15. Hawkins, D. and Peeters, S., Lab. Invest. 24:483, 1971.
16. Weissmann, G., Zurier, R.B., Spieler, P.J. and Goldstein, I.R., J. Exp. Med. 134:149s, 1971.
17. Taichman, N.S., Pruzanski, W. and Ranadive, N.S., Int. Arch. Allergy 43:182, 1972.
18. Sajnani, A.N., Ranadive, N.S. and Movat, H.Z., Life Sci. 14:2427, 1974.
19. Henson, P.M., Johnson, H.B. and Spiegelberg, H.L., J. Immunol. 109:1182, 1972.
20. Goldstein, I.R., Roos, D., Kaplan, H.B. and Weissmann, G., J. Clin. Invest. 56:1155, 1975.
21. Henson, P.M. and Oades, Z.G., J. Clin. Invest. 56:1053, 1975.
22. Johnston, R.B. and Lehmeyer, J.E., J. Clin. Invest. 57:836, 1976.
23. Lawrance, D.A., Weigle, W.O. and Spiegelberg, H.L., J. Clin. Invest. 55: 368, 1975.
24. Spiegelberg, H.L., Lawrance, D.A. and Henson, P.M., in Int'l. Symposium on the Immunoglobulin A System, edited by F.W. Kraus, pp. 67-74, Plenum Pub. Co, New York, 1974.
25. Lefell, M.S. and Spitznagel, J.K., Infect. and Immun. 12:813, 1975.
26. Walker, W.S., J. Immunol. 116:911, 1976.
27. Grey, H.M., Anderson, C.L., Heusser, C.H., Borthistle, B.K., Von Eschen, K. B. and Chiller, J.M., Proceedings Cold Spring Harbour Symposium, In press, 1976.
28. Lamm, M.E., Advan. Immunol. 22:223, 1976.
29. MacLennan, I.C.M., Connell, G.E. and Gotch, F.M., Immunol. 26:303, 1974.
30. Spiegelberg, H.L., J. Clin. Invest. 56:588, 1975.
31. Kazmierowski, J.A., Nisonoff, A., Quie, P.G. and Williams, R.C., J. Immunol. 106:605, 1971.
32. Weissmann, G., Brand, A. and Franklin, E.C., J. Clin. Invest. 53:536, 1974.
33. Ward, H.K. and Enders, J.F., J. Exp. Med. 57:527, 1933.
34. Ecker, E.E. and Lopez-Castro, G., J. Immunol. 55:169, 1947.
35. Nelson, D.S., Complement Ciba Foundation Symposium, edited by G.E.W. Wolstenholme and J. Knight, p. 222, Little, Brown and Co., Boston, 1964.
36. Gigli, I. and Nelson, R.A., Exp. Cell. Res. 51:45, 1968.
37. Logue, G.L., Rosse, W.F. and Adams, J.P., Clin. Immunol. and Immunopathol. 1:398, 1974.
38. Ross, G.D., Polley, M.J., Rabellino, E.M. and Grey, H.M., J. Exp. Med. 138:798, 1973.
39. Eden, A., Miller, G.W. and Nussenzweig, V., J. Clin. Invest. 52:3239, 1973.
40. Henson, P.M. in Biological Activities of Complement, edited by D.G. Ingram, p. 173, S. Karger, Basel, 1972.
41. Stossel, T.P., J. Cell. Biol. 58:346, 1973.
42. Stossel, T.P., Alper, C.A. and Rosen, F.S., J. Exp. Med. 137:690, 1973.
43. Stossel, T.P., Field, R.J., Gitlin, J.D., Alper, C.A. and Rosen, F.S., J. Exp. Med. 141:1329, 1975.
44. Henson, P.M., J. Immunol. 111:300, 1972.
45. Nelson, D.S. and Uhlenbruck, G., Vox. Sang. 12:43, 1967.
46. Dierich, M.P. and Reisfeld, R.A., J. Immunol. 114:1676, 1975.
47. Rothman, I.K., Gelfand, G.A., Fauci, A.S. and Frank, M.M., J. Immunol. 115:1312, 1975.
48. Cooper, N.R., Science 165:396, 1969.

49. Ross, G.D. and Polley, M.J. Fed. Proc. 33:759, 1974.
50. Ross, G.D. and Polley, M.J., J. Exp. Med. 141:1163, 1975.
51. Bokisch, V.A. and Sobel, A.T., J. Exp. Med. 140:1336, 1974.
52. Nilsson, U.R., Miller, M.E. and Wyman, S., J. Immunol. 112:1164, 1974.
53. Miller, M.E. and Nilsson, U.R., Clin. Immunol. and Immunopathol. 2:246, 1974.
54. Hill, J.H. and Ward, P.A., J. Exp. Med. 130:505, 1969.
55. Bokisch, V.A., Muller Eberhard, H.J. and Cochrane, C.G., J. Exp. Med. 129: 1109, 1969.
56. Ward, P.A. and Newman, L.J., J. Immunol. 102:93, 1969.
57. Shin, H.S., Snyderman, R., Friedman, E., Mellors, A. and Mayer, M.M., Science 162:361, 1968.
58. Ward, P.A., Cochrane, C.G. and Muller Eberhard, H.J., Immunol. 11:141, 1966.
59. Lachmann, P.J., Kay, A.B. and Thompson, R.A., Immunol. 19:895, 1970.
60. Ward, P.A., Am. J. Pathol. 77:519, 1974.
61. Till, G., Int. Arch. Allergy, In press, 1976.
62. Keller, H.U., Hess, M.W. and Cottier, H., Seminars in Hematology 12:47, 1975.
63. Cochrane, C.G. and Janoff, A., in The Inflammatory Process, edited by B.W. Zwaifach, R.T., McCluskey and L. Grant, p. 85, Academic Press, New York, 1974.
64. Wilkinson, P.C., Chemotaxis and Inflammation. Churchill Livingston, London, 1974.
65. Fernandez, H., Henson, P.M. and Hugli, T.E., J. Immunol. In press, 1976.
66. Conroy, M.C., Ozols, J. and Lepow, I.H., J. Immunol. 116:1682, 1976.
67. Ward, P.A., in Biological Activities of Complement, edited by D.Ingram, p. 108, Karger, Basel, 1972.
68. Showell, H.J., Freer, R.J., Zigmond, S.H., Schiffmann, E., Aswanikumar, S., Corcoran, B. and Becker, E.L., In press, 1976.
69. Becker, E.L., Fed. Proc. In press, 1976.
70. Becker, E.L., Henson, P.M., Showell, H.J. and Hsu, L.S., J. Immunol. 112: 2047, 1974.
71. Goldstein, I., Hoffstein, S., Gallin, J. and Weissmann, G., Proc. Nat. Acad. Sci. USA 70:2916, 1973.
72. Goldstein, I.M., Feit, F. and Weissmann, G., J. Immunol. 114:516, 1975.
73. Ward, P.A., Lepow, I. and Newman, L.J., Am. J. Pathol. 52:725, 1968.
74. Schiffman, E., Showell, H.J., Corcoran, B.A., Ward, P.A., Smith, E. and Becker, E.L., J. Immunol. 114:1831, 1975.
75. Schiffman, E., Corcoran, B.A. and Wahl, S.M., Proc. Nat. Acad. Sci. USA 72:1059, 1975.
76. Kaplan, A.P., Kay, A.B. and Austen, K.F., J. Exp. Med. 135:81, 1072.
77. Goetzl, E.J. and Austen, K.F., J. Clin. Invest. 53:591, 1974.
78. Kaplan, A.P., Goetzl, E.J. and Austen, K.F., J. Clin. Invest. 52:2591, 1973.
79. Vennerød, A.M. and Laake, K., Thromb. Res. 8:519, 1976.
80. Romeo, D., Zabucchi, G. and Rossi, F., Nature New Biol. 243:111, 1973.
81. Ryan, G.B., Borysenko, J.Z. and Karnovsky, M.J., J. Cell Biol. 62:351, 1974.
82. Oliver, J.M., Zurier, R.B. and Berlin, R.D., Nature Lond. 253:471, 1975.
83. Hawkins, D., J. Immunol. 113:1864, 1974.
84. Berlin, R.D., Nature New Biol. 235:44, 1972.
85. Rossi, F., Zatti, M., Patriarca, P. and Cramer, R., J. Reticuloend. Soc. 9:67, 1971.
86. Tedesco, F., Trani, S., Soranzo, M.R. and Patriarca, P., F.E.B.S. Letters 51: 232, 1975.

87. Zurier, R.B. and Sayadoff, D.M., Inflammation 1:93, 1975.
88. Root, R.K., Metcalf, J., Oshino, N. and Chance, B., J. Clin. Invest. 55:945, 1975.
89. Gale, R.P. and Zighelboim, J., J. Immunol. 114:1047, 1975.
90. Dean, D.A., Wistar, R. and Murrell, K.D., Am. J. Trop. Med. & Hyg. 23:420, 1974.
91. Henson, P.M., Prog. Immunol., vol. 2, edited by L. Brent and J. Holborow, p. 95, North-Holland, Amsterdam, 1974.
92. Hawkins, D., J. Immunol. 107:344, 1971.
93. Becker, E.L. and Showell, H.J., J. Immunol. 112:2055, 1974.
94. Griffin, F.M., Griffin, J.A., Leider, J.E. and Silverstein, S.C., J. Exp. Med. 142:1263, 1975.
95. Scribner, D.J. and Fahrney, D., J. Immunol. 116:892, 1976.
96. Griffin, F.M., Bianco, C. and Silverstein, S.C., J. Exp. Med. 141:1269, 1975.
97. Mantovani, B., Rabinovitch, M. and Nussenzweig, V., J. Exp. Med. 135:780, 1972.
98. Huber, H., Polley, M.J., Linscott, W.D., Fudenberg, H.H. and Muller Eberhard, H.J., Science 162:1281, 1968.
99. Mantovani, B., J. Immunol. 115:15, 1975.
100. Lachmann, P.J., This volume.

This work was supported by USPHS grants GMS 19322 and HL 17786.

Chapter 9

A CHEMOTACTIC RECEPTOR FOR VAL(ALA)-GLY-SER-GLU ON HUMAN

EOSINOPHIL POLYMORPHONUCLEAR LEUKOCYTES

R. Neal Boswell, K. Frank Austen, and E. J. Goetzl
Departments of Medicine,
Harvard Medical School and the Robert B. Brigham Hospital,
Boston, Mass. 02120, USA

Abstract

Preferential eosinophil chemotactic activity is an _in vitro_ and _in vivo_ property of eosinophil chemotactic factor of anaphylaxis (ECF-A), a mixture of two peptides, Val-Gly-Ser-Glu and Ala-Gly-Ser- Glu, isolated from extracts and anaphylactic diffusates of human lung tissue. Purified native and synthetic ECF-A share with the synthetic N-formyl methionyl peptides such features as _in vitro_ activity in nanomolar amounts, high dose inhibition of effect and a requirement for hydrophobic amino acid residues. The capacity of the substituents of ECF-A, Val-Gly-Ser, Ala-Gly-Ser, and Gly-Ser-Glu to modulate eosinophil chemotaxis has permitted a preliminary functional characterization of an eosinophil surface receptor. The activity, specificity, and structural characteristics of the active tetrapeptides suggest that distinct interactions of the peptide with a stereospecific receptor on the eosinophil surface is required for chemotactic movement.

General Features of Leukotaxis

The directed migration of polymorphonuclear (PMN) and mononuclear

leukocytes along a concentration gradient of a chemotactic stimulus can

be reproducibly quantitated _in vitro_ utilizing micropore filter chambers

and other assay systems (1,2,3). The initiation and continuation of

directed migration is dependent on the presence of an adequate

concentration and gradient of the chemotactic stimulus (2,4).

Reversal of the direction of a chemotactic factor gradient reverses

149

the direction of leukocyte migration (5,6) in association with
reorientation of the cytoskeletal assembly toward the higher
concentration (7). Cell directed inhibitors of chemotaxis with or
without chemotactic activity may alter the cell response to a
chemotactic stimulus through such diverse mechanisms as deactivation
(8,9,10), high dose inhibition (10,11,12), or by enhancement or
suppression of critical metabolic pathways (13-17). Neutrophil
immobilizing factor suppresses chemotaxis and random migration of
neutrophils and eosinophils by a direct effect on leukocytes independent
of the specific test chemotactic stimulus (18). Ascorbate stimulates
the hexose monophosphate shunt (HMPS) of neutrophils, monocytes and
eosinophils in parallel with enhancement of their random and directed
migration (19,20). For one series, the synthetic N-formyl methionyl
peptides, there is knowledge of the relationship between structure
and function as chemotactic factors and as releasers of lysosomal
enzymes (21,22). Maximal activity of the synthetic N-formyl methionyl
peptides for neutrophils is dependent on the neutralization of the
positively charged NH_2-terminal amino acid group, and the presence of
nonpolar amino acids in the adjacent position (22).

The chemotactic response of leukocytes to structurally diverse
chemotactic factors is dependent on the presence of divalent cations
(13,23), intact pathways of glucose metabolism (13-15, 17), and the
integrity of intracellular microtubular and microfibrillar structures
(7,24). Manifestations of the interaction of chemotactic factors with
target leukocytes in the absence of a concentration gradient include
activation of PMN leukocyte esterases (8,13,25-30), stimulation of the
PMN leukocyte HMPS (17,20,31), and the release of granular enzymes
(22,32,33).

It has been postulated that receptors for chemotactic factors on the surface of leukocytes can sense both the concentration of the factor and its concentration gradient across the diameter of the leukocyte (2,4). The availability of highly purified natural and synthetic eosinophilotactic factors and synthetic analogues (10,11,34) has permitted the preliminary functional characterization of an eosinophil surface receptor involved in chemotactic activation and deactivation. Any model of a chemotactic factor receptor on leukocytes must account for the structural characteristics which define the cell specificity and potency of the agonist as well as the structural characteristics of specific antagonists which compete with the chemotactic factor. For the purposes of this essay a leukocyte chemotactic factor assessed in vitro by a modification of the Boyden chemotactic assay (1,2) will be defined as an agonist that: induces directed migration of a responsive leukocyte toward an increasing concentration of that factor; does not enhance migration when present in both the stimulus and cell compartments of a Boyden chemotactic chamber so as to eliminate a gradient; and deactivates in a time and dose-dependent manner the chemotactic response of the cell to a subsequent homologous or heterologous chemotactic stimulus.

Eosinophil Chemotactic Factor of Anaphylaxis (ECF-A)

The eosinophil chemotactic factor of anaphylaxis was discovered in 1971 as a mediator released during immediate hypersensitivity reactions in guinea pig (35) and human (36) lung slices. ECF-A was subsequently found to exist totally preformed in rat peritoneal mast cell granules (37) and in mast cell-rich fractions of human lung and nasal polyps (37,38,39). Preliminary physicochemical characterization of ECF-A

obtained by extraction or immunologic activation of human lung slices
revealed a molecular weight of approximately 500 by filtration on
Sephadex G-25 (36). This low molecular weight eosinophilotactic
material was subsequently sequentially purified by Dowex-1, Sephadex
G-10 and paper chromatography (34). Highly purified eosinophilotactic
fractions contained two acidic tetrapeptides of amino acid sequence
Ala-Gly-Ser-Glu and Val-Gly-Ser-Glu. Synthetic peptides of this
structure were prepared and shown to possess ECF-A like activity at
concentrations comparable to purified, native ECF-A.

Both the synthetic tetrapeptides and purified ECF-A are
maximally active in amounts from 0.05 - 1.0 nmole per chemotactic
chamber representing concentrations of $5x10^{-8} - 10^{-6}$M (34). Both
natural and synthetic tetrapeptides rapidly deactivate eosinophils in
quantities as small as 0.1 picomole per $5x10^{6}$ eosinophils and exhibit
a cross-deactivating capacity for the other tetrapeptides and C5a
(10,11,40). Preincubation of $8x10^{6}$ eosinophils in concentrations of
alanyl-tetrapeptide ranging from 10^{-8}M to 10^{-12}M, followed by extensive
washing of the cells and assessment of their chemotactic response to
10^{-7}M valyl-tetrapeptide revealed 75-80% deactivation after a 5 min.
exposure to 10^{-8}or 10^{-10}M alanyl-tetrapeptide. The deactivation of
eosinophils preincubated in 10^{-12}M alanyl-tetrapeptide varied with
different samples of cells from 0-40% after 1-5 min., reached 60-70%
by 40 min. and 80-90% at 1 1/2-2 hours. Thus 1/100 to 1/1000 of the
minimal chemotactic dose of alanyl-tetrapeptide gave brisk and nearly
complete deactivation of eosinophils (10).

The in vivo potency of the tetrapeptides paralleled the in vitro
dose-response relationship. Intraperitoneal administration of
0.1 - 1.0 nmoles of synthetic valyl-tetrapeptide in guinea pigs

elicited an early selective influx of eosinophils, followed by

neutrophils (41).

Functional Features of the Eosinophil Receptor for ECF-A

Hydrophobic amino acid derivatives and peptide analogues or

substituents of the ECF-A tetrapeptides can inhibit the chemotactic

activity of the intact tetrapeptide (10,11,34,40). Valine amide (Val-

amide) and the NH_2-terminal tripeptides Val-Gly-Ser and Ala-Gly-Ser

possess neither chemotactic nor deactivating activity. Both Val-amide

and Val-Gly-Ser suppress the eosinophilotactic activity of 10^{-7}M

valyl-tetrapeptide when present concomitantly in the stimulus

compartment of the Boyden chamber with 50% mean inhibition at 10^{-6}M

and 10^{-7}M, respectively (40). Introduction of the NH_2-terminal

tripeptide into the cell compartment is also inhibitory, but this effect

is completely reversed if the cells are washed before introduction of

the tetrapeptide on the stimulus side (10). Val-amide and Val-Gly-Ser

presumably inhibit the eosinophil chemotactic response by competing

with ECF-A for a hydrophobic binding site on the eosinophil surface,

and thus are reversible in action and without effect on eosinophil

random migration. This capacity of Val-Gly-Ser to inhibit eosinophil

chemotaxis to valyl-tetrapeptide in an equimolar relationship has been

observed in vivo as well (41).

The COOH-terminal tripeptide, Gly-Ser-Glu, is marginally

chemotactic alone and weakly inhibitory when present on the stimulus

side with the valyl-tetrapeptide. However, preincubation of

eosinophils with Gly-Ser-Glu results in a dose-dependent suppression of

their chemotactic response to a reference stimulus of 10^{-7}M valyl-tetrapeptide. This suppression reaches a maximum 80% inhibition after

a 20 min. exposure of 9×10^6 eosinophils to 10^{-6}M COOH-terminal

tripeptide and is irreversible (11,40). The charged carboxy-terminal

glutamic acid produces this irreversible cell-directed effect

presumably at an ionic domain in the eosinophil chemotactic receptor

site involved in activation and deactivation.

Several synthetic peptide analogues have also been studied and

afforded additional insights into the nature of the receptor (40).

While Leu-Gly-Ser-Glu, which has a slightly larger NH_2-terminus, has

a peak activity comparable to the ECF-A tetrapeptides occurring at a

concentration of 10^{-8} - 10^{-7}M, Phe-Gly-Ser-Glu requires a concentration

of 10^{-4}M to achieve a similar cell response. Inversion of the internal

dipeptide to yield Val-Ser-Gly-Glu produces no reduction in activity

while the condensed tripeptide Val-Ser-Glu elicits only half the

cellular response of the tetrapeptide at its most effective concentration

of 10^{-7}M. These preliminary studies are consistent with preferences of

the receptor for hydrophobic residues at the NH_2-terminus and internal

residues which provide an optimal separation between the NH_2- and

COOH-terminal amino acids. Others, working with the synthetic N-formyl

methionyl peptides have also noted that small structural changes are

capable of creating large differences in chemotactic activity (22).

Histamine, which induces directed migration of eosinophils in in

vitro assays that employ thin polycarbonate filters (12) or brief

incubation periods with standard cellulose nitrate filters (10,11)

modulated the eosinophil chemotactic response to the ECF-A tetrapeptides

under standard assay conditions (10). The addition of histamine to the cell

compartment of Boyden chambers at a dose which only minimally stimulated

eosinophil migration, enhanced the chemotactic response of eosinophils

to low concentrations of valyl-tetrapeptide (10). A dose of

histamine in a cell compartment which suppressed spontaneous migration

of eosinophils inhibited directed migration to the valyl-tetrapeptide

(11). Thus, other eosinophil receptors can modulate the cellular

response to the tetrapeptides.

Concluding Comments

The structural requirements for the chemotactic activity of

highly purified natural and synthetic ECF-A tetrapeptides and synthetic

peptide analogues have been assessed in vitro and to a limited extent

in vivo. A model is proposed for a tetrapeptide chemotactic receptor on

eosinophils that is based on functional data derived solely from a

modified micropore filter assay. This has permitted the recognition

of the agonists and some inhibitors which may function by blocking the

interaction of these principles with an eosinophil surface receptor.

Experiments are underway to corroborate this model utilizing both

other functional methods and direct binding assays to assess the

interaction of the tetrapeptide factors with the putative receptors.

The studies completed to date provide support for the hypothesis that

eosinophil directed migration occurs following a primary interaction

between the ECF-A tetrapeptides and a stereospecific receptor on the

eosinophil surface. The activity of the ECF-A tetrapeptides Val-Gly-Ser-

Glu and Ala-Gly-Ser-Glu is suppressed in vitro and in vivo by

equimolar quantities of its NH_2-terminal tripeptide substituent,

presumably by reversible eosinophil membrane receptor competition. The

COOH-terminal tripeptide substituent of ECF-A irreversibly suppresses

directed eosinophil migration to ECF-A possibly by activation and

deactivation that is non-productive of directed migration.

The inhibition of eosinophil migration to ECF-A by both NH_2-
terminal and COOH-terminal tripeptide substituents suggests that
multiple eosinophil surface sites may exist for interaction of "the
receptor" with the chemotactic factors (Fig. 1). Since neither of
the tripeptide substituents exhibit significant intrinsic chemotactic
activity, both the nonpolar NH_2-terminal residues Ala or Val and the
negatively charged COOH-terminal Glu are required for maximal chemotactic
activity. Hydrogen bonding of the serine hydroxyl-group may further
facilitate stabilization of the COOH-terminal glutamic acid at an
activation site in an ionic domain. Lacking the negatively charged
COOH-terminal glutamic acid, the NH_2-terminal tripeptide blocks ECF-A
activity at apparently equimolar concentrations by competing with the

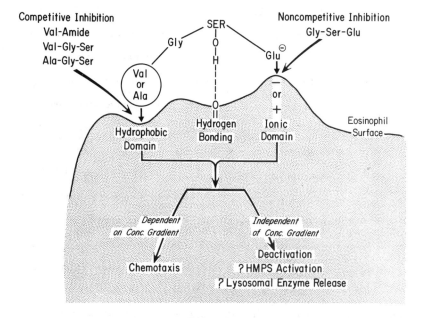

FIGURE 1

Hypothetical model for an eosinophil surface chemotactic receptor for
Val(Ala)-Gly-Ser-Glu.

tetrapeptide for a recognition site in the hydrophobic domain. By

interacting weakly or briefly with a site in the ionic domain, the

COOH-terminal tripeptide at 100-1000 fold the concentration required

for deactivation by the tetrapeptide irreversibly suppresses eosinophil

directed migration possibly by activation - deactivation. The failure

of the COOH-terminal tripeptide to bind to the hydrophobic domain would

prevent it from inducing directed migration and account for the high

concentrations required for irreversible deactivation. A concentration

gradient of the tetrapeptides is necessary for chemotactic migration,

whereas a uniform interaction with the eosinophil receptors leads to

deactivation and possibly other less sensitive cellular functions such

as lysosomal enzyme release.

Acknowledgments

Supported by grants AI-07722 and AI-10356 from the National
Institutes of Health. Dr. Boswell is sponsored by the Air Force
Institute of Technology, Wright-Patterson AFB, Ohio, 45433.
Dr. Goetzl is an Investigator of the Howard Hughes Medical Institute.

References

(1) Boyden, S.V., J. Exp. Med., 115:453, 1962.

(2) Zigmond, S.H., and Hirsch, J.G.,J. Exp. Med., 137:387, 1973.

(3) John, J.T., and Sieber, O.F.Jr., Life Sciences, 18:177, 1975.

(4) Wilkinson, P.C., in Chemotaxis and Inflammation p. 34,
     Churchill Livingstone, Edinburgh, 1974.

(5) Cornely, H.P., Proc. Soc. Exp. Biol. Med., 122:831, 1966.

(6) Wilkinson, P.C., in Chemotaxis and Inflammation p. 38,
     Churchill Livingstone, Edinburgh, 1974.

(7) Malech, H.C., Root, R.K., and Gallin, J.I., Clin. Res., 24:314A, 1976.

(8) Ward, P.A., and Becker, E.L., J. Exp. Med., 127:693, 1968.

(9) Goetzl, E.J., and Austen, K.F., Antibiotics and Chemotherapy, 19:218, 1974.

(10) Goetzl, E.J., and Austen, K.F., Nobel Symposium #33, in press, 1976.

(11) Goetzl, E.J., Am. J. Path., in press, 1976.

(12) Clark, R.A.F., Gallin, J.I., and Kaplan, A.P., J. Exp. Med., 142:1462, 1975.

(13) Becker, E.L., in Biochemistry of the Acute Allergic Reaction, edited by K.F. Austen and E.L. Becker, p. 243, Blackwell, Oxford & Edinburgh, 1971.

(14) Carruthers, B.M., Can. J. Physio. & Pharm., 44:475, 1966.

(15) Carruthers, B.M., Can. J. Physio. & Pharm., 45:269, 1967.

(16) Ward, P.A., J. Exp. Med., 124:209, 1966.

(17) Goetzl, E.J., and Austen, K.F., J. Clin. Invest., 53:591, 1974.

(18) Goetzl, E.J., Gigli, I., Wasserman, S.I., and Austen, K.F., J. Immunol, 111:938, 1973.

(19) Goetzl, E.J., Wasserman, S.I., Gigli, I., and Austen, K.F., J. Clin. Invest., 53:938, 1973.

(20) Goetzl, E.J., Ann. N.Y. Acad. Sci., 256:210, 1975.

(21) Schiffmann, E., Corcoran, R.A., and Wahl, S.M., Proc. Natl. Acad. Sci. USA, 72:1095, 1975.

(22) Showell, H.J., Freer, R.J., Zigmond, S.H., Schiffmann, E., Aswanikuman, S., Corcoran, B., and Becker, E.L., J. Exp. Med., 143:1154, 1976.

(23) Becker, E.L., and Showell, H.J., Z. Immun.-Forsch. Bd., 143:466, 1972.

(24) Zigmond, S.H., and Hirsch, J.G., Exp. Cell Res., 73:383, 1972.

(25) Becker, E.L., and Ward, P.A., J. Exp. Med., 125:1021, 1967.

(26) Ward, P.A., and Becker, E.L., J. Exp. Med., 125:1001, 1967.

(27) Becker, E.L., and Ward, P.A., J. Exp. Med., 129:569, 1969.

(28) Ward, P.A., and Becker, E.L., J. Immunol., 105:1057, 1970.

(29) Becker, E.L., J. Exp. Med., 135:376,]972.

(30) Becker, E.L., in The Phagocytic Cell in Host Resistance, edited by J.A. Bellanti and D.H. Dayton, P. 1, Raven Press, New York, 1975.

(31) Wasserman, S.I., Whitman, D., Goetzl, E.J., and Austen, K.F., Proc. Soc. Exp. Biol. Med., 148:301, 1975.

(32) Becker, E.L., in Mechanisms in Allergy, edited by L. Goodfriend, A.H. Sehon, and R.P. Orange, p. 339, Marcel Decker Inc., New York, 1973.

(33) Goetzl, E.J., Wasserman, S.I., Austen, K.F., in Progress in Immunology II Vol. 4, edited by L. Brent and J. Holborow, p. 41, North Holland Publishing Co., Amsterdam, 1974.

(34) Goetzl, E.J., and Austen, K.F., Proc. Nat. Acad. Sci. USA, 72:4123, 1975.

(35) Kay, A.B, Stechschults, D.J., and Austen, K.F., J. Exp. Med., 133:602, 1971.

(36) Kay, A.B., and Austen, K.F., J. Immunol, 107:899, 1971.

(37) Wasserman, S.I., Goetzl, E.J., and Austen, K.F., J. Immunol., 112:351, 1974.

(38) Kaliner, M.A., Wasserman, S.I., and Austen, K.F., N. Engl. J. Med., 289:277, 1973.

(39) Austen, K.F., Wasserman, S.I., Goetzl, E.J., Nobel Symposium #33, in press, 1976.

(40) Goetzl, E.J., Boswell, R.N, and Austen, K.F., Fed. Proc., 35:515, 1976.

(41) Wasserman, S.I., Boswell, R.N., Drazen, J.M., Goetzl, E.J., and Austen, K.F., J. Allergy Clin. Immunol., 57:190, 1976.

ₒChapter 10

COMPLEMENT RECEPTORS AND CELL ASSOCIATED COMPLEMENT COMPONENTS

Ian McConnell and P.J. Lachmann
MRC Group on Mechanisms in Tumour Immunity, c/o the Laboratory
of Molecular Biology, The Medical School, Hills Road, Cambridge,
England

Abstract

Membrane receptors for activated complement components are widely
distributed amongst tissue cells of most mammalian species. Common amongst
these are receptors for C3b which mediate many of the biological functions
of C3. In addition, the genetic control of certain complement components
is linked to the genes which code for the major histocompatibility complex.
Many of these components are also present on cell surfaces. This suggests
that the function of the complement system and the major histocompatibility
complex may be related.

1. Introduction

Many of the biological functions of the complement system are mediated

by cell membrane receptors which bind activated complement components. This

association between cells and complement components is predominantly a

consequence of complement activation in the cell's microenvironment. There

is, however, a more direct association between complement components and cell

membranes. In several species the genetic control of the serum expression

of certain complement components is linked to genes coding for the major

histocompatibility complex (MHC). Many of these components are also

present on the cell surface. Since the complement system is an important

effector mechanism in humoral immunity its association with a system involved
in cellular immune reactions is intriguing.

2. Complement Reaction Pathways

The components of the complement system and their reaction
mechanisms have been well discussed elsewhere (1, 2) and will thus not be
reviewed here. For the purpose of this review the complement sequence can
be considered as occurring in three major stages.

2.1 The generation of C3 splitting enzymes

C3 is the major complement component in the serum and C3 splitting
enzymes (C3 convertases) are generated by both the classical and alternative pathways of complement activation. The components of these pathways
and their reaction requirements are summarised in Table 1.

2.2 The activation of C3

This is the central 'bulk' reaction of the complement sequence and is
of major biological importance. Classical or alternative pathway C3 convertases cleave a small fragment from C3, C3a (8,900 M.W.), leaving a
larger unstable fragment - nascent C3b. C3a is an anaphylatoxin, mediates
a number of biological reactions and acts as a potent chemotactic factor.
Nascent C3b is unstable and possesses a highly labile binding site which can
form strong hydrophobic interactions with cell surfaces or antigen antibody
precipitates. Unless this happens immediately the molecule undergoes
secondary changes and the hydrophobic binding site is lost giving rise to
C3b. C3b, either in the fluid phase or attached hydrophobically to cells,
can now only bind to certain tissue cells via their C3b receptors.

C3b itself, either bound hydrophobically to a cell surface or free in
solution can produce further splitting of C3 via the C3b feedback cycle. It
does so by combining with Factor B in the presence of Factor D, generating

COMPLEMENT RECEPTORS

"TABLE" I

	Classical	Alternative (or C3b feedback pathway)
Activating agents		
Aggregates of human	IgG1 & 3(&2);IgM	IgA
rabbit	IgG IgM	F(ab')$_2$
guinea pig	IgG2	IgG1
ruminant	IgG1	IgG2
		Inulin
	(Lipid A)	Zymosan
		Endotoxin LPS
		CVF
Factors required to	C1*	Properdin, factor D
generated C3 convertase	C4	C3
	C2	factor B
Total serum requirement	Dilute	Concentrated
Ion requirements	Ca and Mg	Mg

*C1 is a trimolecular complex of C1q, C1r and C1s. Classical pathway activation is initiated when C1q binds to immunoglobulin.

CVF-Bb - the alternative pathway C3 convertase. This splits more native C3 giving rise to further C3b and the cycle is again repeated.

C3b is prevented from producing exhaustive breakdown of C3 via the feedback cycle by the C3b inactivator, sometimes known as conglutinogen activating factor or KAF. KAF acts on C3b, generating C3bi. This is haemolytically inactive, no longer binds to C3b receptors and does not fire the feedback cycle. Subsequent attack of C3bi by tryptic-like enzymes leaves a small fragment, C3d. This is biologically inactive but binds to

C3d receptors on certain cells. The different cell surface binding

mechanism for C3 are shown in Figure 1.

Cobra venom factor (CVF) acts analogously to C3b, combining with factor

B and giving rise to $\overline{\text{CVF-Bb}}$ – another C3 convertase. Since CVF is insusc-

eptible to mammalian KAF it produces massive C3 activation. The analogous

reaction pathways of CVF and C3b are due to the fact that the factor in

cobra venom with this property is cobra C3b (3). Since CVF forms a C3

convertase with Factor B it can be used as a specific probe for this

component.

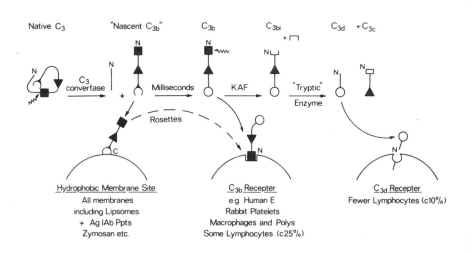

"FIGURE" 1

Binding of C3 split products to cell surfaces.

2.3 The terminal stages of the complement sequence

This is initiated by the splitting of C5 by C5 convertase. A small
fragment, C5a (also an anaphylatoxin) is cleaved leaving the larger C5b
fragment which binds weakly to cell surfaces. It then reacts sequentially
with C6 and C7 to form a trimolecular complex C$\overline{567}$. This binds one
molecule of C8 and six of C9. The C5-9 complex inserts itself into the
lipid bilayer and lysis ensues.

It will be apparent from the above that C4, C3 and C5 undergo
similar cleavage during complement activation. Nascent C4b, C3b and C5b
all bind to cell surfaces albeit with different affinities. The smaller
fragments, especially C3a and C5a are potent anaphylatoxins. C4, C3 and
C5 are also structurally similar. C3 and C5 comprise two polypeptide chains,
one of 140,000 molecular weight (heavy or alpha chain) and one of 70,000
molecular weight (light or beta chain) both held together by disulphide
bridges. C4 also has two chains of similar molecular weight, but in
addition a third smaller gamma chain of 33,000 molecular weight.

3. Complement Receptors

3.1 Receptors for Clq

Receptors for Clq have been reported on both T and B lymphocytes as
well as on certain lymphoblastoid cells (4, 5). Although isolated Clq binds
to these cells, this is unrelated to the presence of surface Ig, for either

Fc or C3b receptors. Cells lacking these other receptors bind Clq. The
biological significance of the Clq receptor is unknown.

Receptors for Clq are also thought to be present on rabbit platelets.
These cells release vasoactive amines on binding collagen and this reaction
can be inhibited by Clq (a collagen-like structure) but not by native Cl (6).

3.2 The C3b receptor

Immune adherence reactions between trypanosomes or microorganisms
treated with antibody and complement and human erythrocytes were the first
indication that mammalian cells had membrane receptors for altered complement
components (7, 8). It was subsequently shown that C3b receptors had a wide-
spread tissue distribution and could be detected on the surface of primate
red cells, many non-primate platelets, polymorphonuclear leukocytes,
macrophages, monocytes, a subpopulation of lymphocytes and the epithelial
cells of the normal renal glomeruli(see reviews 9, 10). The C3b receptors
on many of the above cells also react with C4b. This was first recognised
by Cooper (11) who showed that the immune adherence receptor on human
erythrocytes could also bind C4b. The C3b receptors on lymphocytes and
lymphoblastoid cells similarly bind C4b and their identity is further shown
by the fact that on these cells the two receptors co-cap (12).

The affinity of the C3b receptor for C4b is presumably a reflection
of the structural similarity between C3b and C4b. It is unknown whether the
other member of the series, C5b, can bind to the C3b receptors. This is
difficult to test directly since C5b as such is not stably bound at the
complement fixation site. EC$\overline{567}$, however is stable but immune adherence
negative.

3.3 Physicochemical studies of the C3b receptor

If C3b receptored lymphoid cells are treated with trypsin or reducing
agents (dithiothreitol, mercaptoethanol) receptor activity is lost (13).

Regeneration of the receptor takes 8 - 10 hours. It cannot be excluded that this treatment does not affect some other surface structures necessary for receptor expression. The tentative conclusion reached is that the receptor is protein-like and requires disulphide bridges for activity.

C3b receptor carrying plasma membrane fragments from Raji cells have also been isolated and partially characterised (14). Detergent solubilised membrane fragments were found to have no activity and active membrane fragments were only recovered using membrane disintegrating agents like potassium bromide. When lipid was removed from this material activity was lost suggesting that the C3b receptor was a unique type of membrane receptor in being a lipid-protein complex. This seems unlikely since the assay for activity required multipoint binding, presumably best achieved when C3b receptors were present as a multivalent complex on a membrane fragment. Partial purification of detergent solubilised C3b receptors using a different assay has been described (15). Platelet membranes were solubilised in detergent, fractionated on G200 sephadex and the fractions then used to inhibit histamine release from platelets by C3b coated zymosan particles (ZC3). In this assay there is less requirement for the inhibiting material to be multivalent and hence a better assessment of the molecular weight of the C3b receptor has been achieved. Although a variety of different molecular weight fractions were found to inhibit,an inhibitory 30 - 40,000 molecular weight fraction was consistently found which was also able to bind to ZC3. In the absence of detergent, the fractions readily aggregated into fractions of larger molecular size.

3.4 The C3d receptor

KAF-reacted C3b has little biological activity. It no longer fires the feedback cycle, cannot bind to the C3b receptors on erythrocytes, macrophages, polymorphs, platelets or the renal glomerulus. A small proportion of

normal B lymphocytes and certain lymphoblastoid cells however, do bind
KAF-treated EAC (16, 17).

   This was first recognised by Okada & Nishioka (16) who showed that
Daudi cells preferentially bound EAC prepared with mouse rather than
purified human components.   The latter became adherent for Daudi cells after
KAF treatment.   The fact that the C3d receptors are detected using mouse
serum is due to the fact that most investigators incubate EA in mouse
complement for 30 minutes.   During this prolonged incubation, mouse KAF
acts on the bound C3b generating many C3d sites (18).   Mouse serum rather
than human serum is conventionally used to avoid lysis of EA but this can
be overcome by using human R3 reagents (i.e. normal human serum depleted of
C5 and C6 by zymosan absorption (19)).   With this reagent 70 seconds
incubation with R3 followed by antrypol is sufficient to generate EAC
with a considerable amount of C3b.   EAC incubated with mouse complement
still have sufficient C3b and can nonetheless react with the human C3b
receptor to produce immune adherence reactions (18).

   Native C3 does not bind to the C3b receptor on normal cells.   With
certain lymphoblastoid cells, notably Raji cells it has been claimed that
native C3 can bind to and block the complement receptor (20).   Since this
can be demonstrated only using "purified" C3 and not native C3 present in
serum EDTA it can be asked whether or not the purified C3 contains some
C3b or is altered in some subtle way by the purification procedure.   The
possibility that there is an inhibitor in serum preventing the binding of
native C3 to the Raji cell looks improbable but cannot be wholly excluded.

3.5   Relationship of the C3b receptor on lymphocytes to other surface
      receptors

   The original claim that the C3b receptor was a B cell marker is only
partially correct (21).   C3b receptors are present on all antigen binding

cells (22) (for certain antigens) but absent from the surface of small numbers of immunoglobulin bearing cells, including the precursors of antibody secreting cells (23), as well as the majority of IgM and IgG antibody forming cells (22). This suggests that the C3b receptor, like immunoglobulin (24) is lost from differentiating B cells. Alternatively, activated B cells are like lymphoblastoid cells and spontaneously activate the alternative pathway in vivo generating C3b at the cell surface which blocks the C3b receptor (see 4.5). There is one unsubstantiated report that T cells have C3b receptors (25).

C3b receptors on lymphocytes are distinct from certain other surface receptors. They are neither affected by 'capping' surface immunoglobulin or Fc receptors nor by inhibition of the latter with complement-free immune complexes. Trypsin destroys C3b receptors but has no effect on Fc receptors (26).

There is an association between the B lymphocyte receptor for Epstein Barr (EB) virus and the C3d receptor (27). Blocking of the C3d receptor produces blocking of the EB virus receptor (and vice versa and both receptors co-cap on the lymphocyte surface suggesting that they are identical. These interesting observations may explain why EB virus unlike other viruses (e.g. measles virus) only infects B lymphocytes.

The relationship between histocompatibility antigens and C3b receptors has also been examined. In mice it has been reported that certain anti-H-2 sera raised across I-C, S and D region incompatibilities will block lymphocyte C3b rosettes (28). Susceptibility to blocking is determined by genes mapping in the I-C or S region. Both whole IgG and $F(ab')_2$ antibody blocks. However H-2 identical mice were found to differ with respect to their susceptibility to inhibition. Antisera blocked C3b rosettes in DBA/2-(H-2^d) but not in haplotype identical Balb/c (H-2^d) mice.

Similarly only AKR (H-2^k) but not CBA or C3H (both H-2^k) were inhibited by
the appropriate antisera. This surprising observation was attributed to genes
mapping outside H-2 which were involved in the expression of the C3b recept-
or on the cell surface. In humans antisera to the diallelic, HLA 4a/4b
system were also reported as being able to block C3b receptors in individuals homozygous for either allele (29)

These studies are in contrast to several other observations. Antisera
to the MHC in rats does not block C3b receptors on lymphocytes (22) and
independent studies with the human 4a/4b system have so far not confirmed
the above result (Bright 1976, personal communication), nor is there
correlation between C3b or C3d receptors and HLA phenotype (30) and cells
lacking HLA antigens (Daudi cells) have C3d receptors.

4. Cell Associated Complement Components

The genes controlling the serum expression of certain complement
components, notably C2, C4, Factor B and C8 are now known to be linked to
the genes coding for the MHC (see review by Hobart & Lachmann (31)). Some
of these components are also present on the cell surfaces. By contrast genes
controlling C1, C5, C6 and C7 are unlinked to the MHC and none of these
components are present on cells.

4.1 C4

The fourth component of complement or at least one of its chains is
coded by genes within the MHC. In man C4 deficiency (32) and C4 polymorphism
have been reported as linked to HLA (33). In guinea pigs C4 deficiency is
also linked to the MHC (34) and in mice the Ss protein, a product of the
S region of the H2 complex is now known to be C4 (35, 36).

At a cellular level there is evidence for the existence of C4 on cell
surfaces. Anti-C4 has been reported to block stimulation in a mixed

lymphocyte culture and by an antibody absorption technique, small amounts of

C4, about 1000 molecules per cell have been detected on the surface of

lymphoblastoid cells cultured in the absence of any external source of C4

(37). In mice we have used antisera to the Ss protein to search for C4 on

lymphocytes with no success. However Saunders & Eididin have reported that

anti SS serum detects the Ss component (i.e. C4) on the surface of mouse

fibroblasts (38).

4.2 C8

C8 is one of the terminal lytic components of the classical pathway and

its role in two types of cell mediated cytotoxic reactions has been

investigated. Target erythrocytes coated with antibody and C1 - C7 can

be lysed by lymphocytes faster than the same targets lacking the later

components from C5 onwards (39). In this system the lymphocytes have been

regarded as supplying C8 thus producing lysis. Target erythrocytes

susceptible to lysis by C8 can however be bult up without the participation

of antibody and the early components, by the phenomenon of reactive lysis.

When complexes of C5 and C6 ($\overline{C56}$) are mixed with C7 in the presence of

erythrocytes the nascent $\overline{C567}$ trimolecular complex binds to the cell surfaces.

These cells are now susceptible to lysis with serum C8 and C9. Similarly

these cells can be lysed with human peripheral blood lymphocytes and this

can be inhibited by $F(ab')_2$ anti-human C8. This indicates that C8 is

released by the cells in this special type of cell mediated cytotoxic reaction

(40). Although it has also been reported that whole anti-C8 inhibits

antibody-dependent, cell mediated cytotoxicity this cannot be achieved when

$F(ab')_2$ anti-C8 is used. The inhibition observed with the whole antibody

is presumably due to the presence of small amounts of immune complexes of

C8/anti-C8 which are potent inhibitors of antibody dependent cytotoxicity.

4.3 Factor B

Factor B of the alternative pathway of complement activation was first shown to be associated with normal and neoplastic lymphocytes, particularly those of the B cell lineage (41, 42, 43). These papers fully describe the assay system for detecting factor B on lymphocytes. Briefly, they rely on the ability of lymphocytes (or other cells) to replace factor B in the generation of an alternative pathway C3 convertase with CVF. In this way CVF is used as a specific probe for Factor B on cells.

Factor B has been detected on a variety of different cell types. Initial studies showed that cells of the B cell lineage were predominantly associated with Factor B and that thymus cells and erythrocytes were negative. More recently we have found Factor B to be associated with a large number of different in vitro cultured cell lines such as Daudi cells, HeLa cells, Chang liver cells and several different mouse cell lines. All the cell lines have been grown in the absence of any extrinsic source of Factor B. With cells derived from in vivo sources it is less easy to exclude the possibility that the cell associated Factor B has not been pinocytosed from the serum.

The other major points from our studies on Factor B can be summarised as follows:

1. Membrane associated $\overline{CVF-Bb}$ convertase does not remain attached to the cell surface but is released into the supernatant within 5 minutes of incubation of cells and CVF.

2. The reaction can be inhibited with $F(ab')_2$ anti-Factor B.

3. Generation of the convertase proceeds via the alternative pathway, occurs in C2 and C4 deficient sera and requires Mg^{++} but not Ca^{++}.

4. Cells do not secrete Factor B into the supernatant nor can they be induced to synthesise Factor B following culture with CVF. Mouse macrophages are known to synthesise Factor B (44).

5. Factor B cannot be detected serologically on the cell surface. In the
assay CVF presumably reacts with a CVF reactive site on the surface of the
cells which may itself be associated with Factor B buried in the membrane.

6. In mice a strain difference in the level of Factor B positive cells has
been detected. Balb/c mice (H-2^d) have a significantly greater percentage
(24.8%) of Factor B positive cells in the lymph nodes than CBA strain mice
(H-2^k, mean 9.2%). Preliminary genetic experiments tend to suggest that this
difference in levels is linked to the major histocompatibility complex. This
is of interest in view of the fact that in man Factor B allotypes are linked
to HLA (45).With the single cell assay for Factor B in mice, however, it is
not clear whether the difference in levels detected measures a real difference
in the levels of Factor B positive lymphocytes or some other cell surface
characteristic involved in the assay used.

4.4   Complement activation by cells

Certain substances activate the alternative pathway without involving
antibody. These are zymosan, inulin, bacteria lipopolysaccharides and even
raised concentrations of magnesium ions (see reviews 1 and 2).

Recently it has become apparent that a variety of cell types can
similarly activate the alternative pathway in the absence of antibody.
Rabbit erythrocytes, lymphocytes and thymocytes can activate the alternative
pathway in heterologous, agammaglobulinaemic sera (46). Trypanosomes, (T.
cruzi) can similarly induce activation in normal and agammaglobulinaemic
chicken serum, the parasite being lysed in the process (47). This may
represent an innate resistance mechanism for the elimination of the organism.

We have investigated this phenomenon in an entirely autologous system
using human lymphoblastoid cell lines (LCL) and a variety of human serum
reagents (48). In vitro cultured cell lines derived from Burkitt's lymphoma

or infectious mononucleosis all produce alternative pathway activation. The
reaction occurs in C2 and C4 deficient sera, is magnesium but not calcium
dependent and requires Factor B. Factor D or properdin are not essential.
Antibody is not involved and activation proceeds in normal human serum
absorbed with human cells and hypogammaglobulinaemic human serum.

All of the cell lines tested in the above assay were EBNA positive and
carried the EB virus in their genome. This raises the possibility that
infection of a cell with EB virus confers on LCL the ability to activate the
alternative pathway. To test this possibility we have recently tested
certain LCL which lack EB virus in their genome as well as these same cells
following superinfection with the EB virus. These interesting cell lines
were first described by Klein et al (49).

Both EBNA negative and EBNA positive lines produce activation.
However, there does seem to be a quantitative difference between the EB
negative and EB positive lines, in that considerable C3 conversion can
be produced with very low numbers of EB positive cells but not with similar
concentrations of EB negative cells. So far the most striking difference has
been observed with the Ramos line. Klein et al (50) using complement
consumption as a measure of activation had earlier shown that EB positive
lines produced more complement consumption than EB negative lines. In view
of the intriguing association between EB virus and the complement receptor
it will be interesting to define the difference between these cell lines
with respect to their complement activating properties.

5. Biological Functions of the C3 Receptors

The role played by C3b receptors clearly depends upon the cells on
which they are found. These possible roles will therefore be considered
separately for the different cell types.

5.1 Primate erythrocyte C3b receptor

Since there are no obvious biological functions which are unique to primates by virtue of their having erythrocyte C3b receptors it can be inferred that this receptor plays no necessary biological role. That it may play a facilitating role in bacterial phagocytosis is suggested by the observations that adherence of antibody and complement coated streptococci to the erythrocyte C3b receptor facilitates their phagocytosis by poly-morphonuclear leucocytes in vitro (51).

There are no known erythrocyte functions which can be attributed to the interaction between soluble C3b and the C3b receptor on erythrocytes.

5.2 The platelet C3b receptor

This is another example of a receptor found only on certain species' platelets. Primate and ruminant platelets lack the C3b receptor normally found on most other mammalian species' platelets. Human platelets, however, have an Fc receptor for the Fc part of IgG which can probably act as a 'substitute' for the C3b receptor in in vivo reactions involving platelets.

The immune adherence of rabbit platelets to C3b coated particles, usually zymosan (ZC3), promotes exocytosis of vasoactive amines from the platelet (52). Platelet lysis occurs either in the presence of poly-morphonuclear leucocytes or the late acting complement components. In the latter case platelets bound by immune adherence reactions in the vicinity of a complement activating site undergo bystander lysis. This reaction is believed to operate during normal blood clotting in the rabbit and may explain why in C6 deficient rabbits there is inadequate prothrombin consumption. In C6 deficient humans normal blood clotting occurs even although the platelets are immune adherence negative and in this species other mech-anisms are clearly involved for adequate platelet activation to occur. Furthermore in humans it is likely that the Fc receptor substitutes for the

C3b receptor in immune adherence-like reactions mediated by the C3b receptor
in non-primate species.

5.3 C3b receptors on phagocytic cells

This C3b receptor is of great importance in promoting the adherence of
complement coated microorganisms to phagocytic cells. These adherence
reactions are among the most biologically relevant functions of the
complement system both in protection against microorganisms through
enhanced phagocytosis and in the induction of allergic inflammation.
Adherence of microorganisms to the C3b receptor on polymorphs leads to
rapid phagocytosis. If the particle is non-phagocytosable exocytosis
of the lysosomal content of the polymorph occurs. This is one of the major
mechanisms whereby polymorphonuclear induced inflammation occurs.

Interaction between the C3b and the C3b receptor on the polymorphonuclear
leucocyte may also cause intracellular changes which promote its
cytolytic potential   The presence of C3b on organisms enhances their intracellular killing by a mechanism quite independently of their phagocytosis
(53).

Compared to polymorphonuclear leucocytes the adherence of C3b coated
bacteria to the C3b receptor on macrophages is a less efficient signal for
phagocytosis. It is not true that no phagocytosis occurs by this mechanism
since C3b coated erythrocytes which do not carry antibody are readily
phagocytosed by Kuppfer cells. However attachment of C3b to the C3b receptor
on macrophages can"angrify"macrophages increasing their content of lysosomal
enzymes within the endoplasmic reticulum thereby enhancing the bacteriocidal
properties of the cell. The same signal also promotes the secretion of C3a
(an anaphylatoxin) by macrophages which has been regarded as a specific
cytolytic agent for inter alia tumour cells (54).

A special function may be attributed to the C3b receptor on dendritic
cells within lymph nodes. These cells may be macrophage like but are clearly

non-phagocytic. The attachment of immune complexes to these cells requires both antibody and complement and this type of localisation is an important pre-requisite for the development of germinal centres and B cell memory. This localisation however does not occur in tolerant animals or those chronically depleted of C3 (55). In the latter case antigen (TNP-KLH) fails to localise within germinal centres and B cell memory to this antigen does not develop. In comparable experiments White et al (56) have shown that germinal centre formation in response to antigenic stimulation does not occur in decomplemented chickens. This phenomenon seems likely to lie at the base of the caucus of observations that C3 plays a role in the induction of the immune response (see 5.5).

5.4 C3b receptor on lymphocytes

The role of this receptor has been investigated with somewhat variable results. It has been both claimed (57) and denied (58, 59) that interaction between C3b and the C3b receptor has a mitogenic effect on lymphocytes. Whether or not this is observed is probably dependent on methodology and quantities of C3b used. Failure to find any lymphocyte abnormality in KAF deficient patients who have high levels of continuously circulating C3b suggests that the above phenomenon is clearly not of major importance in vivo.

It has further been claimed that the interaction between the C3b and the C3b receptor on lymphocytes induces the release of a mononuclear cell chemotactic factor (59, 60). This would provide a mechanism for the accumulation of mononuclear cells at the site of allergic reactions.

The possibility has also been broached that the C3b receptor on lymphocytes is important in facilitating the antibody response to antigens. This is done either by promoting bridging between antigen carrying macrophages and lymphocytes or by enhancing the binding of antigen particularly monomeric antigen to the B cell surface via the lymphocyte C3b receptor.

5.5 <u>The role of complement in the induction of the immune response</u>

The substantial literature which has accumulated on this subject in recent years (see review by Pepys (61)) can be summarised as follows. If mice are depleted of C3 in vivo by CVF given at a critical time before immunisation the IgG, IgA and IgE antibody responses to thymus dependent antigens are suppressed. The IgM response to thymus independent antigens is normal. The effect can also be obtained in situations where C3 depletion of the serum is not obtained.

In vitro the situation is more complex. The addition of phospholipase free CVF to complement sufficient cultures has no effect on either the IgG or IgM antibody responses (58). The effect of adding anti-murine C3 to the cultures is variable. With whole IgG anti-C3 the IgM response as opposed to the IgG response to thymus dependent antigens is suppressed (62) but when $F(ab')_2$ anti-C3 was used no inhibition was observed by Waldmann & Lachmann (58). Feldmann & Pepys (62) claim to be able to produce inhibition with $F(ab')_2$ antibody. The reasons for this discrepancy are not wholly apparent. It seems unlikely to be related to the quantity of antibody since measurable quantities of anti-C3 remain even after the end of the culture period. There are possibly more subtle differences such as the degree of priming of B cells used in the culture systems with poorly primed B cells showing a greater requirement for C3. It seems clear however that the inhibition observed with whole IgG anti-C3 requires the Fc part of the molecule suggesting that the inhibition observed is due to the blocking activity of immune complexes present on the macrophage surface rather than to inhibition of C3 function.

As was described above Klaus & Humphrey (55) have provided a possible explanation for the role of C3 in the induction of the immune response. Since CVF is a potent thymus dependent antigen in mice the C3 depletion produced is transient and subsequent injections of CVF are neutralised by antibody.

However in B mice, which fail to make antibody, prolonged C3 depletion can

be achieved by repeated injection of CVF. These mice fail to localise

antigen on the surface of the dendritic reticular cells within lymphoid

tissue and neither germinal centre formation or the development of B cell

memory occurs.

It seems likely therefore that the role of complement in the induction

of the immune response is to facilitate the localisation of antigen on

the surface of dendritic cells via the C3b receptors on these cells.

It is this failure to achieve correct antigen localisation within

the microenvironment of the lymphon which apparently underlies the

failure of C3 depleted mice to mount effective antibody responses. It is

not excluded that the lymphocyte C3b receptor is not similarly involved in

the localisation of antigen or its transport to developing follicles

within lymph nodes. However the lymphocyte C3b receptor plays no necessary

role in in vivo antibody responses since lymphocytes which lack C3b

receptors can produce a normal response (63)

"References"

1.   Müller-Eberhard H.J.   Ann. Rev. Biochem. 44, 697, 1975.

2.   Lachmann P.J.   The Immune System edited by Ian McConnell & M.J. Hobart
      p. 56, Blackwell, Oxford, 1975.

3.   Alper C.A. and Balavitch D. Science 191, 1275, 1976.

4.   Dickler H.B. and Kunkel H.G. J. Exp. Med. 136, 191, 1972.

5.   Bokisch V.A. and Sobel A.T. J. Exp. Med. 140, 1336, 1974.

6.   Suba E.A. and Csako G. J. Immunol. 117, 304, 1976.

7.   Duke H.L. and Wallace J.M. Parasitology 22, 414, 1930.

8.   Nelson R.A. Science 118, 733, 1953.

9.   Nelson R.A. Advanc. Immunol. 3, 131, 1963.

10.   Nussenzweig V. Advanc. Immunol. 19, 217, 1974.

11.   Cooper N.R. Science (Wash. D.C.) 165, 396, 1969.

12.   Ross G.D. and Polley M.J. J. Exp. Med. 141, 1163, 1975.

13.   Dierich M.P., Ferrone S., Pellegrino M.A. and Reisfeld R.A.
      J. Immunol. 113, 940, 1974.

14.   Dierich M.P., and Reisfeld R.A. J. Immunol. 114, 1676, 1975.

15.   Henson P.M. and Neshyba J. J. Immunol. 116,   1976.

16.   Okada H. and Nishioka K. J. Immunol. 111, 309, 1973.

17.   Ross G.D., Polley M.J., Rabellino E.M. and Grey H.M. J. Exp. Med. 138,
      798, 1973.

18.   Okada H. and Okada N. Japanese J. Exp. Med. 44, 301, 1974.

19.   Lachmann P.J., Hobart M.J. and Aston W.P. Handbook of Experimental
      Immunology, edited by D.M. Weir, Chapter 5, Blackwell, Edinburgh, 1973.

20.   Theofilopoulos A.N., Bokisch V.A. and Dixon F.J. J. Exp. Med. 139, 696,
      1974.

21.   Bianco C., Patrick R. and Nussenzweig V. J. Exp. Med. 132, 702, 1970.

22.   McConnell I. and Hurd C.M. Immunology 30, 825, 1976.

23.   Parish C.R. and Hayward J.A. Proc. Roy. Soc. Lond. B. 187, 65, 1974.

24.   McConnell I. Nature New Biol. 233, 177, 1971.

25.   Gyongossy M.I.C., Arnaiz-Villena A., Soteriades-Vlachos C. and
      Playfair J.H.L. Clin. Exp. Immunol. 19, 485, 1975.

26.   Parish C.R. and Hayward J.A. Proc. Roy. Soc. Lond. B. 287, 47, 1974.

27.   Jondal M., Klein G., Oldstone M.B.A., Bokisch V. and Yefenof E.
      Scand. J. Immunol. 5, 401, 1976.

28.   Arnaiz-Villena A., Halloran P., David C.S. and Festenstein H.
      J. Immunogenetics 2, 415, 1975.

29.   Arnaiz-Villena A. and Festeinstein H. Nature (Lond) 258, 734, 1975.

30.   Dierich M.P., Pellegrino M.A., Ferrone S. and Reisfeld R.A. J. Immunol.
      112, 1766 (1974).

31.   Hobart M.J. and Lachmann P.J. Transpl. Rev. 32, 26, 1976.

32.   Rittner C., Hauptmann G., Grosse-Wilde H., Grosshans E., Tongio M.M.
      and Mayer S. Histocompatibility testing p.945, 1975.

33. Teisberg P., Abersson I., Olaisen B., Sedde-Dahl Jr., T. and Norrby E. Nature 264, 253, 1976.

34. Ellman L., Green F. and Frank M. Science 170, 74, 1970.

35. Lachmann P.J., Grennan D., Martin A. and Demant P. Nature 258, 242, 1975.

36. Meo T., Krasteff T. and Shreffler D.C. Proc. Nat. Acad. Sci. 72, 4536, 1975.

37. Ferrone S., Pellegrino M.A. and Cooper M. Science 193, 43, 1976.

38. Saunders D. and Edidin M. J. Immunol. 112, 2210, 1974.

39. Perlmann P., Perlmann H., Muller-Eberhard H.J. and Manni J.A. Science 163, 937, 1969.

40. Perlmann P., Perlmann H. and Lachmann P.J. Scand. J. Immunol. 3, 77, 1974.

41. Halbwachs L., McConnell I. and Lachmann P.J. in International Symposium on Membrane Receptors of Lymphocytes, Edited by M. Seligmann, J.L. Preudhomme and F.M. Kourilsky, North Holland Publishing Co.

42. Halbwachs L. and Lachmann P.J. Scand. J. Immunol. 5, 697, 1976.

43. McConnell I. and Lachmann P.J. Transpl. Rev. 32, 72, 1976.

44. Bentley C., Bitter-Suermann D., Hadding U. and Brade V. Europ. J. Immunol. 6, 393, 1976.

45. Allen F.H. Jr. Vox. Sang. 27, 382, 1974.

46. Platts-Mills T.E. and Ishizaka K. J. Immunol. 113, 348, 1974.

47. Kierszenbaum F., Ivanyi J. and Budzko D.B. Immunology 30, 1, 1976.

48. Budzko D.B., Lachmann P.J. and McConnell I. Cellular Immunology 22, 98, 1976.

49. Klein G., Dombos L., and Gothoskar B. Int. J. Cancer 10, 44, 1972.

50. Klein G., Zeuthen J., Terasaki P., Billing R., Honig R., Jondal M. Westman A. and Clements G. Int. J. Cancer, 1977 (in press)

51. Nelson R.A. Proc. Roy. Soc. Med. 49, 55, 1956.

52. Henson P.M. J. Immunol. 105, 476, 1970.

53. Yamamura M. and Valdimarsson H. Scand. J. Immunol, 1977 (in press).

54. Schorlemmer H-U. and Allison A.C. Immunology 31, 781, 1976.

55. Klaus G.G.B. and Humphrey J.H. Immunology, 1977 (in press).

56. White R.G., Henderson D.C., Eslami M.B. and Nielsen K.H. Immunology 28, 1, 1975.

57.   Hartmann K.U. and Bokisch V.A. J. Exp. Med. 142, 600, 1975.

58.   Waldmann H. and Lachmann P.J. Europ. J. Immunol. 5, 185, 1975.

59.   Koopman W.J., Sandberg A.L., Wahl S.M. and Mergenhagen S.E.
       J. Immunol. 116,   1976.

60.   Sandberg A.L., Wahl S.M., and Mergenhagen S.E. J. Immunol. 115,
       139, 1975.

61.   Pepys M.B. Transpl. Rev. 32, 93, 1976.

62.   Feldmann M. and Pepys M.B. Nature 249, 159, 1974.

63.   Mason D.W. J. Exp. Med. 143, 1111, 1976.

Chapter 11

PROPERTIES OF THE Fc RECEPTOR ON MACROPHAGES

AND MONOCYTES

Keith J. Dorrington
Department of Biochemistry and Institute
of Immunology, University of Toronto,
Toronto, Canada M5S 1A8

Abstract

Macrophages and monocytes possess a surface receptor specific for the
Fc region of certain subclasses of IgG. The binding site on IgG is localized
within the $C\gamma3$ homology regions of the heavy chains. The intrinsic affinity
of the receptor for IgG ranges from 10^6 to 10^8 M^{-1} depending on species and
subclass of IgG. The most definitive studies on mouse macrophages indicate
that IgG2a rapidly associates and dissociates from the receptor. The overall reaction is exothermic; increasing temperature lowers the intrinsic
affinity. The Fc receptor, in common with many other membrane components,
may be capped by polyvalent ligands under permissive conditions and capping
is inhibited by azide. Data on the chemistry of the receptor is both sparse
and conflicting. Sensitivity of the receptor to proteolytic enzymes has
been clearly demonstrated for mouse macrophages although rabbit and guinea-
pig cells appeared to carry resistant receptors. IgG-binding may be inhibited
if cells are treated with phospholipases and certain group-specific reagents.
Evidence is reviewed indicating that the Fc receptors found on various cell
types are different. The macrophage Fc receptor appears to play a role in
mediating phagocytosis and in non-immune cytotoxicity. Whether the receptor
serves only to concentrate sensitized target cells at the cell surface or
whether occupation of the receptors results in modulation of effector cell
function remains to be determined.

Introduction

The importance of the macrophage in homeostasis is incontroversial (see,

for example, the recent monograph edited by Nelson (1)). It plays a central

role in host defense against infection, in inflammation and in the induction

of specific immunity. Although the macrophage has the ability to distinguish

self from non-self by mechanisms which do not involve the intervention of

antibody, the enhancement of phagocytosis of red cells and bacteria when the
latter were coated with antibody (opsonized) was recognized early in this
century (early data has been summarized by Bloom (2)). Boyden and Sorkin
(3,4) pioneered the study of cytophilic antibodies which were distinct from
reaginic (or IgE) antibodies. Since that time an enormous literature has
accumulated regarding the Fc receptor on macrophages and monocytes. Despite
these efforts our understanding of the chemical nature and in vivo functions
of this receptor remains rudimentary. Studies on the specificity of the
receptor at the level of the immunoglobulin ligand are well advanced but detailed chemical data is still lacking. I have not attempted the unrewarding
task of comprehensively reviewing the literature on cytophilic antibody since
this has already been done to varying degrees (e.g. 5) and because much of
the data is repetitive or confirmatory.

Before discussing the Fc receptor it is important to define what is
implied by the term 'receptor'. In the present context I define the Fc receptor as a distinct molecular entity present on the plasma membrane of the
macrophage and the monocyte with the following properties. (i) Saturability.
There are a finite number of receptors per cell, all of which can be occupied
at high free-ligand concentrations. (ii) Specificity. The receptor shows
specificity for only certain classes or subclasses of immunoglobulin. (iii)
High affinity. The specific ligands are bound with high energy implying a
close stereochemical relationship to the receptor. These various aspects of
the Fc receptor are developed below.

Many of the studies to be reviewed have employed the 'rosette' test to
assay for the presence of the Fc receptor. Red cells are coated with immunoglobulin either specifically, using antibody specific for a red-cell determinant, or passively following chemical modification of the red cell surface
(e.g. with chromic chloride or tannic acid). Sensitized red cells will bind
to macrophages (or other cells carrying the receptor) forming a characteristic
rosette as shown in Figure 1. Recently more sensitive assays involving the
binding of radiolabelled or fluorescent immunoglobulin have been developed
which more readily yield quantitative data.

Properties of Fc receptor

Specificity The Fc receptor shows a marked specifity for IgG although
there are a few studies which indicate that some other Ig classes may bind
(see below). In species where subclasses of IgG have been defined there is

an asymmetric distribution of binding activity among the subclasses. For example, in man the receptor binds monomeric IgG1 and IgG3 but not IgG2 or IgG4 (e.g. 6). It is interesting that a similar distribution of classical complement-fixing activity exists among the subclasses although the two effector functions are mediated by different regions of Fc (7). The same situation exists in the mouse where IgG2a and IgG2b both interact with the Fc receptor and activate complement whereas IgG1 mediates neither of these effector functions. In the guinea pig also, these two activities are restricted to the IgG2 subclass. The biological significance, if any, of this combination of functions is unclear.

Early studies showed that the effector site for macrophage/monocyte binding is localized exclusively in the Fc region (8). More recently attempts have been made to localize the binding site in subfragments of Fc. These have been stimulated, at least in part, by Edelman's proposal (9) that the homology regions apparent in the primary structure of heavy and light chains have evolved to perform unique biological functions. There is now persuasive evidence (reviewed in Ref. 10) that each homology region is folded into a compact globular 'domain'. Fc is composed of two pairs of domains; the amino-terminal Cγ2 domains and carboxy-terminal Cγ3 domains. The residues mediating the interaction with the Fc receptor and other effector functions (e.g. complement fixation) are presumably exposed on the surface of the domains. Well characterized fragments derived from the Cγ3 region have been available for some time and have proved useful in defining the ligand for the Fc receptor.

Yasmeen et al (7,11) were the first to demonstrate that the binding site for the Fc receptor was localized in the Cγ3 region. Two experimental approaches were used in a heterologous system (guinea-pig peritoneal macrophages and human IgG1 fragments). In a direct quantitative rosette assay Cγ3-fragments were coated onto tanned SRBC and shown to be approximately one third as active, on a molar basis, in promoting rosette formation as red cells coated with the parent intact IgG. This apparent reduction in activity may have been due to a proportion of the low molecular weight fragments being bound to the SRBC in a fashion which sterically blocked the binding site for macrophages. This possibility was supported by the results of an inhibition assay, in which the ability of unaggregated Cγ3 fragment to compete with IgG1 coated onto tanned SRBC for the Fc receptor was assessed. In fact in this assay the Cγ3 fragment appeared to have twice the molar binding activity of intact IgG when it was used as the competing species. The reason for this

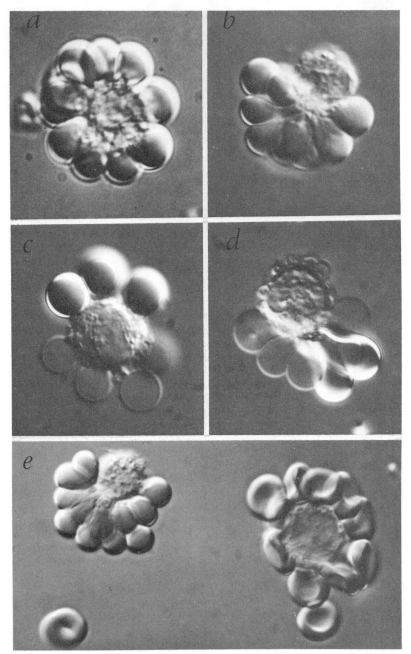

enhanced activity is not clear. A well-defined fragment corresponding to
the Cγ2 domain (12) was completely without activity in either the direct or
indirect assay. The involvement of Cγ3 was confirmed by Okafor et al (13)
who showed inhibition of rosette formation between human monocytes and anti-
Rh$_o$ (D)-coated human RBC by pFc' fragments derived from IgG1 and IgG3 human
myeloma proteins. Corresponding fragments derived from IgG2 and IgG4
proteins were not active as anticipated from earlier data on the subclass
specificity of the Fc receptor (6). The failure of Abramson et al (14) to
demonstrate inhibition by pFc' in an apparently identical assay is difficult
to reconcile with the above data.

From the structure of the smallest Cγ3 fragment shown to be active in
the rosette inhibition assay (the Fc' fragment of Turner and Bennich (15))
the binding site must be located somewhere between Gln 342 and His 433 (Eu
numbering of the γ chain. A recent report by Ciccimarra et al (16) places
the site in a ten-residue stretch between Tyr 407 and Arg 416. A fragment
corresponding to this peptide was obtained from Fc fragments of pooled human
IgG by a somewhat fortuitous combination of papain, acid and cyanogen bromide
cleavages. The results they obtained with this peptide in a rosette
inhibition assay were frankly surprising. Comparable inhibition (~ 90%)
was seen with 0.6 nmol of the peptide and 280 nmol of IgG, i.e. the peptide
is about 460 times more active than the parent molecule! Further surprises
come if one locates the homologous sequences in the high-resolution X-ray
model of any of the domains in the Fab fragment; these sequences are buried
in the interior of the domains. Although comparable crystallographic data

Figure 1

Interference contrast photomicrographs of adherent human blood monocyte
EA(anti-RH$_o$(D)) rosettes in serum. Non-rosetting EA appear as normal bi-
concave discs and serve as internal standards of size (ca. 7 microns) and
morphology. a, b, non-capped and capped (respectively) trypan blue-negative
EA rosettes. Both cells have induced a tense spherocytosis in bound EA. The
extreme cap in b shows morula-like rosetting EA tightly convergent upon a
single focus of cell membrane; c, d, live non-capped and capped blood monocyte rosettes bearing EA both as spherocytes and as haemolysed ghosts. In
d, capping has polarized to that region antipodal to the characteristic re-
niform nucleus; e, adjacent live (left) and dead (right) human peritoneal
macrophage EA rosettes in serum. The cell on the left has induced a spherocytic change in its capped EA. The cell on the right showed nuclear staining
by trypan blue; its rosetting EA remained randomly distributed and of normal
morphology.

is not available for Fc domains it is reasonable to predict that the homology
in sequence will yield similar three-dimensional folding patterns. These
data also lead to the conclusion that the binding site is linear rather
than conformational i.e. is formed from a linear array of amino acids rather
than one-dimensionally distant residues brought into close apposition by
tertiary folding. These workers also found that certain other peptides enhanced, rather than inhibited, rosette formation. This effect was attributed
to the presence of glycine; a contention supported by the apparent ability of
enormous concentrations ($> 10^{-2}$ M) of glycine-containing dipeptides to enhance
rosette formation. The logic behind their suggestion that this data indicates
a role for glycine in Fc-receptor activity is not entirely clear since, if
this were so, one might expect that the peptides would show an affinity for
Cγ3 and block its ability to interact with the Fc receptor.

     The subtle structural features which determine whether or not an
immunoglobulin will exhibit affinity for the Fc receptor are illustrated by
considering human IgG1 and IgG4. Comparison of the sequences of these two
subclasses in the Cγ3 region between residues Gln 342 and His 433 shows only
two amino acid interchanges (17). One is a conservative replacement (Arg in
IgG4 for Lys in IgG1 at position 409) whereas the other is not (Gln for Arg
at position 355). It is tempting to speculate the one or both of these
residues in IgG1 is responsible for the binding specificity. This idea is
supported by the presence of the same residues in rabbit IgG (18) and guinea-
pig IgG2 (19) both of which are cytophilic. Unfortunately the sequence of
cytophilic mouse IgG2a has a deletion at position 355 (20) which tends to
implicate Lys 409 as the crucial residue. This residue is contained in the
peptide studied by Ciccimarra et al (16) discussed critically above.

     Problems associated with the correlation of primary structure and
effector function have also been encountered for complement fixation in the
Cγ2 region; e.g. why does IgG1 fix Clq and IgG4 not? This particular enigma
was partially resolved by Isenman et al (21) who found Fc fragments from the
two subclasses to be equally active in fixing Clq and suggested that quaternary interactions between Fab and Cγ2 in IgG4 blocked the binding site. A
similar mechanism cannot be invoked with respect to Fc receptor binding since
Okafor et al (12) have shown that Cγ3 fragments prepared from human IgG subclass proteins showed the same distribution of cytophilic activity as the
the parent proteins.

     Although the bulk of the available experimental data suggests that the
Fc receptor is specific for IgG, sporadic reports indicate that binding of

other classes may occur in some systems. Thus, while there is almost general agreement that pentameric (19S) IgM does not interact with the receptor (e.g. 22), Rhodes (23) has data suggesting that monomeric (7S) IgM$_S$-coated SRBCs can form rosettes with guinea-pig splenic macrophages. The rosette-forming cells were identified as macrophages by a variety of tests including magnetic depletion following ingestion of carbonyl iron. Binding occurred in the absence of complement and lymphocytes were inactive. No significant rosette formation occurred with the parent 19S IgM antibody. Rhodes suggests that the putative receptor for monomeric IgM may be involved in the collaborative function that macrophages are known to exercise in the immune response. He picks up Feldman's proposal (24) that T cells may release an IgMs-antigen complex which binds to macrophages in an array appropriate for presentation to B cells. It would be interesting to see this observation extended, particularly to determine if IgMs and IgG were binding to the same receptor and, if so, whether the affinities were similar.

Although IgE is known to be cytophilic for mast cells and basophils (see chapter by Froese in this volume) the only evidence that IgE may also bind to macrophages has come from some studies by Capron et al (25). Adherent peritoneal cells from normal Fisher rats (identified as macrophages by electron microscopy) were found to bind strongly to the schistosomules of the helminth Schistosoma mansoni in the presence of serum from rats immune to the parasite. Adherence of the macrophages was inhibited if the immune serum was heated at 56° for 3h and if the serum was absorbed with anti-rat IgE, either in the fluid phase or with the anti-IgE coupled to Sepharose. Absorption with anti-rat IgG, IgM and IgA did not affect adherence. These and a number of other experiments provided convincing evidence that rat macrophages carry a receptor for IgE. It would be interesting to determine if IgE binds to the same receptors as cytophilic IgG and, since IgE is homocytotropic for mast cells, whether it binds to heterologous macrophages. Stimulated by the above observation we have tried and failed to inhibit rosette formation between human peritoneal macrophages and anti-Rh$_o$(D)-sensitized red cells with a human IgE myeloma protein (Romans, Bennich and Dorrington-unpublished data). Whether this was due to a very low intrinsic affinity of monomeric IgE for the IgG receptor or the presence of two classes of receptor remains to be determined.

Using a variety of assays, both human serum and secretory IgA have been shown to have no measureable affinity for the Fc receptor on macrophages (26).

Number and affinity of receptors. Several studies have been published
in which the number and intrinsic affinity of Fc receptors for monomeric IgG
on macrophages have been measured, the most definitive of which has come
from Unkeless and Eisen (27). They studied the binding of $[^{125}I]$-labelled
mouse myeloma proteins to normal and thioglycollate-stimulated mouse
peritoneal macrophages and the macrophage-like cell line (P388D$_1$) described
by Koren et al (28). The assay system used allowed the measurement of
equilibrium constants as well as rates of association and dissociation. IgA,
IgM and IgG1 were not bound significantly, IgG2a was bound with high affinity
(K_A = 1.1 x 10^8 M^{-1} at 4°) and IgG2b showed a twenty-fold lower affinity.
All three types of cell used showed similar binding characteristics and the
linearity of the Scatchard plots indicated a single class of binding site.
Binding was exothermal, as the temperature was lowered the affinity increased
(2.2 x 10^7 M^{-1} at 37° and 1.1 x 10^8 M^{-1} at 4°) and the dissociation rate
constant decreased (0.44 min^{-1} at 37° and 0.11 min^{-1} at 4°). The association
and dissociation rate constants for IgG2a binding were both large indicating
that in vivo there is a dynamic equilibrium between bound and free Ig. This
is likely to be functionally important feature of the receptor since a free
concentration of only 7 µg IgG2a/ml would result in half the receptors
being occupied. The serum concentration of IgG2a is presumably much higher
than this value and the receptors normally saturated. Mouse IgG2a is bound
with higher affinity than other Ig's are to their homologous receptors. In
fact mouse IgG2b is bound with an affinity in line with the other species;
about 10^6 M^{-1} for rabbit (29) and guinea-pig (30) macrophages. Teleo-
logically one might expect the affinity to be related to the serum concentration of ligand. For example, with an affinity of 10^6 M^{-1} the majority
of receptors would be occupied at a IgG concentration of 8 x 10^{-5}M (i.e. 12
mg/ml). Since the concentration of any specific antibody would be much
lower than this, the formation of antibody-antigen complexes with high
avidity for Fc receptors arising from their polyvalency, would be necessary
to ensure effective presentation of the antigen to the macrophage. Problems
would arise if either the receptor affinity was very high or the dissociation
rate very slow. Under these circumstances antibody-antigen complexes (or
antibody-coated target cells) would compete inefficiently with a large excess
of monomeric IgG for receptors and the number of unoccupied receptors would
be very small.

Rabbit (29) and guinea-pig (30) macrophages possess approximately 10^6
receptors/cell while mouse cells carry about 10^5 (27). Stimulated

('activated') macrophages have more receptors probably because of their increased size and surface area. From the number of receptors/cell, the surface area of an unstimulated mouse macrophage (1000 μm^2 (27)) and the cross-sectional area of Fc (1.7 x 10^{-5} μm^2 (8)) one can calculate that about 0.2% of the cell surface would be occupied by IgG if all the receptors were filled.

An intriguing feature of Unkeless and Eisen's data was the finding that normal and thioglycollate-stimulated macrophages apparently possessed twice as many binding sites for IgG2a than for IgG2b (27). If this difference is outside experimental error (as seems likely) it raises the possibility that more than one population of receptor exists on the cell surface. Receptor heterogeneity was not apparent from the Scatchard analysis of the binding data for either subclass. Preliminary experiments indicated mutual interference in the binding of IgG2a and IgG2b but whether this was to due to competition for the same receptors or steric hinderance was not evaluated. The latter seems unlikely in view of the small proportion of cell surface accounted for by receptors unless, of course, the two types of receptor were linked. Clearly there are contradictions here which deserve further study (31). One can envisage situations in which subclass-specific receptors could be functionally significant. For example, if human IgG1 and IgG3 share the same receptor, the much higher (about tenfold) serum concentration of IgG1 would result in the majority of receptors being occupied by this subclass. A separate receptor for IgG3 would insure that antibodies of this subclass had appropriate access to macrophage effector mechanisms.

Receptor mobility. Recently, Singer and Nicolson (32) have formulated a fluid mosaic model to synthesize the rapidly accumulating and occasionally conflicting data relating to the structure and biological functions of mammaliam membranes. In this model, which is based primarily on thermodynamic considerations rather than morphological data, cell membranes are viewed as two-dimensional solutions of globular and amphipathic integral proteins randomly distributed throughout a discontinous phospolipid matrix. Important evidence for this model has come from observations that specific receptors can be induced to form local clusters or 'patches' separated by unlabelled regions of membrane when cells are incubated with the appropriate ligand. Under permissive conditions (e.g. at 37°) these patches coalesce into a 'cap' over one pole of the cell. This phenomenon has been described, using a variety of techniques, for a number of cell surface components (for access to literature see Ref. 33).

Romans et al (33) have demonstrated a translational mobility and polar distribution of the Fc receptor on human blood monocytes and peritoneal macrophages using anti-Rh$_o$ (D)-sensitized red cells (EA) in a rosette technique (Fig. 1). Capping, defined as binding 6 or more EA confined to the cell half-perimeter, proceeded in a time-and temperature-dependent fashion; 80% live rosettes formed caps after approximately 40-60 min at 37°. Virtually no caps formed at 4°. The presence of serum appeared to promote rosette capping compared to PBS; about 70% of the caps observed at 37° were extreme caps in which bound EA appeared as a morula contiguous with the adherent cell. Capping was reversibly inhibited by azide although inhibition was only about 80% complete even at 0.1M, a concentration tenfold higher than that giving complete inhibition of antigen-receptor capping on lymphocytes. Re-exposure of capped rosettes to EA failed to increase either the proportion of cells forming rosettes or the fullness of such rosettes indicating that a critical number of receptors had been capped.

Capping of the Fc receptor on macrophages has also been demonstrated by Thrasher et al (34) using a different technique. Guinea pig alveolar macrophages were incubated with cytophilic rabbit antibody and then stained with fluoresceine-labelled goat anti-rabbit globulin. Cells incubated at 4° or room temperature showed continuous or peripheral staining which formed into a cap within 30 min of incubation at 37°.

Chemistry of receptor. Over the past ten years several attempts have been made to determine the gross chemical features of the Fc receptor using proteolytic enzymes, phospholipases and a variety of agents capable of modifying protein functional groups.

Fc receptors on guinea-pig and rabbit macrophages appear to be resistant to trypsin, chymotrypsin, papain, ficin and Pronase (35). In fact several studies suggest that enzyme-treated cells bind more cytophilic antibody than untreated control cells. Arend and Mannik (29), for example, found that macrophages from male rabbits when treated with trypsin or pronase bound IgG to the same number of receptors as control cells but the average affinity increased two-fold. Cells from females showed an increased number of receptors with no change in affinity after treatment. In their studies on mouse macrophages, Unkeless and Eisen (27) clearly demonstrated that the Fc receptor was trypsin sensitive. In fact their assay for binding affinity was based on this property. Binding of Ig2a and Ig2b was almost completely abrogated following a 15 min incubation with trypsin. However, about 60% of the Ig-binding activity was restored after incubating the trypsinized cells for 12h in serum-free medium. In the presence of cycloheximide (100 ug/ml), receptor regeneration was completely inhibited. Less complete inhibition was observed with actinomycin D. As the authors suggest these data are consistent with either an intracellular pool of receptor, the incorporation of which into the membrane requires other protein synthesis (e.g. glycosidases), or the presence of a fairly stable messenger RNA for the receptor, whose translation is blocked by cycloheximide.

Thus we are faced with conflicting data on the protein nature of the receptor. The issue is further confused by other data obtained in the mouse suggesting that the receptors which bind cytophilic antibody produced early in the immune response to SRBC (36) or oxazalone (34,37) were trypsin sensitive while those binding antibody from late antisera were resistant. I find these observations difficult to rationalize especially since Ig binding to macrophages is mediated via a site in the constant-region of the H chain and, as such, should be independent of antibody combining site specificity or affinity. These observations cannot be related to existence of two classes of receptor, one specific for IgG2a and the other for IgG2b, with these two subclasses being unequally represented in early and late antibody populations since the receptors for the two subclasses are both trypsin sensitive (27).

In the studies showing resistance of the receptor to proteolytic enzymes a susceptibility to phospholipase has been demonstrated (35). This does not necessarily mean that the receptor is phospholipid in nature since any perturbation of the fluid bilayer could adversely affect the activity of the constituent membrane proteins.

Evidence suggesting the requirement for free sulphydryl and amino groups in receptor binding has come from chemical modification data. However, these experiments were performed under conditions where the reagents used (e.g. iodoacetamide) were capable of reacting with functional groups other than those predicted.

Clearly we know almost nothing about the chemistry of the Fc receptor. It seems likely that the current intense interest in membrane-associated molecules and rapid development of techniques for their isolation and characterization will lead to definitive studies in this area. There is no shortage of interesting questions to be answered: Are Fc-receptors structurally related to other surface markers of importance in the immune response (38). Is β_2-microglobulin associated with the receptor? Are Fc receptors on different cell types structurally related?

Comparison of Fc receptors on different cell types. Fc receptors are
present on a wide variety of different cell types other than macrophages and
monocytes, i.e., mast cells, basophils, neutrophils, platelets, K cells, B
lymphocytes, 'activated' T lymphocytes as well as on yolk sac and placental
membranes (for references see review by Kerbel and Davies (39)). These
receptors are only operationally similar; i.e., they show affinity for sites
in the Fc portion of certain immunoglobulin classes. There is no a priori
reason to suppose that they will be structurally similar. In fact, it may
be worthwhile to briefly summarize the evidence that the several Fc receptors
exhibit different ligand specificities and binding affinities.

The receptor on mast cells and basophils is the most fastidious showing
affinity for IgE only of the same, or closely related species (hence homocytotropic antibody). It also shows the highest affinity ($> 10^9$ M^{-1}) being
2 - 3 orders of magnitude greater than that shown by the macrophage receptor
for monomeric IgG. The localization of the binding site within Fc or IgE
is not known but it has been suggested that more than one domain may be
involved (40). The receptors on mast cells for IgE and macrophages for
IgG are porbably unique among blood cells in showing high affinity for monomeric immunoglobulin. Thus in vivo both cells types will carry Ig on their
surface even in the absence of antigen. In contrast, receptor occupancy by
IgG is likely to be low on neutrophils, platelets and B cells, because of
the low intrinsic affinity, unless a polyvalent array of Fc regions is
presented in the form of antibody-antigen complexes. Almost unavoidably
this enhanced avidity exhibited by complexed (or even heat-treated) antibody
has been attributed to a conformational change within Fc. Such changes,
although frequently invoked to account for such phenomena, have been difficult
to demonstrate (41). A minimal hypothesis requires only that the high
avidity of complexes arises from the cooperative association of several
intrinsically weak interactions (41).

Data presented in an earlier section clearly indicated that the macrophage Fc receptor is specific for a site on the Cγ3 domain of Fc. Recent
studies on the Fc receptor on the human placental syncitiotrophoblast and
on the cell (or cells) responsible for antibody-dependent cell-mediated
cytotoxicity (ADCC) suggest that both Cγ2 and Cγ3 domains may be required.
McNabb et al (42), demonstrated saturable, specific binding of IgG (order
of affinity; IgG1 = IgG3 > IgG4 > IgG2) to receptors on placental membrane
preparations. Fc was found to bind with the same affinity and to the same
number of sites as the parent IgG. However, well-characterized fragments

corresponding to the $C\gamma2$ and $C\gamma3$ domains were completely without activity
in both direct and inhibition assays. This was interpreted to mean either
that the binding site is formed from regions on both domains or that the
site is present on one domain but only active in the presence of the second.
These authors suggest that the placental receptor may have evolved to accomo-
date the IgG molecule rather than vice versa as predicted by Edelman's
domain hypothesis (7). A similar type of situation exists with respect to
the mediation of ADCC. The killing of antibody-coated target cells by K
cells cannot be inhibited by $C\gamma3$ fragments nor by a fragment, Facb, which
represents an IgG molecule lacking the $C\gamma3$ domains (43). Michaelsen et al
(44) have provided evidence suggesting that either close association of the
paired $C\gamma2$ domains or quaternary interactions between $C\gamma2$ and $C\gamma3$ are
necessary for binding to the effector cell. It is possible that IgG binds
through a site in $C\gamma3$ but that a site on $C\gamma2$ is responsible for initiating
the events leading to cytotoxicity that but both activities depend upon the
native, quaternary structure of Fc.

Biological Functions of the Fc Receptor

The macrophage is truly a cell for all seasons; it shows an insatiable
appetite for particulate matter ranging from bacteria and red cells to latex
beads; kills a variety of target cells by a non-phagocytic, complement-in-
dependent mechanism; releases a plethora of enzymes, colony-stimulating factors
and synthesizes at least two of the complement proteins (C2 and C4). In
addition the macrophage plays an often obligatory role in the afferent as
well as the efferent arm of the immune response. This functional diversity
requires either a very versatile cell or some degree of differentation within
the macrophage population. There is clear evidence that macrophages can be
'activated' in vivo and in vitro by a variety of chemical stimuli. Activation is accompanied by metabolic, secretory and morphological changes as
well as an increase in Fc receptor activity (45). The question of interest
here concerns what role, if any, does the Fc receptor play in these diverse
macrophage functions?

Antibody-dependent cytotoxicity. In a variety of in vitro systems it
has been demonstrated that normal lymphoid cells can destroy target cells
sensitized with antibody specific for target-cell surface antigens in the
absence of complement (46). The 'classical' K cell of ADCC appears to be
non-adherent, does not possess surface immunoglobulin nor T-cell markers.

Resolution of the confusion that surrounds this phenomenon is aided, in no small degree, by recognizing that more than one cell type can mediate cytotoxicity (47,48). Macrophages and blood monocytes can certainly kill a variety of sensitized target cells. Some of the features of this cytotoxicity can be explored by reference to the EA rosette system as illustrated in Figure 1. When bound via the Fc receptor to macrophages or monocytes the red cells undergo a spherocytic change and ultimately lyse (33) (see especially Fig. 1a). Viable mononuclear cells are required for this change since dead rosettes carry EA of normal morphology (Fig. 1e). Receptor capping is not a prerequisite for spherocytosis (compare Fig. 1a and b). It seems reasonable to suppose that this morphological change reflects altered membrane permeability to ions. The Fc receptor could play two distinct roles in this phenomenum:

1.   It may serve to concentrate EA in the vicinity of the macrophage facilitating contact between the membranes of the two cell types. Target-cell damage may follow as the result of the action of a membrane-bound effector molecule (e.g. phospholipase or protease) or release of a soluble mediator. In this model Fc receptors play a passive role.

2.   Occupation of the receptor by Fc actively triggers the release of activation of effector molecules perhaps via a second messenger (cGMP?). Although at this time it is not possible to select between these two possibilities there are one or two observations which may be relevant.

Goldman and Cooper (49) have studied binding of red cells to live macrophages mediated by sub-agglutinating doses of concanavalin A (Con A) and found that the red cells maintained their normal morphology. Close association of the membranes of the two cell types occurred and the red cells were ultimately ingested. This observation tends to suggest an active role for the Fc receptor in the induction of osmotic fragility. However one cannot exclude an inhibition of this effector mechanism by bound Con A.

The molecular mechanism of killing is unknown. In the EA-rosette system the spherocytosis and lysis are probably mediated by a molecule(s) bound to the macrophage membrane rather than a mechanism involving soluble factors since unbound red cells lying near to a rosette retain normal morphology, unless the effector molecule has a short half-life. This observation would tend to exclude anyone of the wide variety of enzymes which may be released from the macrophage as the effector molecule. The lack of requirement for Ca^{2+} in ADCC (50) suggests that stimulus-secretion mechanisms are not involved Preliminary studies on murine K cell cytotoxicity suggest that phospholipase

A may mediate the kill phase (51). Inhibitors and natural substrate of this enzyme also inhibited killing but did not interfere with the initial binding (i.e. rosette formation). If killing results from this rather simple mechanism then the Fc receptor might serve a passive role in bringing the enzyme and substrate into close apposition. The inhibition of cytotoxicity by agents which interfere with microfilament function and which increase intracellular cAMP levels, as well as augumentation by agents which increase cGMP levels, suggest either that killing is a more complex phenomenum or that these agents modulate membrane fluidity and consequently effect enzyme-substrate interaction. The finding that membrane changes were not apparently induced in EA bound to dead macrophages suggest that if a simple enzymic mechanism is involved the enzyme must have a short half-life or its activity is dependent upon native membrane structure. Thus, although Fc receptor - IgG interactions are not dependent of cell viability, the putative membrane plasticity necessary to ensure enzyme-target cell interaction is likely to be lost on dead cells. The macrophage EA-rosette system would seem to have much to offer as a model in which to study factors involved in cytotoxicity.

Phagocytosis. The ability of macrophages to damage foreign or abnormal cells by a phagocytic mechanism as well as by an extracellular mechanism (previous section) are probably closely linked. For example, in the EA rosette system a proportion of the red cell spherocytes generated via the extracellular cytotoxicity will ultimately be phagocytosed and further degraded by lysozomal enzymes. However, while recognition of an antibody-coated target cell by Fc receptors is an absolute requirement for extracellular lysis, phagocytosis may proceed in the absence of antibody. Presumable some mechanism must exist whereby the macrophage recognizes another cell as abnormal; surface glycoproteins may provide such a mechanism. There certainly seems to be good evidence that malignant cells are characterized by surface glycoprotein abnormalities (52). Although antibody is not an obligatory requirement for phagocytosis, its presence greatly enhances this process (i.e. the classical opsonization phenomenon). The most logical analysis of this enhancement must invoke the involvement of the Fc receptor in generating close association between the membrane and the particle being ingested and perhaps maintaining this association for a critical time period. However, in a series of papers, van Oss and his collegues (see ref. 53) have analysed phagocytosis (mainly by neutrophils) as a non-specific surface phenomenum. Coating of a particle with antibody, in such a way that the Fc regions protrude from the particle's surface is thought to form a more hydrophobic outer surface which

facilitates phagocytosis. The crucial factors in whether a particle is engulfed or not are the contact angle between the particle (which is subject
to random Brownian motion) and the surface of the phagocyte, and the inter-
facial free energy. This model does not require the existence of an Fc receptor as defined in the Introduction. While a non-specific mechanism may
be involved in the phagocytosis of latex beads, carbon particles, etc.,
this does not seem likely for sensitized cells and antibody-antigen complexes
at least not for the macrophage. The data summarized in the previous section
strongly supports the concept of a finite number of high-affinity receptors
showing specificity for the Fc of certain immunoglobulin classes on the macrophage membrane.

Secretion. The macrophage is a secretory cell (54). In culture, for
example, macrophages secrete a variety of macromolecular products including
lysozyme, several neutral proteases including plasminogen activators, collagenase and elastase. The level of enzyme secretion varies markedly with
the functional state of the cell. Lysozyme is continually secreted but plasmin
ogen activator and other neutral proteases are inducible enzymes associated
with stimulation and phagocytosis. The involvement of Fc receptors in the
secretion process stimulated by antibody-antigen complexes or heat aggregated
IgG is probably similar to their involvement in phagocytosis per se. There
is no reason at this time to suppose that occupation of Fc receptors makes
any unique contribution to the induction of enzyme secretion.

Acknowledgements

I am endebted to Drs. D.G. Romans, T. McNabb and S.R. Loube for extensive
discussions and for reading the manuscript. Work described from the author's
laboratory was supported by the Medical Research Council of Canada.

References

1. Nelson, D.S. (Editor). Immunobiology of Macrophage. Academic Press;
      New York, 1976
2. Bloom W., Arch. Path & Lab. Med., 3:608, 1927.
3. Boyden, S.V. and Sorkin, E., Immunology, 3:272, 1960.
4. Boyden, S.V., Immunology, 7:474, 1964.
5. Tizard, I.R., Bact. Rev., 35:365, 1971.

6. Hay, D.C., Torrigiani, and Roitt, I.M., Europ. J. Immunol., 2:257, 1972.

7. Yasmeen, D., Ellerson, J.R., Dorrington, K.J. and Painter, R.H., J. Immunol., 116:518, 1976.

8. Liew, F.Y., Immunology, 20:817, 1971.

9. Edelman, G.M., Biochemistry, 9:3197, 1970.

10. Cathou, R.E. and Dorrington, K.J., in Biological Macromolecules, Subunits in Biological Systems, edited by G.D. Fasman and S.N. Timasheff, p. 91, Marcel Dekker, New York, 1975.

11. Yasmeen, D., Ellerson, J.R., Dorrington,K.J. and Painter,R.H.,J. Immunol., 110:1706, 1973.

12. Ellerson, J.R., Yasmeen, D., Painter, R.H. and Dorrington, K.J., J. Immunol., 116:510, 1976.

13. Okafor, G.O., Turner, M.W. and Hay, F.C., Nature, 248:228, 1974.

14. Abramson, N., Gelfand, E.W., Jandl, J.H. and Rosen, F.S., J. Exp. Med. 132:1207, 1970.

15. Turner, M.W. and Bennich, H., Biochem. J., 107:171, 1968.

16. Ciccimarra, F., Rosen., F.S. and Merler, E., Proc. Nat. Acad. Sci. 72:2081, 1975.

17. Dayhoff, M.O., Atlas of Protein Sequence and Structure, vol. 5, p. D376, National Biomedical Research Foundation, Washington, 1972.

18. Hill, R.L., Delaney, R., Fellows, R.E. and Lebovitz, Proc. Nat. Acad. Sci., 56:1762, 1966.

19. Tracey, O.E. and Cebra, J.J., Biochemistry, 13:4804, 1974.

20. Bourgois, A., Fougereau, M. and Rocca-Serra, J., Europ. J. Biochem. 43:423, 1974.

21. Isenman, D., Dorrington, K.J. and Painter, R.H., J. Immunol., 114:1726, 1975.

22. Ralph, P., Richard, J. and Cohn, M., J. Immunol., 114:898, 1975.

23. Rhodes, J., Nature, 243:527, 1973.

24. Feldmann, M., J. Exp. Med., 136:737, 1972.

25. Capron, A., Dessaint, J-P., Capron, M. and Bazin, H., Nature, 253:474, 1975.

26. Huber, H., Douglas, S.D., Huber, C. and Goldberg, L.S., Int. Arch. Allergy, 41:262, 1971.

27. Unkeless, J.C. and Eisen, H.N., J. Exp. Med. 142:1520, 1975.

28. Koren, H.S., Handwerger, B.S. and Wunderlich, J.R., J. Immunol. 114:894, 1975.

29. Arend, W.P. and Mannik, M., J. Immunol., 110:1455, 1973.
30. Leslie, R.G.Q. and Cohen, S., Immunology, 27:577, 1974.
31. Walker, W.S., J. Immunol., 116:911, 1976. This author has provided evidence that separate receptors for IgG2a and IgG2b exist on another mouse macrophage-like cell line (1C-21).
32. Singer, S.J. and Nicolson, G.L., Science, 175:720, 1972.
33. Romans, D.G., Pinteric. L., Falk, R.E. and Dorrington, K.J., J. Immunol., 116:1473, 1976.
34. Thrasher, S., Bigazzi, P.E., Yoshida, T. and Cohen, S., Immunol. Comm., 4:219, 1975.
35. Davey, M.J. and Asherson, G.L., Immunology, 12:13, 1967.
36. Kossard, S. and Nelson, D.S., Aust. J. Exp. Biol. Med. Sci., 46:63, 1968.
37. Askenase, P.W. and Hayden, B.J., Immunology, 27:563, 1974.
38. The possible relationship between Fc receptors and Ia antigens is discussed by Halloran and Schirrmacher in this volume.
39. Kerbel, R.S., and Davies, A.J.S., Cell, 3:105, 1974.
40. Dorrington, K.J. and Bennich, H., J. Biol. Chem., 248:8378, 1973.
41. Metzger, H., Adv. Immunol., 18:169, 1974.
42. McNabb, T., Koh, T.Y., Dorrington, K.J. and Painter, R.H., J. Immunol., 117: in press, 1976.
43. Connell, G.E. and Porter, R.R., Biochem. J., 123:53, 1971.
44. Michaelsen, T.E., Wisloff, F. and Natvig, J.B., Scand. J. Immunol., 4:71, 1975.
45. Rhodes, J., J. Immunol., 114:976, 1975.
46. MacLennan, I.C.M., Transplant. Rev., 13:67, 1972.
47. Scornik, J.C. and Cosenza, H., J. Immunol., 113:1527, 1974.
48. Gelfand, E.W., Resch, K. and Prester, M., Europ. J. Immunol., 2:419, 1972.
49. Goldman, R. and Cooper, R.A., Exp. Cell Res., 95:223, 1975.
50. Goldstein, P. and Gomperts, B.D., J. Immunol., 114:1264, 1975.
51. Frye, L. and Friou, G., in Proc. 10th Leucocyte Cult. Conf., edited by V.P. Eijsvoogel, in press, Academic Press, New York. 1976.
52. Rapin, A.M.C. and Burger, M.M., Adv. Cancer Res., 20:1, 1974.
53. van Oss, C.J., Gillman, C.F. and Neumann, A.W., Immunol. Comm., 3:77,1974
54. Gordon, S., Unkeless, J.C. and Cohn, Z.A., in Immune Recognition, edited by A.S. Rosenthal, p. 589, Academic Press, New York, 1975.

Chapter 12

Fc RECEPTORS AND Ia ANTIGENS

P. Halloran and V. Schirrmacher
Department of Medicine, Mt. Sinai Hospital,
Toronto, Canada, and the Department of Cellular Immunology,
Institute of Immunology and Genetics,
Deutsches Krebsforschungszentrum, 69 Heidelberg, W. Germany

Abstract

Antibody against Ia antigens inhibits the ability of B lymphocytes to bind aggregated immunoglobulin and to form EA rosettes. The explanation suggested for this phenomenon has been that Ia antigens are identical to or closely associated with Fc receptors. But a variety of observations preclude acceptance of this explanation since inhibition is also demonstrable with antibody against a wide variety of B cell surface components, including H-2K, H-2D, β_2 microglobulin, immunoglobulin, Ly 4.2. Furthermore, Fc receptor function can be separated from Ia antigens by capping or by isolation of membrane components. It seems likely that the mechanism of inhibition of Fc receptors by antibody against Ia antigens is part of a broader spectrum of effects induced when antibody binds to cell membrane antigens.

Introduction

The discovery that serum factors, both heat labile and heat stable, could

opsonize bacteria for subsequent phagocytosis played an important role in resolving

the controversy between the proponents of the central role of the phagocytes and

the proponents of the central role of humoral factors in achieving immunity (1,2).

Today, it is clear that the capacity of certain cell membranes to bind immunoglobulin

via its Fc portion, particularly after Fc "activation" by antigen binding, plays an

important role as an interface between many cellular and humoral events. This Fc

binding capacity is conventionally attributed to cell membrane "Fc receptors". The

relationship of Fc receptors to various cell surface components, and particularly the

postulated relationship of B lymphocyte Fc receptors to antigens controlled by the I

region genes of the mouse H-2 system (3), has stimulated great interest (4,5).

The purpose of this review is to survey briefly what is known about Fc receptors and about Ia antigens and to consider the current work on potential relationships between Fc receptors and Ia antigens against this background.

The Nature of "Fc Receptors"

The "Fc receptor" is a concept which describes those cell membrane components which are capable of combining with Fc portions of immunoglobulin molecules. Whether a particular component is or is not an "Fc receptor", and whether or not a particular cell type has "Fc receptors", may well be dependent on the sensitivity of the assay for the detection of cell membrane-immunoglobulin interactions. For example, although a wide variety of investigators have concluded that B lymphocytes usually have Fc receptors, there remains considerable controversy as to whether or not Fc receptors exist on T lymphocytes and on thymocytes. In this context it must be remembered that Weissman et al. (6) demonstrated that artificial phospholipid membranes (liposomes) could bind immunoglobulin Fc with considerable evidence of subclass specificity, acting as though they had "Fc receptors". One must conclude that at least some interactions of immunoglobulin with cell membranes could be due to hydrophobic interactions of Fc with cell membrane phospholipids. Clearly, the term "Fc receptor" refers to a binding event, which can occur with greater or lesser degrees of avidity to different cell types and does not necessarily imply that discrete structures (such as protein "locks" into which the Fc "keys" must fit) will account for all the binding activity. Thus the question "Do thymocytes have Fc receptors?" should be rephrased: "What is the relative avidity with which thymocytes bind immunoglobulin in a particular assay?". Furthermore, a change in the avidity of binding of immunoglobulin by a particular cell type may reflect either a change in the number of binding sites or a change in the avidity with which the individual binding sites of the cell bind immunoglobulin; conventional assays cannot distinguish between these two types of change.

The Fc receptor system which received the earliest attention was the phagocytosis of opsonized bacteria or red cells (for discussion of early results see 7). We can also note in passing that studies of "reagins" have always involved Fc receptors of mast cells and basophils, although the nature of the specialized receptors on mast cells for the Fc portion of IgE will be considered elsewhere in these reviews. Studies of cytophilic antibodies which were distinct from reagins were initiated by Boyden and Sorkin (8,9). It now seems likely that the receptor involved in such studies is identical to the macrophage Fc receptor involved in phagocytosis of opsonized material and in other conventional Fc receptor binding assays (for discussion see 10,11).

The ability of lymphocytes to bind immunoglobulin Fc was first described by Uhr (12) using antibody-coated bacteria, and the term "Fc receptor" first arose in the inhibition of reverse immune cytoadherence system (RICA) (13). Currently, lymphocyte Fc receptors are usually detected by the binding of immunoglobulin-coated material to cell membranes, in assays such as (i) EA rosette formation (the binding of antibody-coated erythrocytes to Fc receptor bearing cells); and (ii) the binding of radiolabelled or fluorescent immune complexes (15) or aggregated immunoglobulin (16,17) to the cell membranes. Another widely studied Fc receptor dependent assay is antibody-dependent cellular cytotoxicity (ADCC). This assay resembles phagocytosis in that Fc binding is detected by the effects of the Fc receptor bearing cell on the antibody-coated target cell (for reviews see 18,19). Theoretically, any assay in which cell behaviour is altered in some detectable way by aggregated IgG or immune complexes can be used as an assay of Fc receptors.

In general the characteristics of the Fc receptor binding detected in each of these systems have been reasonably consistent. Thus pre-treatment of the cells with proteolytic enzymes, such as trypsin and chymotrypsin, fails to reduce their Fc binding ability (20,21) (with the exception of pronase treatment which does reduce Fc binding ability) (22; confirmed by V. Schirrmacher, unpublished). The binding of Fc is sensitive to inhibition by oxidizing agents, by agents which bind to free sulfhydryl groups, by phospholipases A and C, and by lysolecithin (20,21). Most Fc receptor systems are sensitive to inhibition by whole IgG or by its Fc portion, by aggregated IgG, and by immune complexes.

The tissue distribution and biological roles of Fc receptors were recently
reviewed (23). Fc receptors are widely distributed on normal cells (macrophages,
monocytes, neutrophils, mast cells, platelets, B lymphocytes, perhaps some T
lymphocytes and thymocytes, activated T cells, placenta, yolk sac, and effector
cells in antibody–dependent cell–mediated cytotoxicity) and on a variety of human,
murine and other mammalian neoplasms. The in vivo functions attributed to Fc
receptors include phagocytosis (for review see 24), follicular localization of
immune complexes (14), antibody–dependent cell–mediated cytotoxicity (ADCC) (18,19),
placental transfer of immunoglobulin (25), feedback inhibition of antibody synthesis
and immune complex disease.

When antibody binds to cell surface antigens on the Fc receptor bearing cell,
inhibition frequently occurs. For example, macrophage phagocytosis was inhibited
by antibody directed against the macrophage surface but this inhibition was dependent
on the intact Fc portion of the anti–macrophage antibody (30). Similarly antibody
directed against antigens in a cytotoxic effector cell population inhibited ADCC
(31,32,33) but only when the Fc portion of the immunoglobulin was intact (34, for
discussion see below). EA rosettes formed with rat lymphocytes were inhibited by
antibody against the AgB histocompatibility antigens of the lymphocytes (35, see also
below) and the absorption of antibody–coated erythrocytes to Fc receptor bearing cell
in human tumours was inhibited by heterologous antibody against these tumours (36).
Heterologous anti–mouse lymphocyte serum inhibited Fc binding in the RICA inhibition
technique (14). Some of these inhibitory activities probably reflected
competition for binding between the antibody in the indicator system and the
inhibitory antibody, but an Fc independent mechanism must have been operating
in at least some cases in Fc receptor inhibition by antibody, since inhibition
was also demonstrable with the $F(ab')_2$ fragments. In general, the more sensitive
Fc receptor systems(those in which the system detects small amounts of antibody,
such as phagocytosis or ADCC) have been inhibited by Fc dependent mechanisms;
whereas the systems in which larger amounts of immunoglobulin are required, such
as EA rosettes and aggregated IgG binding, have been inhibited by Fc independent

mechanisms. In short, the mechanisms of inhibition of Fc receptors by antibody
directed against the Fc receptor bearing cells are complex, multiple, and not
completely understood. It is against this background that we must view the results
obtained with anti Ia antisera inhibition of B lymphocyte Fc receptors.

Ia Antigens

The major histocompatibility complex of the mouse, the H-2 region, represents
a series of closely linked genes whose gene products were detected initially by
serologic and transplantation techniques (37) and more recently by other techniques
(for reviews see 38,39). A conventional representation of the genes mapping in
this complex are given in Table I.

The present discussion will focus on the antigens which are detected on
nucleated cells by serologic techniques, ie. the H-2K, H-2D and Ia antigens. The
H-2K and H-2D antigens are glycoproteins with molecular weights of approximately
45,000 and are very widely distributed in mouse tissues. The Ia antigens (42,43,

TABLE   I

The Mouse H-2 Complex

Properties	Mapping of the Corresponding Gene(s)				
	K	I	S	G	D
Serologically-detected antigens	+	+ (Ia antigens)		+	+
Antigens detected by transplantation and cell-mediated lysis	+	+			+
Strong MLC or graft vs. host determinants		+			
Immune response (Ir) genes		+			
Ss protein levels and S1p antigen			+		

44,45,46,47,48) are also glycoproteins, but with molecular weights of about 30,000

and are probably controlled by at least two separate genetic sub-regions (38,40,41).

The tissue distribution of Ia antigens is more restricted than that of the H-2K and

H-2D antigens: Ia antigens are found on B lymphocytes, perhaps on some T lymphocytes

and thymocytes, (49,50),as well as on some macrophages, fetal liver cells, epithelial

cells and spermatozoa; but not on brain, liver, red cells or platelets (41,51,52).

It is interesting to speculate on the relationship of genes controlling Ia antigens

to the other genes mapping in the I region, controlling such characteristics as

specific immunologic responsiveness to particular antigens (51), lymphocyte

activating determinants (53,54), and antigens detected in transplantation (H-2I)

(55) and in graft enhancement (56,57,58). Ia antigens may have a role in

cooperation between various cells in the initiation of the immune response,

including macrophages, T lymphocytes and B lymphocytes (59,60,61). Ia antigens

have apparently been conserved in evolution, occurring in the human (62,63) and

rat (56,58) as well as in mouse. The possible significance of Ia antigens in

human transplantation offers an exciting new area for clinical research.

The Inhibition of Aggregated IgG Binding to B Lymphocytes by Antibody Against Ia Antigens

When human IgG, conjugated with fluorescein (Fl-HGG), is incubated with mouse

spleen lymphocytes, a subpopulation of cells which seems to be chiefly B lymphocytes

becomes fluorescent, due to binding of the immunoglobulin to the cell membrane.

This assay has been used to study B lymphocyte Fc receptors (22). When H-2 anti-

sera were pre-incubated with spleen cells, subsequent binding of Fl-HGG to the B

lymphocytes was almost totally inhibited (3). The inhibitory activity was directed

chiefly against antigens in the I region, and not against those in the K or D regions

antibody against antigens in the K and D regions failed to inhibit Fl-HGG binding.

Pepsin digestion of the antibody, which removed the Fc portions, still left some

inhibitory activity in the Fab preparations, although much of the activity was lost

in these preparations. These observations were interpreted as reflecting "identity

or close association" of the B lymphocyte Fc receptors and Ia antigens (3).

Subsequent studies of F1-HGG binding have demonstrated that the ability to inhibit F1-HGG binding is not a unique property of antibody against Ia antigens, but is manifested by antibody against some non H-2 antigen or antigens as well (64). Using a serum A/J anti B10 antibody directed against some non H-2 antigen or antigens were shown to inhibit F1-HGG binding of cells from a strain H-2 identical with A/J (B10.A). Such a reaction could only be attributed to antibody against non H-2 antigens. This serum was raised across differences in Ly 4.2, a non H-2 specificity found mainly on B lymphocytes (65,66,67,68). Dickler and his co-authors speculated that this antiserum was directed against a non H-2 antigen whose relationship to the non H-2 immune response genes was similar to the relationship of Ia antigens to the H-2 linked immune response genes. There is as yet no evidence to support this interesting suggestion.

Using human lymphocytes inhibition of F1-HGG binding has been demonstrated with anti HL-A antisera. At least some of the inhibitory activity in such sera seemed to be due to antibody directed against human analogues of Ia antigens, but the possibility that antibody against the conventional HL-A.A and HL-A.B antigens was also capable of inhibiting F1-HGG binding was not excluded in these experiments (69).

From the outset, certain difficulties were inherent in the thesis that Ia antigens were identical to or closely associated with Fc receptors. Firstly, if Ia antigens were Fc receptors only anti Ia should inhibit Fc receptors, and no inhibition should be demonstrable with antibody against H-2K or H-2D antigens. Although initial experiments suggested this (3), recent unpublished studies have revealed that some H-2K or H-2D antisera with no known anti Ia activity were inhibitory, albeit much less inhibitory than were anti Ia sera of comparable cytotoxic activity (David Sachs, personal communication). This latter observation makes identity or close association between Ia antigens and Fc receptors more difficult to envisage. Secondly if each Ia antigen were a separate protein acting as an Fc receptor, then antibody against any one group of Ia specificities on lymphocyte cell surface should inhibit only a portion of the Fc receptors. For example, antibody against antigens controlled by the I-A subregion should inhibit only those Fc receptors controlled by that subregion, and not those controlled by

I-B or I-C subregions. Yet antibody against Ia antigens in one I subregion inhibited

all of the Fc receptors of the B lymphocytes (3). Similarly in the F_1 animal

antibody against either parent inhibited all of the Fc receptors (3). Once again,

if each Ia antigen were a separate Fc receptor, only 50% inhibition of Fc receptors

would be expected. In order to accommodate these findings Sachs and Dickler (29)

have produced an interesting but complicated model of the relationship between Ia

antigens and Fc receptors. Based on experiments of Abbas et al. (70) who reported

that immunoglobulins occur in patches or microaggregates on the cell surface,

Sachs and Dickler proposed that Ia antigens might occur in similar patches on the

cell surface, each of which served as an Fc receptor. Antibody against any one Ia

specificity would thus inhibit all the Fc receptors. Further experiments by

Abbas et·al. have suggested that Ia antigens might be distributed in such micro-

clusters (89), and thus the possibility that Fc binding could be a property of the

microclusters which contain Ia antigens remains open. Further experimental study

of the uniformity of distribution of Ia and other H-2 antigens on the cell surface

will be needed before the possibility raised by Sachs and Dickler to account for

their findings can be either confirmed or refuted.

Evidence tending to confirm the existence of a relationship between Ia antigens

and Fc receptors was presented by Basten et al. (71). In their system, which

involved the binding of I^{125} labelled fowl gammaglobulin or labelled immune complexes

to B lymphocytes, H-2 antisera inhibited Fc binding. The activity in the H-2 antiser

was absorbed by spleen cells but not by thymocytes. This evidence is suggestive of

of relationship between Ia antigens and Fc receptors but unfortunately the crucial

experiment (in which anti H-2K or D antisera of known potency but devoid of any

anti Ia activity were tested in parallel for their blocking ability) was not done.

Until strong anti H-2 sera devoid of anti Ia activity are shown to have no inhibition

in this system, the inhibiton of Fc binding observed with anti Ia in this system

cannot be taken as evidence of a unique association between Ia antigens and Fc

receptors.

In the rat, antibody against alloantigens associated with with AgB complex

and probably analogous to Ia antigens of the mouse, have been shown to inhibit

EA rosettes, whereas eluates of antibody against antigens, presumably analogous

to the K and D antigens of the mouse (antisera absorbed to and eluted   from

platelets), did not inhibit (61). This was interpreted as indicating that in

the rat, like the mouse, Ia antigens were associated with B lymphocyte Fc receptors.

However, heterologous antibody against rat immunoglobulin also completely blocked

Fc receptor binding, so that blocking of EA rosettes was not unique to antibody

against Ia antigens. This finding calls into question the validity of the

interpretation of Fc receptor inhibition as reflecting a specific association of

Ia antigens with Fc receptors. Also, the inability of eluates to inhibit Fc

receptor binding may be due to the potency or affinity of these preparations;

when similar eluates of mouse anti H-2 sera were tested by ourselves, the Fc

receptor inhibitory activity was much weaker in relationship to cytotoxic titres

than the Fc receptor inhibitory activity of anti Ia sera, but was nevertheless

present and specific (72).

In summary, the case for identity or close association of Ia antigens with

Fc receptors at present rests primarily on the observation that antibody against

Ia antigens(and the $F(ab')_2$ of such antibody) inhibits Fc receptors.

Evidence Which Fails to Support the Thesis that Ia Antigens are Identical or Closely Associated with Fc Receptors of B Lymphocytes

We undertook to study the mechanisms of interaction between antibody against

cell surface antigens and Fc receptors of cytotoxic effector cells involved in

antibody dependent cell-mediated cytotoxicity (ADCC). Antisera directed against

a wide variety of antigens in the effector cell population inhibited cytotoxicity

in this system (32,33). But the mechanism of inhibition was not due to specific

binding of the antibody combining site to antigens associated with the effector

cell Fc receptor, because the inhibition was totally dependent on intact Fc

fragments of the antibody. Thus the pepsin digested preparations $(F(ab')_2$

completely lost their ability to inhibit specifically ADCC, whereas the IgG
preparations of the same antisera were powerfully and specifically inhibitory
(34,73). This dependence on intact Fc portions was observed even when antisera
containing very strong anti Ia activity were used. Hence the sole mechanism by
which antibody against Ia or other cell membrane antigens inhibited Fc receptors
of cytotoxic effector cells was by competition between the Fc portions of these
antibodies and the Fc portions of the antibody on the target cell. Since the
$F(ab')_2$ fragments of anti Ia cannot inhibit the Fc receptors of cytotoxic effector
cells, the Fc receptor of the cytotoxic effector cell is not identical to or
closely associated with Ia antigens (73).

Since no relationship between Fc receptors and Ia antigens was apparent in
the ADCC system, we undertook to assess directly the immunogenetic significance of
the inhibition of B lymphocyte Fc receptors by anti Ia, using a system of EA
rosettes in which antibody-coated chicken erythrocytes (EA) bind to Fc receptor-
bearing mouse spleen cells or lymph node cells. In such a system, most of the
rosette-forming cells are B lymphocytes, although occasional macrophages and
possibly some T lymphocytes may be forming rosettes as well, as has been reported
in other Fc receptor systems (17,74,75,76,77). About 50% of the spleen cells
and 25% of lymph node cells formed EA rosettes.

In this EA rosette system, inhibition occurred with antibody against the
whole H-2 complex or against Ia antigens alone (73) thus resembling initially the
results previously obtained in the inhibition of Fl-HGG binding by Dickler and Sachs.
But when alloantisera raised against Ia negative tumours were used, strong inhibition
was also observed (78), suggesting that antibody against alloantigens other than Ia
might be capable of mediating EA rosette inhibition. Accordingly, an assortment
of antisera directed against K and D region antigens was selected and tested in
EA rosette inhibition against cells from strains differing only at the K or D
regions. These studies, particularly those with antisera lacking Ia reactivity,
led to the conclusion that antisera recognizing only K or D region antigens could
inhibit EA rosettes, as well as antisera recognizing only Ia antigens (78).

Furthermore, whether the inhibition was due to anti Ia, anti H-2D, or anti H-2K, the Fc portion of the antibody did not seem to be necessary, since the pepsin-digested IgG (ie. $F(ab')_2$ fragments) was almost as potent as the undigested IgG (78). The titre of the antisera in EA rosette inhibition was not as high in some anti H-2K sera as in the anti Ia sera (relative to the cytotoxic titres) but the titres of certain H-2D sera were as high as the anti Ia (72).

Antiserum against antigens unrelated to the H-2 complex blocked EA rosettes. One antiserum identifying a non H-2 antigen possibly identical to Ly 4.2 was strongly inhibitory (73), as has been shown for a similar antiserum in Fl-HGG binding (see above). Antibody against mouse immunoglobulin (and its $F(ab')_2$ fragments) strongly inhibited EA rosette formation (73) although previous studies in EA rosettes of anti-immunoglobulin had been equivocal (79) and anti-immunoglobulin had failed to inhibit Fl-HGG binding (22). The reason for this discrepancy may have been related to the use in some of these studies of anti-immunoglobulin which was capable of binding to the Fc portions of the antibody on the EA or the Fl-HGG respectively. Such a binding was shown by us to permit "mixed agglutination" (11) independent of Fc receptors but capable of simulating Fc receptor-mediated binding, producing the impression that anti-immunoglobulin did not inhibit Fc receptors, whereas in reality the inhibition of Fc receptor function was complete (80).

Antibody against MBLA also inhibited B lymphocyte EA rosettes, although the existence of anti Ia activity in such a serum has not been ruled out. Other investigators have tested antibody against β_2 microglobulin in both EA rosettes (81) and in aggregated IgG binding (82) and have demonstrated that it inhibited Fc receptor binding, independently of its Fc portion.

The inhibition of B lymphocyte Fc receptors by the IgG and $F(ab')_2$ of anti Ia, anti H-2D, anti H-2K, anti Ly 4.2, anti mouse immunoglobulin, anti MBLA, and anti β_2 microglobulin is thus not restricted to one particular type of antigen; but the inhibition is specific nonetheless, in that antibody against irrelevant H-2 or Ia antigens or against T cell antigens (eg. Thy 1.2) failed to inhibit B cell Ea rosettes (73).

To further elucidate the significance of antiserum-induced inhibition of B cell Fc receptors, we assessed the effect of antisera against other varieties of receptor-bearing cells. When antisera against H-2K, H-2D, and Ia antigens were tested against mouse bone marrow cells (which are predominantly Ia negative) (51), P815Y tumour cells (Ia negative), or macrophages (partially Ia negative) (52) the anti H-2K or D sera were much more potently inhibitory than were the anti Ia sera (72,73,80). This result suggests that the inhibition of B cell Fc receptors by antibody against B cell surface antigens is not due to a unique association of B cell Fc receptors with Ia antigens but is part of a more general mechanism whereby antisera against surface antigens of Fc binding cells reduce or inhibit their Fc binding affinity.

Such a conclusion finds support in other types of investigations involving Fc receptors and Ia antigens. For example, identity or close association of Ia antigens and Fc receptors might be expected to result in co-capping of these surface markers. However, experiments attempting to demonstrate co-capping of Ia antigens and Fc receptors have shown that Ia antigens remain detectable on the cell surface after Fc receptors have been capped (83,84). The reverse experiment, in which anti Ia is capped, is not instructive since the anti Ia pre-treatment inhibits Fc receptor function. Another approach to the study of possible associations of Fc receptors with Ia antigens has come from attempts to isolate the B cell Fc receptor. Rask et al. (85) isolated an aggregated IgG-binding glycoprotein with a molecular weight of about 65,000,which these workers felt was possibly the B lymphocyte Fc receptor. No evidence of identity or close association of this protein with Ia antigens has been found, since this "Fc receptor" could cap separately from Ia, was resistant to trypsin digestion which destroyed most Ia antigens, and was removed by an aggregated IgG column which allowed Ia antigens to pass freely. Ia determinants could not be found on this protein.

Studies by Krammer et al. (86) of the distribution of Ia antigens and Fc receptors on various populations of activated T lymphocytes did not reveal any discernable relationship between these surface markers. For example, activated thoracic duct lymphocytes were 60% Ia positive but were Fc receptor negative. Such T cells provide a contrast to the studies with P815Y mentioned above: the latter is Fc receptor positive Ia negative, whereas the former are Ia positive Fc receptor negative.

Perspective and Conclusions

The observation that anti Ia inhibits B lymphocyte Fl-HGG binding much more strongly than do H-2 antisera is interesting and provocative, and has stimulated much useful research. The fact that simple identity or physical attachment between B lymphocyte Fc receptor binding sites and the proteins on which Ia antigens are expressed now seems to be unlikely does not preclude the possibility that Fl-HGG binding inhibition may prove useful in studies of Ia-like antigens in various species, including man. But the explanation for the manner in which various antisera inhibit Fl-HGG or EA-rosette assays independently of the Fc portions of these antibodies continues to be elusive, and invites speculation.

We favour a model based on the assumption that the physiologic state of the cell membrane contributes to the affinity of that membrane for immunoglobulin Fc. Reasonable justification for this assumption exists, in that the activation of cells has been shown to alter their binding affinity (87) and the physiologic state of an animal seems to determine the Fc binding ability of spleen cells in at least one experimental situation (88). Antibody to cell membranes might induce local membrane "perturbation", as suggested by Unanue et al. (83), which would indirectly influence the Fc binding sites. Whether or not "perturbation" occurred when an antibody bound to a surface antigen might depend on the characteristics of that antigen, such as its chemical composition, density of distribution, uniformity or patchiness of distribution, etc. Thus the binding of antibody to certain antigens might induce a higher degree of "perturbation" than the binding of antibody to other antigens. Furthermore, certain assays which detected binding to components other than that which is conventionally detected in Fc receptor assays,

as has been suggested may occur with the F1-HGG binding assay (89), might be more
resistant to "perturbation" than assays which did not detect such binding. Thus
quantitative or even apparently qualitative differences in the degree of inhibition
observed after pre-treatment of cells with various antisera need not reflect
direct effects of the antibody on Fc receptor function (ie. covalent association
of the antigen with Fc receptors). Instead, the Fc receptor inhibition would be
indirect, through the intermediate events in the membrane produced by the antibody-
antigen reaction.

   In summary, present data do not support any covalent association between Ia
antigens and the majority of B lymphocyte Fc receptors. Furthermore, even if Ia
antigens do have some topographic relationship with Fc receptors, the present
evidence indicates that a variety of other types of antigen (H-2D, H-2K, Ly 4.2,
immunoglobulin) could have a similar relationship with Fc receptors. The key to
understanding the phenomena which occur when an antibody inhibits Fc receptor
function is to establish the nature of the Fc receptor and the forces involved in
immunoglobulin-Fc receptor binding; and to elucidate the nature of the cell membrane
changes which follow antibody-antigen interaction on the cell surface and how such
changes could influence Fc receptor function. The resolution of these problems
will be beneficial to studies of the immunology of receptors, the functions of
surface antigens, and the physiology of cell membranes in general.

REFERENCES

1.   Wright, A.E. and Douglas, S.R., Proc. R. Soc. B., 72:364, 1904.

2.   Neufeld, F. and Rimpau, R., Dtsch. med. Wschr., 11:1458, 1904.

3.   Dickler, H. and Sachs, D., J. Exp. Med., 140:779, 1974.

4.   Kerbel, R.S., Nature, 255:576, 1975.

5.   Kerbel, R.S., Nature, 257:180, 1975.

6.   Weissman, G., Brand, A. and Franklin, C., J. Clin. Invest., 53:536, 1974.

7.   Bloom, W., Arch. Path. & Lab. Med., 3:608, 1927.

8.   Boyden, S.V. and Sorkin, E.,   Immunology,   3:272, 1960.

9.   Boyden, S.V.,   Immunology,   7:474, 1964.

10. Berkin, A. and Benacerraf, B.,   J. Exp. Med.,   123:119, 1966.

11. Coombs, R.R.A. and Franks, D.,   Prog. Allerg.,   13:174, 1969.

12. Uhr, J.W.,   Proc. Nat. Acad. Sci.,   54:1599, 1965.

13. Paraskevas, F., Lee, S.-T., Orr, K.B. and Israels, L.G.,   J. Immunol.,
     108:1319, 1972.

14. Paraskevas, F., Orr, K.B., Anderson, E.D., Lee, S.-T., and Israels, L.G.,
     J. Immunol.,   108:1729, 1972.

15. Basten, A., Miller, J.F.A., Sprent, J. and Pyre, J.,   J. Exp. Med.,
     135:610, 1972.

16. Dickler, H. and Kunkel, H.G.,   J. Exp. Med.,   136:191, 1972.

17. Anderson, C.L. and Grey, H.M.,   J. Exp. Med.,   139:1175, 1974.

18. MacLennan, I.C.M.,   Transplant. Rev.,   13:67, 1972.

19. Perlmann, P., Perlmann, H. and Wigzell, H.,   Transplant. Rev.,   13:91, 1972.

20. Howard, J.G. and Benacerraf, B.,   Brit. J. Exp. Path.,   47:193, 1966.

21. Davey, M.J. and Asherson, G.L.,   Immunology,   12:13, 1967.

22. Dickler, H.,   J. Exp. Med.,   140:508, 1974.

23. Kerbel, R.S. and Davies, A.J.S.,   Cell,   3:105, 1974.

24. Stossel, T.P.,   New Eng. J. Med.,   290:717, 1974.

25. Elson, J., Jenkinson, E.J. and Billington, W.D.,   Nature,   255:712, 1975.

26. Perlmann, P., Perlmann, H. and Biberfield, P.,   J. Immunol.,   108:558, 1972.

27. Pollack, S.,   Int. J. Cancer,   11:138, 1973.

28. Greenberg, A.H. and Shen, L.,   Nature (New Biol.),   245:282, 1973.

29. Sachs, D.H. and Dickler, H.B.,   Transplant. Rev.,   23:159, 1975.

30. Holland, P., Holland N.H. and Cohen, Z.A.,   J. Exp. Med.,   135:458, 1972.

31. Perlmann, P. and Holm, G.,   in Advances in Immunology II, p. 117,
     Academic Press Inc., New York, 1969.

32. Halloran, P. and Festenstein, H.,   Nature,   250:52, 1974.

33. Halloran, P., Schirrmacher, V. and Festenstein, H., J. Exp. Med.,
     140:1348, 1974.

34. Schirrmacher, V., Halloran, P., Ross, E. and Festenstein, H., Cell. Immunol.,
     16:362, 1975.

35. Parish, C.R. and Hayward, J.A., Proc. R. Soc. Lond. B., 187:47, 1974.

36. Tønder, A. and Thunøld, S., Scand. J. Immunol., 2:207, 1973.

37. Gorer, P.A., J. Path. Bacteriol., 47:231, 1938.

38. Shreffler, D.C. and David, C.S., Adv. Immunol., 20:125, 1975.

39. Klein, J., Biology of the Mouse Histocompatibility-2 Complex.
     Springer-Verlag New York Inc., 1975.

40. Cullen, S.E., David, C.S., Shreffler, D.C. and Nathanson, S.G.,
     Proc. Nat. Acad. Sci. (Wash), 71:648, 1974.

41. Delovitch, T.L. and McDevitt, H.O., Immunogenetics, 2:39, 1975.

42. David, C.S., Frelinger, J.A. and Shreffler, D.C., Transplantation,
     17:122, 1973.

43. David, C.S., Shreffler, D.C. and Frelinger, J.A., Proc. Nat. Acad. Sci.,
     70:2509, 1973.

44. Hauptfeld, V., Klein, D. and Klein, J., Science, 181:167, 1973.

45. Sachs, D.H. and Cone, J.L., J. Exp. Med., 138:1289, 1973.

46. Götze, D., Reisfeld, R.A. and Klein, J., J. Exp. Med., 138:1003, 1973.

47. Davies, D.A.L. and Hess, M., Nature, 250:228, 1974.

48. Hämmerling, G.J., Deak, B.D., Mauve, G., Hämmerling, U. and McDevitt, H.O.,
     Immunogenetics, 1:68, 1974.

49. Frelinger, J.A., Niederhuber, J.E., David, C.S. and Shreffler, D.C.,
     J. Exp. Med., 140:1273, 1974.

50. Götze, D., Immunogenetics, 1:495, 1975.

51. McDevitt, H.O., Bechtol, K.B., Hammerling, G.J., Lonai, P. and Delovitch, T.L.,
     in The Immune System: Genes, Receptors, and Signals, edited by E.E. Sercarz,
     A.R. Williamson and C.F. Fox, p. 597, Academic Press, New York, 1974.

52.   Hammerling, G.J., Mauve, G., Goldberg, E. and McDevitt, H.O.,
      Immunogenetics, 1:428, 1975.

53.   Meo, T., Vives, G., Rijnbeck, A.M., Miggiano, V.C., Nabholz, M. and
      Shreffler, D.C., Transplant. Proc., 5:1339, 1973.

54.   Fathman, C.G., Handwerger, B.S. and Sachs, D.H., J. Exp. Med., 140:853, 1974.

55.   Klein, J., Bach, F.H., Festenstein, H., McDevitt, H.O., Shreffler, D.C.,
      Snell, G.D. and Stimpfling, J.H., Immunogenetics, 1:184, 1974.

56.   Davies, D.A.L. and Alkins, B.J., Nature, 247:294, 1974.

57.   Archer, J.R., Smith, D.A., Davies, D.A.L. and Staines, N.A.,J. Immunogenetics,
      1:315, 1974.

58.   Soulillou, J.-P., Carpenter, C.B., d'Apice, A.J.F. and Strom, T.B.,
      J. Exp. Med., 143:405, 1976.

59.   Armerding, D., Sachs, D.H. and Katz, D.H., J. Exp. Med., 140:1717, 1974.

60.   Munro, A.J. and Taussig, M.J., Nature, 256:103, 1975.

61.   Katz, D.H., Greaves, M., Dorf, M. and Dimuzio, H., J. Exp. Med.,
      141:263, 1975.

62.   Winchester, R.J., Wemet, P., Dupont, B. and Kunkel, H.G., in Membrane
      Receptors of Lymphocytes, edited by M. Seligmann, J.L. Preud'homme and
      F.M. Kourilsky, p. 323, North-Holland Pub. Co., 1975.

63.   Mann, D.L., Abelson, L., Henkart, P., Harris, S.D. and Amos, D.B.,
      Proc. Nat. Acad. Sci. (USA), 72:5103, 1975.

64.   Dickler, H.B., Cone, J.L., Kubicek, M.T. and Sachs, D.H., J. Exp. Med.,
      142:796, 1975.

65.   Snell, G.D., Cherry, M. and McKenzie, I.F.C., Proc. Nat. Acad. Sci. (Wash),
      70:1108, 1973.

66.   McKenzie, I.F.C. and Plate, J.M.D., Cell. Immunol., 14:376, 1974.

67.   McKenzie, I.F.C., J. Immunol., 114:848, 1975.

68.   McKenzie, I.F.C., J. Immunol., 114:856, 1975.

69.   Dickler, H.B., Arbeit, R.D. and Sachs, D.H., in <u>Membrane Receptors of</u>
       <u>Lymphocytes</u> , edited by M. Seligmann, J.L. Preud'homme and F.M. Kourilsky,
       p. 259, North-Holland Pub. Co., 1975.

70.   Abbas, A.K., Ault, K.A., Karnovsky, M.J. and Unanue, E.R., J. Immunol.,
       <u>114</u>:1197, 1975.

71.   Basten, A., Miller, J.F.A. and Abraham, R., J. Exp. Med., <u>141</u>:547, 1975.

72.   Halloran, P., Schirrmacher, V., David, C.S. and Staines, N.A.,
       J. Immunogenetics, in press, 1976.

73.   Schirrmacher, V., Halloran, P. and David, C.S., J. Exp. Med., <u>141</u>:1201, 1975.

74.   Yoshida, T.O. and Andersson, B., Scand. J. Immunol., <u>1</u>:401, 1972.

75.   Soteriades-Vlachos, C., Gyongyossy, M.I.C. and Playfair, J.H.L., Clin. exp.
       Immunol., 18:187, 1974.

76.   Santana, V. and Turk, J.L., Immunology, 28:1173, 1975.

77.   Stout, R.D. and Herzenberg, L.A., J. Exp. Med., 142:611, 1975.

78.   Halloran, P., Schirrmacher, V. and David, C.S., Immunogenetics, <u>2</u>:349, 1975.

79.   Hallberg, T., Gurner, B.W. and Coombs, R.R., Int. Arch. Allergy, <u>44</u>:500, 1973.

80.   Schirrmacher, V. and Halloran, P., in <u>Membrane Receptors of Lymphocytes</u>,
       edited by M. Seligmann, J.L. Preud'homme and F.M. Kourilsky, p. 267,
       North-Holland Pub. Co., 1975.

81.   Vincent, C., Robert, M. and Revillard, J.P., Cell. Immunol., <u>18</u>:152, 1975.

82.   Grey, H.M., Anderson, C.L., Heusser, C.H. and Kurnick, J.T., in <u>Membrane</u>
       <u>Receptors of Lymphocytes</u>, edited by M. Seligmann, J.L. Preud'homme and
       F.M. Kourilsky, p. 185, North-Holland Pub. Co., 1975.

83.   Unanue, E.R., Ault, K.A., Schreiner, G.F. and Sidman, C.L., in <u>Membrane</u>
       <u>Receptors of Lymphocytes</u>, edited by M. Seligmann, J.L. Preud'homme and
       F.M. Kourilsky, p. 363, North-Holland Pub. Co., 1975.

84.   Goding, J.W., Nossal, G.J.V., Shreffler, D.C. and Marchalonis, J.J.,
       J. Immunogenetics, <u>2</u>:9, 1975.

85.   Rask, L., Klareskob, L., Ostberg, L. and Peterson, P.A.,   Nature,

       257:231, 1975.

86.   Krammer, P.H., Hudson, L. and Sprent, J.,   J. Exp. Med.,   142:1403, 1975.

87.   Rhodes, J.,   J. Immunol.,   114:976, 1975.

88.   Rubin, B. and Hertel-Wulff, B.,   Scand. J. Immunol.,   4:451, 1975.

89.   Abbas, A.K., Dorf, M.E., Karnovsky, M.J. and Unanue, E.R.,   J. Immunol.,

       116:371, 1976.

Chapter 13

INHIBITION OF B LYMPHOCYTE ACTIVATION BY INTERACTION WITH Fc RECEPTORS

J.L. Ryan and P.A. Henkart
Immunology Branch, National Cancer Institute,
National Institutes of Health, Bethesda, Maryland   20014

Abstract

Much early work indicated that specific antibody can play an inhibitory role in the immune response. This inhibitory activity was found to be dependent on an intact Fc portion of the antibody used, both in vivo and in vitro. Thus a role for lymphocyte Fc receptors in regulation of the immune response was suggested. However, soluble antigen-antibody complexes, which bind to Fc receptors, do not appear to inhibit B cell activation. Recent experiments have demonstrated that antigen-antibody complexes immobilized on a surface strikingly inhibit LPS induced B cell mitogenesis and polyclonal antibody synthesis. Mechanisms for Fc receptor-mediated inhibition of B cell activation have been considered, and a model proposed to explain many of these findings as well as allow for antigen specific inhibition.

I.   Introduction

A. Purpose and Scope

Surface receptors on lymphocytes have proven vital in defining

lymphocyte subpopulations with different functional roles in the immune

response. The precise physiological role of these surface receptors has

provided the central problem in much recent research, but has been difficult

to establish. For example, it appears evident that the B cell surface

immunoglobulin binds antigen, but the direct consequence of this binding

is unclear and a considerable controversy exists on the question of whether

such binding itself results in a signal being transmitted to the cell

interior (1). In this review we will concentrate on the function of

B lymphocyte Fc receptors; specifically we will consider the evidence that

221

these receptors play an inhibitory role in regulating B cell activation.

Most of the experiments cited have been done using murine spleen cell

systems.

    B. The Fc Receptor

        Fc receptors are defined by their ability to bind the Fc portion

of IgG immunoglobulin which has been complexed with antigen, or has been

aggregated by heat or other means (2,3). Single IgG molecules are bound

to the lymphocyte Fc receptor with a markedly lower affinity (2). Fc

receptors are detected chiefly by monitoring the binding of antigen-antibody

complexes by autoradiography or immunofluorescence (4,5), by binding of

fluorescent labelled heat aggregated IgG (3), or by forming rosettes with

antibody coated erythrocytes (EA rosettes) (6). These receptors do not

bind to IgG $F(ab')_2$ fragments which lack the Fc portion of the antibody

molecule, nor to intact antibodies of other classes (2). They have been

found on B lymphocytes in the mouse (2,7), human (3), and other species and

have recently been described on normal T lymphocytes in the mouse (8). Fc

receptors have also been found on many other types of cells including

macrophages. It seems quite possible at this point that there may be dif-

ferent types of Fc receptors on different cells. A recent review of lympho-

cyte Fc receptors is recommended for a detailed discussion (9).

    II. Antibody Mediated Suppression of the Immune Response

        In a variety of older experiments it was convincingly shown that passive-

ly administered antibody could specifically suppress an antibody response

in an animal given antigen. The early work in this area has been reviewed

by Uhr and Möller (10). These experiments have given rise to the suggestion

that one essential aspect of the in vivo regulation of antibody production

is a negative feedback system in which circulating antibody molecules

specifically inhibit the differentiation of new antibody secreting cells.

More recently, the studies of Sinclair and his colleagues have been of critical importance in this area because they have established the requirement for an intact Fc portion of the antibody molecule to suppress immune responses both in vivo and in vitro (11,12,13). For example, six week old mice were immunized with sheep erythrocytes and varying doses of $F(ab')_2$ or IgG anti-SRBC antibodies. When the spleens were assayed for anti-SRBC plaques 5 days later, it was found that the IgG antibody was much more effective in decreasing the response than the $F(ab')_2$ antibody (12). It was also found that only IgG and not $F(ab')_2$ fragments could inhibit a response which was already underway. Thus they suggested that the inhibitory activity that was seen in the animal treated with antigen plus $F(ab')_2$ antibody was likely due to increased antigen destruction and that the intact IgG was acting by inactivating the antibody-forming cell precursor population by interaction with antigen-antibody complexes.

To explain these data, Sinclair and Chan proposed a tripartite inactivation model featuring an interaction between antigen, antibody, and antigen-sensitive cell in which a negative signal is delivered via the Fc portion of the antibody molecule (12).

Studies showing antigen-specific suppression of in vitro immune responses have also been reported where the suppression was dependent on an intact Fc fragment in the antibody molecule (13,14,15,16). These studies indicate that the spleen cells being cultured contain all the elements necessary to achieve the suppression and eliminate many mechanisms possible in the in vivo experiments such as alterations in renal tubular function (17). Kappler et al. (16) have demonstrated that there is an antigen specific, determinant nonspecific suppression of the immune response of mouse spleen cells to heterologous erythrocytes which was not mediated by $F(ab')_2$ antibodies. They suggested that Fc receptors may be

functioning to bind antigen-antibody complexes rendering the antigen
inaccessible to B cell precursors.

III. Inhibition of B Lymphocyte Activation by Antibody

B lymphocyte Fc receptors seem an obvious candidate for mediating the
suppressive effects discussed above. One major conceptual difficulty is
that these receptors recognize the Fc portion of complexed antibody molecules regardless of their antigen binding specificity while the suppressive
effects are antigen specific. This problem will be more fully discussed
below. Two other lines of evidence have led to speculation regarding the
role of Fc receptors in regulation of the immune response. First, was the
demonstration that there is a close relationship on the lymphocyte cell
surface between the Fc receptor and Ia antigens (18). Since Ia antigens are
coded for by genes which either are the same as or closely linked to immune
response genes, it seemed possible that the Fc receptor may play a role in
regulating the immune response (19). Secondly, it appears that an association between Fc receptors and membrane immunoglobulin can be induced by
binding of ligands to the latter molecules (20,21). Although many questions
remain regarding the structure of Fc receptors and their surface relationships with other cell markers, this evidence also suggests that the Fc
receptors may play an important role in the immune response.

A. Study of B Lymphocyte Activation

In order to establish that the B lymphocyte Fc receptor is capable
of inhibiting B cell activation it is useful to study systems in which the
latter process is being measured and in which the effect of other cell
types can be minimized. This is important because of the presence of Fc
receptors on macrophages and some T cells. B cells can be activated in a
polyclonal fashion by a great many compounds including bacterial lipopolysaccharide, purified protein derivative of M. tuberculosis, and other poly-

anions (22). Such polyclonal activation stimulates B cells of all antigenic specificities to undergo an increase in DNA synthesis, divide, and differentiate into antibody-forming cells. DNA synthesis (mitogenesis) is conveniently measured by incorporation of radiolabelled DNA precursors while differentiation is usually measured using the plaque assay developed by Jerne and Nordin (23). The polyclonal activation of B cells by compounds such as LPS appears to mimic the natural activation of specific B cell clones by antigen and has been studied as a convenient model for the natural event. The advantage of using these systems is that it is possible to focus on B cells in isolation from other cells which are not required for the response being studied. One disadvantage is that antigen specificity is lost.

B.   Lack of Activation of B Cells by Complexes

There was some early success in activating human lymphocytes using soluble and particulate antigen-antibody complexes (24). Attempts to extend this work, however, have met with little success. Möller and Coutinho (25) investigated the ability of both the Fc receptor and the C'3 receptor to act as activating mechanisms for mouse splenic B cells. They found that heterologous erythrocytes coated with 7S rabbit antibody did not activate mouse spleen cells for polyclonal antibody activity. Other experiments showed that soluble complexes were also unable to induce any B cell activation under a variety of conditions. These results have been corroborated in other laboratories. Ramasamy (26) used soluble immune complexes of ferritin-anti-ferritin in an attempt to activate nude spleen cell cultures. No increase in DNA synthesis was observed. Ryan et al. (27,28) attempted to stimulate DNA synthesis in human lymphocyte cultures using immobilized antigen-antibody complexes without success and further showed that neither soluble complexes nor immobilized complexes stimulated DNA synthesis in mouse spleen cell cultures. Thus with the exception of Möller's early results (24), all attempts to activate B cells via interaction with Fc

receptors have been unsuccessful.

C.   Inhibition of B Cell Activation by Complexes

Given the failure of attempts to stimulate B cells by binding of

ligand to the Fc receptor, the question of whether such binding can prevent

activation of the cell by other means has been asked. Moller and Coutinho

(25) found that EA rosetted lymphocytes would not suppress the polyclonal

antibody synthesis induced by LPS. Furthermore, they demonstrated that

soluble antigen-antibody complexes over a broad range of concentrations

exerted no inhibitory effect on the polyclonal antibody response of nude

spleen cells to LPS. More recently, Ramasamy (26) investigated the ability

of soluble antigen-antibody complexes and aggregated immunoglobulin to

inhibit LPS induced mitogenesis in nude spleen cell culture. Under no con-

ditions did the binding of complexed immunoglobulin interfere with LPS

induced B cell activation.

In contrast to these reports the results of Ryan et al. (28,29)

have shown that under the appropriate conditions antigen-antibody complexes

can be strongly suppressive to B lymphocyte activation. They confirmed the

earlier results of Moller and Coutinho (25) showing that soluble complexes

did not lead to an inhibition of B cell activation, but further showed that

if cells were cultured on a surface coated with immobilized antigen-antibody

complexes (30), B cell activation as measured by both mitogenesis and pro-

duction of antibody forming cells was markedly inhibited. The observed

inhibition was attributed to the Fc receptor since similar immobilized

complexes prepared from IgA antibody or $F(ab')_2$ antibody were inactive in

suppressing either function. The inhibitory effect of immobilized com-

plexes was observed for all B cell activators tested including LPS,

pneumococcal polysaccharide SIII, poly I:C, PPD, and 8 BrcGMP. The ability

of complexes to inhibit LPS induced DNA synthesis was later confirmed by

Stout and Herzenberg (31). These investigators used large molecular weight complexes to mediate the inhibition.

D. Coordinated Interactions Between sIg and the Fc Receptor

The search for a physiological role for surface immunoglobulin has a more detailed history than that for the Fc receptor. In a parallel fashion it has been found that, with certain notable exceptions (32,33), anti-immunoglobulin antibodies that combine with lymphocyte surface immunoglobulin do not lead to activation of B cells (26,34). Attempts to block B cell activation have been successful, but at present remain controversial. Andersson et al. (34) reported that B cell activation induced by LPS, PPD, and FCS and measured by plaque-forming cell production was inhibited by anti-immunoglobulin (κ, λ, or μ)or its $F(ab')_2$ fragments. The same sera, however, failed to block LPS induced mitogenesis. Conversely, Schrader (35), found that anti-immunoglobulin sera (polyvalent anti-IgG) did indeed suppress LPS induced mitogenesis. Very recently, Sidman (36) has confirmed and extended the findings of Schrader (35). He found that anti-immunoglobulin inhibited LPS mitogensis, but that $F(ab')_2$ fragments were unable to mediate this inhibition. These two reports (35,36) are in conflict with the first (34). The latter data suggests that a coordinated binding between surface immunoglobulin and Fc receptors or aggregation of Fc receptors induced by the binding of antibody to surface immunoglobulin may be required to inhibit B cell activation.

E. Role of the T Cell Fc Receptor

Since T-B collaboration is a critical event for most antigen-induced B cell activation (37), and Fc receptors have been described on a subpopulation of T cells (8), a possible role for T cell Fc receptors regulating B cell activation must be considered. It has been proposed that antigen-antibody complexes may block Fc receptor positive T cells and prevent their interaction with other cellular species (31). This inference is based on the ability of complexes to partially inhibit the Concanavalin A response of

spleen cells (28,31). This inhibition of the Con A response was found to be

dependent on an intact Fc fragment of the antibody used to create the com-

plexes (28). Further evidence for the role of T cell Fc receptors in

regulating B cell activation has been provided by Fridman and his associates

(38,39). They have found that alloantigen activated T cells release a

factor which combines with the Fc fragment of antigen complexed IgG. It

has been suggested that this factor (immunoglobulin-binding factor, IBF)

may be identical to the T cell Fc receptor. IBF has been shown to inhibit

in vitro 19S antibody formation to both T-dependent and T-independent anti-

gens (40). IBF has further been shown to be capable of suppressing 7S anti-

body response and of being neutralized by aggregated mouse IgG (39). It is

not clear at this time whether IBF reacts directly with B cells. Thus

there is increasing evidence that T cell Fc receptors may play a role in

inhibiting B cell activation.

IV. Mechanism of Inhibition of B Cell Activation by the Fc Receptor

Many questions remain unanswered regarding the molecular events

involved in the inhibition of B cell activation by the Fc receptor. It is

perhaps too much to expect to understand how B cell activation is inhibited

while the most basic aspects of the molecular mechanism of the activation

itself are still highly controversial. Nevertheless, a number of questions

must be considered regarding the relevant experiments in this area. First,

why is effective inhibition observed only with immobilized or very large

molecular weight complexes but not with soluble complexes? While there is

precedent for B cells responding to immobilized, but not soluble stimuli

(41,42), the mechanism remains unclear. Furthermore it is most relevant to

question whether the observed nonspecific inhibition of B cell activation by Fc receptors can indeed explain the antibody mediated suppression of the immune response which is antigen specific. We have proposed (28) that both of these problems could be explained by a variant of the tripartite model proposed by Chan and Sinclair (12). The model assumes that a B lymphocyte may become blocked from being activated by a coordinated binding of the antigen to the sIg receptor and an IgG antibody from the medium bound to this antigen by another determinant. The Fc portion of this complexed antibody would then bind to the lymphocyte surface by interaction with Fc receptors to complete the inactivating signal. This is diagrammed below (Fig. 1).

Several aspects of the model deserve comment. 1) Antigen specificity is provided for by sIg which may operate not only as a passive focussing device, but as a necessary part of the inhibitory event. 2) The cross-linking of membrane receptors which occurs may be a crucial aspect of the inhibition and may be mimicked by the immobilized complexes which were observed to inhibit effectively. 3) The recent experimental data of Sidman (36) would certainly support this model.

Regardless of the correctness of this model there are a number of different levels at which inhibition of B lymphocyte activation via Fc receptors may take place.

1) Blockage of the surface binding event(s) which trigger the cell to become activated. Ryan et al. (28), have argued from indirect evidence that this is not the case. They were unable to demonstrate blocking of Fc receptors by any of a series of B cell activators as measured by competition for binding with that aggregated immunoglobulin. In the more general case of antigen and T cell stimulation of the B cell, this mechanism remains a possibility.

2) The activation could be blocked by an uncoupling of the linkage between surface receptor binding and cytoplasmic events. One possibility

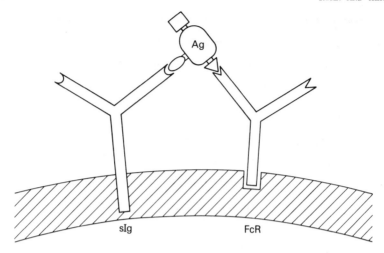

Figure 1:   Model for B cell inactivation by the Fc receptor. Antigen is
            bound to the B cell surface by sIg. An IgG antibody from the
            circulation binds to the same antigen molecule, and the Fc
            portion of this antibody binds to the Fc receptor. The cell
            could be inactivated by a "one signal model" involving the Fc
            receptor alone, or by a "two signal model" which requires
            signals from both sIg and the Fc receptor.

along these lines would be a "freezing" of surface membrane mobility as
envisioned by Edelman (43). However, no experimental evidence yet supports
this idea.

A related mechanism would involve immobilizing B cell surface components. The concept of a redistribution of surface components being
critical for B cell activation has been fully discussed (44,45). Melchers
and Andersson (46) have presented evidence that stimulation of B cells by
LPS results in the aggregation of surface IgM. This aggregation occurs
rapidly and lasts for more than 10 hours. They suggest that this redistribution of sIg is a critical event for activation. While the role, if any,

of Fc receptors in augmenting or inhibiting surface redistribution is
unclear, the possibility exists that the inhibitory event induced by
immobilized or large molecular weight complexes may be related to prevention
of the movement of surface molecules such that a triggering event cannot
take place.

3) The inhibition is a result of a central inhibitory signal generated
by binding to the Fc receptor under appropriate conditions (perhaps binding
also with sIg). The nature of such central (cytoplasmic) signals remains
unknown, but one attractive suggestion is the proposal that cell division
is controlled by the intracellular ratio of cGMP to cAMP (47). In support
of this idea, Watson (49) found that the intracellular level of cGMP
increases rapidly after the addition of LPS to cultures peaking at about 15
minutes. During this time the cAMP levels remained relatively constant.
This increased ratio of cGMP to cAMP has been proposed to be the signal for
B cell activation. The finding that cGMP derivatives are mitogenic for mouse
B lymphocytes when added exogenously while cAMP derivatives are inhibitory
(49) is in accord with this concept. Since it was found that immobilized
antigen-antibody complexes also inhibit the mitogenic effect of 8 BrcGMP
(29), it may be argued that these inhibitory effects are occurring via a
central inhibitory signal. Recent results in our laboratory (50) indicate
that the interaction of mouse spleen cells with immobilized complexes
results in approximately a 50% increase in the intracellular level of cAMP
within 30 minutes. This may represent a central "off" signal delivered to
the cell via Fc receptors.

Studies on the functional role of the lymphocyte Fc receptor in the
immune response are still in their infancy. The data reviewed here lend
promise that further investigation will be rewarding and continue to provide some of the most exciting problems in cellular immunology.

References

1. Coutinho, A. and Möller, G., Scand. J. Immunol., 3:133, 1974.

2. Basten, A., Sprent, J. and Miller, J.F.A.P., Nature, New Biol., 235:178, 1972.

3. Dickler, H.B. and Kunkel, H.G., J. Exp. Med., 136:191, 1972.

4. Basten, A., Miller, J.F.A.P., Sprent, J. and Pye, J., J. Exp. Med., 135:610, 1972.

5. Arbeit, R.D., Henkart, P.A. and Dickler, H.B., in In Vitro Methods in Cell Mediated Immunity, edited by B. Bloom and J. David, vol. II, in press.

6. Yoshida, T.O. and Andersson, B., Scand. J. Immunol., 1:401, 1972.

7. Paraskevas, F., Lee, S.T., Orr, K.B. and Israels, L.C., J. Immunol., 108:1319, 1972.

8. Stout, R.P. and Herzenberg, L.A., J. Exp. Med., 142:611, 1975.

9. Dickler, H.B., Adv. Immunol., in press, 1976.

10. Uhr, J.W. and Möller, G., Adv. Immunol., 8:81, 1968.

11. Sinclair, N.R., St.C., J. Exp. Med., 129:1183, 1969.

12. Chan, P.L. and Sinclair, N.R., St.C., Immunol., 21:967, 1971.

13. Lees, R.K. and Sinclair, N.R., St.C., Immunol., 24:735, 1973.

14. Wason, W.M. and Fitch, F.W., J. Immunol., 110:1427, 1973.

15. Abrahams, S., Phillips, R.A. and Miller, R.G., J. Exp. Med., 137:870, 1973.

16. Kappler, J.W., Van der Haven, A., Dharmarajan, V. and Hoffman, M., J. Immunol., 111:1228, 1973.

17. Spiegelberg, H. and Weigle, W.O., J. Exp. Med., 123:999, 1966.

18. Dickler, H.B. and Sachs, D.H., J. Exp. Med., 140:779, 1974.

19. Sachs, D.H. and Dickler, H.B., Transplant. Rev., 23:159, 1975.

20. Forni, L. and Pernis, B., in <u>Membrane Receptor of Lymphocytes</u>, edited
    by M. Seligmann, J.L. Preud'homme and F.M. Kourilsky, p. 193, North
    Holland, Amsterdam, 1975.

21. Unanue, E.R. and Abbas, A.K., in <u>Membrane Receptor of Lymphocytes</u>,
    edited by M. Seligmann, J.L. Preud'homme and F.M. Kourilsky, p. 281,
    North Holland, Amsterdam, 1975.

22. Coutinho, A. and Möller, G., Nature, New Biol., 245:12, 1973.

23. Jerne, N.K. and Nordin, A.A., Science, 140:405, 1963.

24. Möller, G., Clin. Exp. Immunol., 4:65, 1969.

25. Möller, G. and Coutinho, A., J. Exp. Med., 141:647, 1975.

26. Ramasamy, R., Immunol., 30:559, 1976.

27. Ryan, J.L. and Henkart, P.A., unpublished results.

28. Ryan, J.L., Arbeit, R.D., Dickler, H.B. and Henkart, P.A., J. Exp. Med.,
    142:814, 1975.

29. Ryan, J.L. and Henkart, P.A., submitted for publication.

30. Henkart, P.A. and Alexander, E., manuscript in preparation.

31. Stout, R.D. and Herzenberg, L.A., J. Exp. Med., 142:1041, 1975.

32. Sell, S., Rowe, D.S. and Gell, P.G.H., J. Exp. Med. 122:823, 1965.

33. Weber, W.T., Transplantation, 24:113, 1975.

34. Andersson, J., Bullock, W.W. and Melchers, F., Eur. J. Immunol.,
    4:715, 1974.

35. Schrader, J.W., J. Immunol., 115:323, 1975.

36. Sidman, C.L., Fed. Proc. Abst., 35:820, 1976.

37. Miller, J.F.A.P., Mitchell, G.F., Davies, A.J.S., Claman, H.N.,
    Chaperon, E.A. and Taylor, R.B., Transplant. Rev., 1:1, 1969.

38. Fridman, W. and Goldstein, P., Cell. Immunol., 11:442, 1974.

39. Gisler, R.H. and Fridman, W., Cell. Immunol., 23:99, 1976.

40. Gisler, R.H. and Fridman, W., J. Exp. Med., 142:507, 1975.

41. Greaves, M.F. and Bauminger, S., Nature, New Biol., 235:67, 1972.

42. Feldmann, M., Greaves, M., Parker, P.C. and Rittenberg, M.B., Eur. J. Immunol., 4:591, 1974.

43. Wang, J.L., McClaine, D.A. and Edelman, G.M., Proc. Nat. Acad. Science, 72:1917, 1975.

44. Greaves, M. and Janossy, G., Transplant. Rev., 11:87, 1972.

45. Unanue, E. and Karnovsky, M., Transplant. Rev., 14:184, 1973.

46. Melchers, F. and Andersson, J., Transplant. Rev., 14:76, 1974.

47. Goldberg, N.D., Haddox, M.K., Nicol, S.E., Glass, D.B., Sanford, C.H., Kuehl, F.A. and Estensen, R., Adv. Cyclic. Nucl. Res., 5:307, 1975.

48. Watson, J., J. Exp. Med., 141:97, 1975.

49. Weinstein, Y., Segal, S. and Melmon, K.L., J. Immunol., 115:112, 1975.

50. Ryan, J.L. and Henkart, P.A., unpublished results.

Chapter 14

ANTISERA TO MAST CELLS AND THE RECEPTOR
FOR IgE

A. Froese
MRC Group for Allergy Research
Department of Immunology
Faculty of Medicine, University of Manitoba
Winnipeg, Manitoba, R3E 0W3, Canada

Abstract

Recent studies on the receptor for IgE on rat mast cells and rat basophilic leukemia cells have established that this receptor has a molecular
weight of about 60,000 daltons and is, at least in part, a protein. However, so far no antisera, monospecific for this receptor, are available.
Only antisera to mast cells have been described. Such antisera are capable
of releasing histamine from mast cells, but only in the presence of complement. In most instances such antisera were not specific for mast cells
only, but in some cases they were rendered specific by absorption with
lymph node cells or liver cells. One report demonstrates that such an absorbed antiserum was capable of inhibiting the binding of IgE to mast cells.
Moreover, it precipitated several mast cell surface components. One of
these possibly was the receptor for IgE.

It is now established beyond any doubt that it is the interaction

between homocytotropic antibodies (IgE and some subclasses of IgG) and tar-

get cells such as mast cells and basophils, which mediates the release of

pharmacological substances (histamine, SRS-A) from these cells (1-4). More-

over, the mechanism by which these mediators are released is fairly well

understood (5,6). The initial triggering signal has its origin in the in-

teraction of the homocytotropic antibody with a polyvalent antigen (allergen)

(7), a divalent antibody (3) or even a divalent hapten (7,8). However, very

little is known about the subsequent events which take place on or in the

plasma membrane of the target cell and which are involved in the trans-

mission of the initial signal to the interior of the cell.

In order to gain a better and detailed understanding of these events a characterization of the nature and properties of the receptor for IgE on the plasma membrane of the target cell would seem to be very important. To achieve such a goal, the availability of antibodies specific to such a receptor would be of considerable advantage. Such antibodies could be used to study the antigenic properties of the receptor for IgE both on the cell and in its isolated form; they could be employed in the isolation of the receptor and in attempts to initiate the mediator release from target cells, in order to gain more insight into the triggering mechanism.

Detailed investigations on the receptor for IgE and thus on anti-bodies to this receptor have been made possible in recent years by the dis-covery of monoclonal IgE in man (9) and in the rat (10), by the availability of rat basophilic leukemia cells (RBL) (11) and by the fact that the bind-ing constant for the interaction of the target cell receptor with IgE is rather high (12-14).

At present, very little information is available on antibodies to the receptor for IgE. All related studies have dealt only with antibodies to rat mast cells (RMC) (15-17) and their effect on mast cells. Only in one investigation (18) has an attempt been made to establish whether or not antisera to RMC do contain antibodies, directed against the receptor for IgE. This lack of information is not really surprising, since attempts to isolate this receptor and characterize it at the molecular level have been made only during the past three years.

The Receptor for IgE

Even though attempts to characterize the receptor for IgE on the app-ropriate target cells have been made only in recent years, the presence of this receptor was implied ever since it was realized that homocytotropic antibodies do sensitize target cells (19). The presence of IgE on target

cells was first demonstrated by radioautography (20). Its association
with the plasma membrane of human basophils (21) and of rat mast cells (22)
was shown more clearly by electron microscopy. The presence of receptors
on mouse mast cells was made visible by rosette formation (23). It was
also shown that on human basophils, membrane bound IgE can be made to cap
(24) which indicates that the receptors for IgE are free to move about on
the surface of these cells.

Bach and Brashler (25) were the first to demonstrate that the integrity of the rat mast cell (RMC) is not required for IgE to bind to the
plasma membrane of this cell. König and Ishizaka (26,27) sonicated rat peritoneal mast cells and observed that the cellular component capable of binding IgE remained in the supernatant after centrifugation at 20,000 x g.
Analysis of the supernatant by chromatography on Sepharose 6B indicated
that the subcellular component which was able to block PCA activity, was
eluted ahead of Blue Dextran 2,000 and appeared in the same volume as KLH.
The protein eluted in this peak was associated with 5' nucleotidase, an
enzyme known to be bound to the plasma membranes. These results demonstrated that the receptor for IgE was associated with a membrane component,
having an apparent molecular weight in excess of 2×10^6 daltons. Similar
results were obtained by Carson et al. (28) who cultured rat basophilic
leukemia cells (RBL) at 4°C and observed a subcellular component which was
shed into the medium and which was capable of binding IgE. When complexed
with ^{125}I-labelled rat monoclonal IgE, this cell surface component was eluted in the void volume upon chromatography on Sepharose 6B. These subcellular membrane components and those described by König and Ishizaka (26),
because of their size, must be considered as membrane fragments rather than
individual receptor moelcules or receptor complexes. The fragments obtained by König and Ishizaka (26) appeared to be heterogeneous with respect to charge since, when subjected to DEAE-cellulose chromatography, com-

ponents capable of blocking PCA were eluted at widely different ionic

strengths. When re-run on a column of Sepharose 6B, these components also

exhibited heterogeneity in terms of size, being generally smaller than

before DEAE-cellulose chromatography. However, all components still eluted

ahead of Blue Dextran 2,000. Thus, in spite of some dissociation upon DEAE-

cellulose chromatography, the components with receptor activity, because of

their large size, most likely did not represent individual receptor mole-

cules, and may still have been associated with membrane lipids. Recently,

Metzger et al. (29) have demonstrated that the receptors for IgE on sub-

cellular, particulate fractions, obtained from RBL cells by sonication,

have binding parameters for IgE which are very similar to those on intact

RBL cells.

Conrad et al. (30) used the nonionic detergent Nonidet-P-40 (NP-40)

to solubilize both rat peritoneal mast cells (RMC) and rat basophilic leuk-

emia (RBL) cells. These cells had been incubated with ^{125}I-IgE and washed

free of excess IgE. When such solubilized cells (either RMC or RBL cells)

were passed over a column of Agarose 1.5m equilibrated with NP-40, a single

radioactive peak was eluted in a volume larger than that required for the

elution of IgM but smaller than that required for IgE. A further compari-

son with polymeric IgA suggested that the IgE-receptor complex might have

an apparent molecular weight of $3.5 - 5.5 \times 10^5$ daltons. Subtracting the

molecular weight of IgE, these authors proposed that the receptor itself

might have a molecular weight in the range of $2 - 4 \times 10^5$ daltons.

In a further attempt to characterize the receptor on both RMC and RBL

cells, Conrad and Froese (31) prepared IgE-receptor complexes from cells

which had been surface iodinated using the lactoperoxidase catalyzed iodi-

nation technique (32). They precipitated the complexes, consisting of

[125]I-labelled cell surface molecules and "cold" rat IgE, by a sandwich sys-

tem involving rabbit anti-IgE antibodies and goat anti-rabbit immunoglobulin serum. The washed precipitates were dissolved in buffer containing sodium dodecyl sulfate (SDS) and urea, and analyzed by SDS polyacrylamide gel electrophoresis.. In the presence of SDS, complexes held together by non-covalent bonds are known to dissociate. Thus, when complexes obtained from either cell type were analyzed, there was only one peak which could be identified as the receptor for IgE or a component thereof. The molecular weight of this component, determined by SDS-polyacrylamide gel electrophoresis, was found to be approximately 62,000 daltons. In the case of SDS gel patterns from RMC, two peaks were observed in addition to the receptor peak. One of these was identified as IgE, present on the cells at the time of isolation, and the second as a cell surface component capable of interacting with immune complexes. This latter peak was the only one seen when normal rabbit serum instead of anti-IgE was used in order to precipitate cell surface components.

In view of the fact that in the case of IgE-receptor complexes in the presence of NP-40, the receptor appeared to contribute about $2 - 4 \times 10^5$ daltons to the molecular weight of the complex (30), Conrad and Froese (31) postulated two possible models for the receptor for IgE. They suggested that it consists either of several identical subunits, each having a molecular weight of about 60,000 daltons, or of one such component non-covalently linked to an entirely different molecule which was not detected on polyacrylamide gels since, because of its location in the membrane, it was not radioactively labelled. More recent evidence (33) demonstrates that the apparent molecular weight of the receptor in NP-40 may be somewhat lower than originally estimated. This observation coupled with the fact that NP-40 is most likely bound to hydrophobic portions of the receptor, suggests that the possibility that the receptor consists of only one component of 60,000 daltons, must also be considered. It should be pointed out that any size, determined for either receptor or IgE-receptor complexes in the

presence of NP-40, may represent only a lower limit, since it is conceivable
that the detergent has the capacity to disrupt larger receptor complexes
present on the intact plasma membrane of either RMC or RBL cells.

Antisera to Mast Cells

As pointed out above, there are only a few reports in the literature
concerned with antisera to mast cells (15-18,34,35). Of these, most deal
with antisera obtained by immunizing rabbits with purified preparations
of rat peritoneal mast cells (RMC) (15-18); one is concerned with studies
on an antiserum to peritoneal cells, enriched with respect to mast cells
(32), and one reports on human autoantibodies, capable of reacting with
granules of human and rat tissue mast cells (35). This last study, since
it does not involve antibodies to the mast cell plasma membrane, is really
outside the context of this review.

Preparation of antisera: All antisera were elicited in rabbits, either
by injecting suspensions of mast cells intravenously (15,34) or by emulsi-
fying the cells in complete Fruend's Adjuvant and injecting the mixture
subcutaneously (16-18). Yiu and Froese (18) also attempted to immunize
rabbits by the intravenous route, but did not succeed in obtaining high-
titered antisera.

Detection of antibodies to mast cells: When discussing methods which
can be employed to monitor an immune response to RMC, emphasis will be
placed on those techniques which appear to have been used in what might be
considered a routine screening manner.

Perhaps the simplest technique has been used by Smith and Lewis (15),
who layered disrupted mast cells over the antiserum and observed the formation of a precipitin ring. While this is certainly a simple test, it may
not be sensitive enough to detect the presence of antibodies in some antisera. Valentine et al. (16) made extensive use of histamine release in

the presence of complement. This test, while somewhat tedious to perform, was able to detect anti-RMC activity at a dilution of 1:500. Yiu and Froese (18) used an intradermal skin test, which may be considered as a kind of "reverse PCA" test, to screen their sera for anti-RMC activity. This test, although relatively easy to carry out once optimal conditions are established, is subject to variability, mainly due to variation in the sensitivity of one rat skin to another. Valentine et al. (16) also used the intradermal skin test to characterize their antiserum. In their hands this test appeared to be somewhat less sensitive than the complement mediated release of histamine.

Specificity of anti-RMC sera for mast cells: When considering this topic, it should be understood that such a specificity cannot easily be established in absolute terms. Mast cells can be expected to carry on their surface a variety of proteins, glycoproteins and carbohydrates. Some of these (e.g. transplantation antigens) may be shared with many other types of cells, some may be shared with a few other cell types, while some may, indeed, only be present on mast cells. Superimposed on this there may be partial identities of antigenic determinants between mast cells and other cells of the same animal or of the same inbred strain of animals. Any definition of specificity on an anti-RMC serum will thus depend on the type of cells which are used as controls.

When studying the properties of anti-RMC sera, in order to render their sera specific or to gain some information as to their specificity, investigators either absorbed them with unrelated cells (16,18,34) and/or used control antisera obtained by immunization of rabbits with some other cell types (15,18).

Smith and Lewis (15) compared the activity of an anti-RMC serum to that of an antiserum to rat liver cells. When living mesentery was exposed to either serum and observed under the microscope, the mast cells in this tis-

sue were seen to rupture. However, only when anti-RMC serum was used was
the rupture preceded by oscillations of the cytoplasmic granules and a grad-
ual swelling of the cells. These authors (15) also injected their antisera
into the peritoneum of rats and examined the histamine content of various
tissues for extended periods of time after the injection. The anti-RMC
serum had an effect on the histamine content of mesentery, scrotum and ven-
tral skin, while anti-liver serum seemed to affect mesentery tissue only.
Both sera caused severe damage to the tissues of the peritoneal cavity.
On the basis of this evidence it is difficult to say that the anti-RMC
serum was more selective for RMC than was the antiserum to liver cells.
The difference observed between the two sera could have been quantitative
rather than qualitative, and could have depended largely on the concentration of antibodies to rat antigens present in each of the antisera. The
lower antibody content of the anti-liver serum may also have been reflected
in the fact that this serum failed to give a positive ring test with disrupted mast cells.

Valentine et al. (16) and Yiu and Froese (18) attempted to render
their antisera specific for RMC by absorbing them with cells from tissues
considered to be largely devoid of mast cells. The former used lymph node
cells and the latter liver cells. In addition, Yiu and Froese (18) absorbed their serum with normal rat immunoglobulins and rat IgE. Both groups
of investigators observed that the absorbed anti-RMC sera retained their
cytotoxicity for RMC as measured by histamine release (16) or ^{51}Cr release
(18). In the latter report (18), it was also demonstrated that the absorbed
antiserum was no longer cytotoxic to lymph node cells. A rabbit antiserum
to rat sarcoma cells lost its capacity for the complement mediated release
of ^{51}Cr from either RMC or lymph node cells after absorption with liver
cells. It is conceivable that the absorbed antiserum of Valentine et al.
(16) contained some antibodies to ε-chain; the presence of any significant

concentration of antibodies to rat IgG these authors could rule out by demonstrating that the addition of normal rat serum did not reduce the capacity of their serum to release histamine from RMC in the presence of complement.

Hogarth-Scott and Bingley (34) absorbed their antiserum to rat peritoneal cells with erythrocytes. However, it is very unlikely that this
treatment had rendered their antiserum specific for RMC since others (16,
18) had shown that erythrocytes are not very effective in removing antibodies to non-RMC antigens from anti-RMC sera. It is difficult to ascribe
the prolongation and enhancement of Nippostrongylus brasiliensis infection
in rats, observed by Hogarth-Scott and Bingley (34) as a consequence of the
administration of their antiserum, to antibodies of any particular specificity. It is conceivable that the observed effect was due to the presence
of antibodies to either RMC, lymphocytes, immunoglobulins, or all of these
combined.

Finally, it should be stressed again that anti-RMC sera, exhaustively
absorbed with some other cell type, may still retain the capacity to interact with cells other than RMC. In this connection, the observations of
Thiede et al. (36,37) are of some interest. These investigators had obtained a rabbit antiserum to rat peritoneal macrophages which was also capable of reacting with tissue mast cells. Absorption of this antiserum with
rat erythrocytes and thymocytes did not abolish the cytotoxicity of this
serum for macrophages or tissue mast cells (37). These findings, along with
some other observations, led the authors to suggest that tissue mast cells
and macrophages both originate from monocytes.

Antibodies to the receptor for IgE: As pointed out above, there are
at present no published studies which deal directly with this topic. Only
in one study (18) was an attempt made to answer the question as to whether
or not antibodies to the receptor for IgE are present in an antiserum pro-

duced by a rabbit in response to immunization with mast cells. Thus, an analysis in this context of the existing literature cannot be executed without some speculation. The major recourse has to be a survey of the properties of anti-RMC sera.

For an antiserum to RMC to contain anti-receptor antibodies such an antiserum has to interact with the surface of mast cells. That this was actually the case was demonstrated by Valentine et al. (16), using an immunofluorescence technique and by Yiu and Froese (18), who precipitated radiolabelled surface antigens with their antiserum.

The most extensively studied property of anti-RMC sera has been their capacity to release histamine both in vivo and in vitro. The reactions observed upon injection of anti-RMC sera into the skin of rats are similar to PCA reactions induced by antigen or reverse PCA reactions induced by anti-IgE (16,18). Yet, it must be assumed that the in vivo histamine release caused by anti-RMC sera is due to a cytotoxic reaction since it has been shown that the in vitro histamine release is complement dependent (16, 38). Complement mediated lysis was also shown using the ^{51}Cr release method (18); the microscopic examination of tissue exposed to anti-RMC serum demonstrated what appeared to be complete rupture of cells (15). Moreover, no histamine was released from RMC when these cells were treated with anti-RMC serum in the absence of complement (16), nor did such a treatment of RMC cause the release of SRS-A (17). These results are in contrast to those obtained with antisera to human IgE (39) and to Fab fragments of rat-IgG (40), which were found to release histamine in a non-cytotoxic fashion from human basophils and rat peritoneal mast cells, respectively. Similar observations were made when mouse peritoneal mast cells were treated with an antiserum to mouse γ-globulin (41). In view of these results, it is tempting to speculate that the anti-RMC sera did not contain any antibodies to the receptor for IgE. Such a speculation requires the tacit assumption that antibodies to the receptor for IgE, similar to anti-IgE, will indeed

cause the release of mediators in a non-cytotoxic fashion. However, there

is as yet no evidence to support this assumption. Even the non-cytotoxic

release of mediators by an anti-RMC serum would not necessarily represent

proof for the presence of antibodies for the receptor for IgE. Thus, sev-

eral macromolecules which may not interact with either IgE or its receptor

are capable of causing mast cells to degranulate and to release histamine

in the absence of complement. These include α-chymotrypsin (42), poly-

mixin B coupled to Sepharose 4B (43) and dextran (44). However, there are

others which are considered to trigger mast cells in a fashion similar to

that of allergens and anti-IgE. For example, concanavalin A is thought to

release histamine from rat mast cells by interacting with the carbohydrate

of IgE (45), and murine alloantibodies of the IgG_1 subclass may initiate

the degranulation of mouse mast cells by linking alloantigens and an Fc

receptor on the same cell (46). This Fc receptor could be identical with

the receptor for IgE. It should be pointed out that even if antibodies to

the receptor for IgE could initiate the non-cytotoxic release of mediators,

their interaction with the receptor on the mast cell surface could be steri-

cally blocked by antibodies to other cell surface components. Antibodies

to several cell surface antigens were shown to be present in an anti-RMC

serum, even after this serum had been absorbed with liver cells (18).

The foregoing discussion was presented in order to illustrate that,

based on the histamine release data, it is rather difficult to arrive at

a definite conclusion as to the presence or absence of antibodies to the

IgE receptor in anti-RMC sera.

As pointed out above, there exists, at present, only one report (18)

which describes an attempt to demonstrate the presence of antibodies to

the receptor for IgE in anti-RMC sera. Thus, Yiu and Froese (18) were able

to show that their anti-RMC serum was capable of inhibiting the binding of monoclonal ^{125}I-IgE to RMC. This property could not be abolished by absorption of the serum with rat liver cells and rat immunoglobulins including IgE. Unfortunately, this observation cannot be taken as absolute evidence that the anti-RMC serum contained antibodies capable of interacting with the receptor for IgE. An antibody, bound by an antigen adjacent to the receptor, could have provided enough steric hindrance to block the interaction between IgE and its receptor. That this can happen was demonstrated by the fact that an anti-sarcoma serum could also inhibit the binding of IgE. However, in the case of this serum, the antibodies responsible for the "blocking" activity could be completely absorbed with liver cells (18).

The same authors (18) also characterized the ^{125}I-labelled surface antigens, precipitated by their anti-RMC serum. An SDS polyacrylamide gel electrophoretic pattern of the antigens precipitated by an absorbed anti-RMC serum in shown in Fig. 1. Amont these antigens, there was one which had a mobility identical to a cell surface component identified previously as the receptor for IgE or at least a component thereof (31). This antigen, as well as the receptor for IgE precipitated according to Conrad and Froese (31), are shown near fraction 23 of Fig. 1. However, it has to be pointed out that the fact that the two peaks are in identical positions does not necessarily mean that they represent identical molecules. Thus, the results of Yiu and Froese (18) suggest very strongly that antibodies to the receptor for IgE are present in anti-RMC sera, but they cannot prove it beyond any doubt. The electrophoretic pattern seen in Fig. 1 demonstrates that even an absorbed anti-RMC serum still contains antibodies to several cell surface antigens. It is conceivable that some of these antigens were part of a larger "receptor complex" which was, perhaps, disrupted by the NP-40. Such surface molecules would not have been detected by the precipitation system used by Conrad and Froese (31) and referred to earlier.

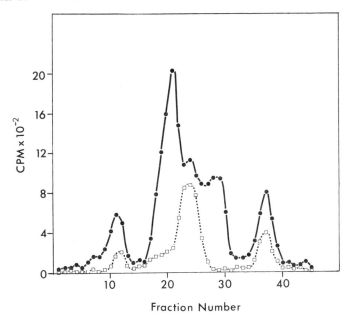

FIGURE 1

Analysis of surface components by SDS-polyacrylamide gel electro-phoresis. Cell surface antigens (^{125}I-labelled) precipitated by an anti-RMC serum, absorbed with liver cells and rat immunoglobulins (●————●), are compared to RMC surface components (^{131}I-labelled) precipitated by ε-specific rabbit anti-IgE serum, from cells which had been pretreated with cold IgE (□-----□). [From: Yiu and Froese (18)]

This precipitation system can only detect surface molecules bound by IgE in the presence of NP-40.

More recently, Conrad and Froese (33) purified receptor from rat baso-philic leukemia (RBL) cells, using an affinity column which consisted of IgE coupled to CL-Sepharose 4B. When rabbits were immunized with this rec-eptor preparation, they produced an antiserum which in turn blocked the binding of IgE to RBL cells and which also precipitated a major surface component having an apparent molecular weight similar to that of the recep-tor for IgE.

Concluding Remarks

While the information currently available on antisera to mast cells and particularly on antisera to the receptor(s) for homocytotropic antibodies on these cells is neither plentiful nor detailed, this situation can be expected to improve during the next few years. With increased knowledge about the receptor for IgE and about the surface antigens of mast and other cells, it should be possible to obtain better antisera to the receptor and to study the properties of such antisera more extensively. Such antisera should, in turn, facilitate the characterization of the receptor for IgE on mast cells and basophils and allow a comparison of the properties of these receptors with receptors for immunoglobulins found on other cell types, such as the Fc receptors on lymphocytes. These latter receptors appear to be, at least as far as size is concerned, similar to the receptor for IgE (47).

Antisera to the receptor for IgE may also help elucidate further the triggering mechanism of histamine release from mast cells. In particular, they may aid to resolve the question as to whether or not two different receptors as suggested by Stanworth (48) are required on mast cells for the IgE mediated release of histamine. Finally, the availability of antisera to the receptor for IgE and to other mast cell surface antigens may be of some advantage in studies on the spatial relationship of these antigens and receptors on the mast cell plasma membrane.

Acknowledgement

The author is indebted to Drs. B.G. Carter, D.H. Conrad and A.H. Sehon and Ms. S.H. Yiu for helpful discussions and suggestions.

References

1. Ishizaka, K. and Ishizaka, T., J. Allergy, $\underline{42}$:330, 1968.

2. Stechschulte, D.J., Orange, R.P. and Austen, K.F., J. Immunol., $\underline{105}$:1082, 1970.

3. Ishizaka, T., Ishizaka, K., Orange, R.P. and Austen, K.F., J. Immunol., $\underline{104}$:335, 1970.

4. Bloch, K.J. and Ohman, J.L. Jr. in Biochemistry of the Acute Allergic Reactions, edited by K.F. Austen and E.L. Becker, p. 45, Blackwell Scientific Publications, Oxford, 1971.

5. Becker, E.L. and Henson, P.M., Adv. Immunol., $\underline{17}$:93, 1973.

6. Austen, K.F., Fed. Proc., $\underline{33}$:2256, 1974.

7. Levine, B.B., J. Immunol., $\underline{94}$:111, 1965.

8. Mossmann, H., Meyer-Delius, M., Vortisch, U., Kickhöfen, B. and Hammer, D.K., J. Exp. Med., $\underline{140}$:1469, 1974.

9. Johansson, S.G.O. and Bennich, H., Immunology, $\underline{13}$:381, 1967.

10. Bazin, H., Querinjean, P., Beckers, A., Heremans, J.F. and Dessy, F., Immunology, $\underline{26}$:713, 1974.

11. Eccleston, E., Leonard, B.J., Lowe, J.S. and Welford, H.J., Nature, $\underline{244}$:73, 1973.

12. Ishizaka, T., Soto, C.S. and Ishizaka, K., J. Immunol., $\underline{111}$:500, 1973.

13. Kulczycki, A. Jr. and Metzger, H., J. Exp. Med., $\underline{140}$:1676, 1974.

14. Conrad, D.H., Bazin, H., Sehon, A.H. and Froese, A., J. Immunol., $\underline{114}$:1688, 1975.

15. Smith, D.E. and Lewis, Y.S., J. Exp. Med., $\underline{113}$:683, 1961.

16. Valentine, M.D., Bloch, K.J. and Austen, K.F., J. Immunol., $\underline{99}$:98, 1967.

17. Orange, R.P., Valentine, M.D. and Austen, K.F., J. Exp. Med., $\underline{127}$:767, 1968.

18. Yiu, S.H. and Froese, A., J. Immunol., in press.

19. Bloch, K.J., Progr. Allergy, 10:84, 1967.

20. Ishizaka, K., Tomioka, H. and Ishizaka, T., J. Immunol., 105:1459, 1970.

21. Sullivan, A.L., Grimley, P.M. and Metzger, H., J. Exp. Med., 134:1403, 1971.

22. Lawson, D., Fewtrell, C., Gomperts, B. and Raff, M.C., J. Exp. Med., 142:391, 1975.

23. Ovary, Z., in Mechanisms in Allergy, edited by L. Goodfriend, A.H. Sehon and R.P. Orange, p. 285, Marcel Dekker, Inc., New York, 1973.

24. Becker, K.E., Ishizaka, T., Metzger, H., Ishizaka, K. and Grimley, P.M., J. Exp. Med., 138:394, 1973.

25. Bach, M.K. and Brashler, J.R., J. Immunol., 111:324, 1973.

26. König, W. and Ishizaka, K., J. Immunol., 113:1237, 1974.

27. König, W. and Ishizaka, K., Immunochemistry, 13:345, 1976.

28. Carson, D.A., Kulczycki, A. and Metzger, H., J. Immunol., 114:158, 1975.

29. Metzger, H., Budman, D. and Lucky, P., Immunochemistry, 13:417, 1976.

30. Conrad, D.H., Berczi, I. and Froese, A., Immunochemistry, 13:329, 1976.

31. Conrad, D.H. and Froese, A., J. Immunol., 116:319, 1976.

32. Marchalonis, J.J., Cone, R.E. and Santer, V., Biochem. J., 124:921, 1971.

33. Conrad, D.H. and Froese, A., to be published.

34. Hogarth-Scott, R.S. and Bingley, J.B., Immunology, 21:87, 1971.

35. Rizetto, M. and Doniach, D., Clin. Exp. Immunol., 14:327, 1973.

36. Thiede, A., Müller-Hermelink, H.-K., Sonntag, H.G., Müller-Ruchholz, W., and Leder, L.-D., Klin. Wschr., 49:435, 1971.

37. Thiede, A., Müller-Hermelink, H.-K., Sonntag, H.G., Müller-Ruchholz, W., and Leder, L.-D., Res. Exp. Med., 157:198, 1972.

38. Kempf, R.A., Gigli, I. and Austen, K.F., Transpl. Proc., 11:676, 1969.

39. Ishizaka, T., Tomioka, H. and Ishizaka, K., J. Immunol., 106:705, 1971.

40. Kaliner, K. and Austen, K.F., J. Immunol., 112:664, 1974.

41. Prouvost-Danon, A., Peixoto, J. and Queiroz-Javierre, M., Immunology, 18:749, 1970.

42. Lagunoff, D., Chi, E.Y. and Wan, H., Biochem. Pharmac., 24:1573, 1975.

43. Morrison, D.C., Roser, J.F., Cochrane, C.G. and Henson, P.M., J. Immunol., 114:966, 1975.

44. Baxter, J.H., J. Immunol., 111:1470, 1973.

45. Keller, R., Clin. Exp. Immunol., 13:139, 1973.

46. Daëron, M., Duc, H.T., Kanellopoulos, J., LeBouteiller, Ph., Kinsky, R., and Voisin, G.A., Cell. Immunol., 20:133, 1975.

47. Rask, L., Klareskog, L., Östberg, L. and Petersen, P.A., Nature, 257:231, 1975.

48. Stanworth, D.R. in Mechanisms in Allergy, edited by L. Goodfriend, A.H. Sehon and R.P. Orange, p. 177, Marcel Dekker, Inc., New York, 1973.

Chapter 15

MEMBRANE RECEPTORS IN THE SPECIFIC TRANSFER OF IMMUNOGLOBULINS FROM MOTHER TO YOUNG

Max Schlamowitz
Department of Biochemistry
The University of Texas System Cancer Center
M. D. Anderson Hospital and Tumor Institute
Houston, Texas 77030

Abstract

Transfer of immunity from mother to young takes place prenatally, postnatally, or both depending on the animal species. Where prenatal transfer occurs the fetal tissues across which the immunoglobulins pass are the yolk sac or the hemochorioplacenta. Postnatal transfer is effected via the gut of the newborn. Transfer mechanisms are discussed and evaluated as is the evidence from both in vivo and in vitro studies that support the view that membrane receptors specific for IgG and its Fc fragment mediate the transfer process. This is followed by a brief discussion of the properties of the receptors.

Introduction

The recognition of the ability of the maternal organism to confer passive immunity to its young stems principally from the observation of Erlich that newborn mice of mothers immunized with toxins acquired a measure of immunity. He inferred that some transmission of immunity took place in utero but that the greater part was conferred through the milk (1). This phenomenon of transferable immunity to the young, prenatally, in utero, and/or shortly after birth, through milk, has been confirmed for all mammalian species studied to date and for avian species as well. The transfer of immunity from mother to young prenatally or shortly after birth should not be taken to mean that the newborn is immunologically incompetent; rather, the newborn has only relatively limited immunologic capability, principally reflecting a lack of antigenic stimulation as a result of having been seque-stered in an essentially germ-free environment during the prenatal phase of its development (2). Considerable evidence has been accumulated to show

253

that fetal lambs and humans have the capacity to respond to immunogenic
stimuli for considerable periods before birth (2-5).

Prenatal capacity to respond immunologically is not surprising in light
of the demonstrated ontogenic maturation of lymphocytes bearing surface IgM,
IgG and IgA early in gestation. Also, somewhat later in gestation, lymphocytic cells, in vitro, are able to proceed to terminal transformation into
IgG-containing plasma cells when stimulated with antigen (6).

The full potential for immunologic response characteristic of the adult
is developed gradually in the young animal and may not be fully achieved
until a considerable period of time after maternal passive immunity has
waned. The pre- and postnatal ontogenic relationship of the appearance of
immunoglobulins in the human is described by Alford (7). It is of interest,
in the context of passive transfer of immunoglobulins from mother to young,
that in the rat, rabbit and human, where extensive studies have been carried
out, transfer appears to be restricted to the IgG class. In essence then,
the complement of IgG antibodies seen in the serum of the newborn animal
mainly reflects its mother's previous antigenic experience.

In the early postnatal period there is a conservation of newly synthesized fetal immunoglobulin and maternoimmunoglobulin, i.e., a sparing
action which prolongs the duration of passive immunity. Such sparing action, or conservation, is indicated by longer half-lives of immunoglobulins
in newborns of several species studied (8,9).

Immunoglobulin Transfer: Time and Principal Routes

Stemming from the principles enunciated by Grosser (10,11) for the
classification of eutherian mammalian placentae, the role of the placenta
in selecting and regulating the transfer of substances from mother to fetus
was viewed as that of an ultrafilter. Needham (12) postulated that a reciprocal relation existed between the number of layers of tissue that separate
maternal from fetal circulation and the size of molecules or colloidal
particles which could pass the "placental barrier." The recognition that
selection and transfer of materials across cells is an active process makes
this static physical model much less acceptable.

The time when immunoglobulin is transferred from mother to young varies
in different animal species. In some, passive immunity is acquired before
birth (guinea pig, gray squirrel, man, monkey, rabbit); in others, after
birth (cow, goat, horse, pig, sheep); and in yet others, during both the

pre- and postnatal periods (cat, dog, hedgehog, mouse, rat). The routes
of transfer also may differ, e.g., the chorio-allantoic placenta, the yolk
sac membrane, or the gut, depending on the species. These points are summarized in TABLE I.

Brambell's treatise on the subject of maternofetal and maternoneonatal
transfer remains the single most comprehensive one on the subject (13) although more recent treatments, updating one or more aspects of the subject,
have appeared by Gitlin and Gitlin (17), Hemmings (16), Waldmann and Jones
(18), and Wild (19,20) and by contributors to the recent symposium edited
by Hemmings (21).

For rabbit and guinea pig the fetal yolk sac is the principal organ
for transfer of immunoglobulins. In these animals, as in rats and mice,
the inverted bilaminar yolk sac, an extensive highly villated and well
vascularized organ, is developed and becomes a functional placenta in the
latter half of pregnancy. To Brunschwig (22) and Everett (23) must go
credit for suggesting that the vascularized yolk sac in rodents constitutes

TABLE I

Time and Principal Routes of Transfer of IgG from
Mother to Young in Mammals[a]

Animal Order	Animal	Transfer		Approximate Duration of Transfer
		Prenatal	Postnatal	
ungulata	cow	−	+	1 day
	goat	−	+	1-2 days
	horse	−	+	1-2 days
	pig	−	+	1-2 days
	sheep	−	+	1-2 days
carnivora	cat	+	+	1-2 days
	dog	+	+	1-2 days
	ferret	+	+	8 days
rodentia	mouse	+	+	16 days
	rat	+	+	21 days
	guinea pig	+	−	
lagomorpha	rabbit	+[b]	−	
primata	man	+	−	
	monkey	+	−	

[a]compiled from data of Brambell (13), Waldmann and Strober (14), and Jeffcott (15).

[b]a recent report (16) that there is transfer across the chorio-allantoic
membrane as well as across the yolk sac in rabbit needs independent confirmation.

a physiological placenta and an important "organ of exchange." This view of the yolk sac as a functional placenta came to be generally accepted following Mossman's all-embracing redefinition of the term (24). As defined by Mossman, "the normal mammalian placenta is an apposition or fusion of fetal membranes to the uterine mucosa (of the maternal organism) for physiological exchange." However, the single most decisive experiment demonstrating that the yolk sac membrane is a functional placenta was provided by Brambell. In his review of 1958 (25), Brambell describes the experiment in which it was shown that ligating the vitelline circulation of the rabbit yolk sac completely arrested the transmission of antibodies to the fetus, whether from the uterine lumen or maternal circulation. Analogous studies, with similar findings were carried out on the guinea pig by Barnes (26). The studies of King and Enders (27) with the yolk sac membrane of guinea pig show that the junctional complexes between the endodermal cells of the fetal yolk sac membrane prevent fenestral passage of immunoglobulin molecules from the maternal uterine lumen to the intercellular spaces of the fetus.

The neonatal gut, composed of an endodermal epithelium and a vascularized mesoderm is morphologically and ontogenically analogous to the fetal yolk sac. The chorionic syncytiotrophoblast, the placental tissue across which transfer occurs in man and monkey, is a tissue of ectodermal origin, but one endowed with microvilli, a microvillar glycocalyx coat, and active endocytotic activity not unlike the yolk sac endoderm (19,28-32).

As seen in TABLE I, among the ungulates transfer occurs via the gut within the first day or two after birth, then ceases. The source of immunoglobulins in the newborn of these species is the maternal colostrum. Numerous studies (33-38) of normal suckling or force-fed animals have shown that proteins and biologically inert macromolecules, e.g., polyvinylpyrollidone (PVP), are taken up nonselectively, implying an absence of specific receptors in the gut of these species. Experiments have shown that the immunoglobulins of horse and cow are poorly transferred across the yolk sac membrane of rabbit in vivo (25) and bind poorly to the formalinized yolk sac membranes in vitro, FIGURE 1. It is interesting to speculate that only the immunoglobulins from species where selective transfer occurs are richly endowed with structural moieties or conformations that adapt them for recognition by cell surface receptors.

Inasmuch as the transfer of immunoglobulins in the ungulates is non-selective, we shall not dwell on the passive immunization process in these

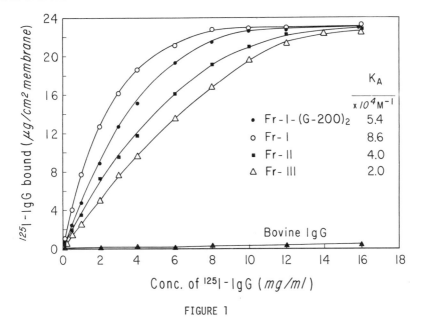

FIGURE 1

Binding of bovine IgG and fractions of rabbit IgG to formalin-fixed rabbit fetal yolk sac membrane. Adapted from Tsay and Schlamowitz (39).

animals. Similarly, although it is known that immunoglobulin transfer in carnivores takes place both pre- and postnatally, additional information with these animals is limited. Therefore, we will devote our main attention to studies of transfer in lagomorphs (rabbit), rodents (guinea pig, mouse, rat) and primates (human).

Mechanisms for Transfer of Immunoglobulins

Hypothetical schemes for the mechanism of selective transfer of immunoglobulins have been based on studies of the yolk sac membranes of rabbit and guinea pig, the intestinal epithelia of newborn mice and rats, and to a lesser extent from studies of primates where passage of immunoglobulins is across the discoid hemochorial placenta.

Three basic mechanisms for the specificity of transfer have been postulated by, in chronological order, Brambell (40), Wild (20), and Hemmings (41). Any mechanism proposed for the transfer process must be able to account for its specificity and for its capacity to transfer large amounts of protein from mother to fetus, or mother to neonate, in a relatively short period of time. To satisfy the latter requirement all of the proposed mechanisms share the view that cellular uptake of proteins is by pinocytosis. Brambell and Wild further share the view that specific receptors are involved in the initial endocytotic uptake of the immunoglobulins.

Although both Brambell's and Wild's hypotheses implicate specific cell surface receptors as necessary elements for the selective transfer process, they do so in different ways. The similarities and differences between the two hypotheses are best seen with reference to FIGURE 2 in which rabbit IgG (IgG_R) is, for illustrative purposes, an example of a transferable protein and bovine IgG (IgG_B), an example of one that is not specifically transferred.

In the Brambell scheme, specific receptors for IgG_R are uniformly distributed over the absorptive cell's microvillar surface and pinocytotic vesiculation occludes both specifically bound and excess unbound IgG_R as well as IgG_B from the luminal contents. The vesicles or phagosomes in the course of their passage through the cell, fuse with lysosomes, whose catheptic enzymes digest all but the receptor-bound, protected protein. This surviving protein is then released from the cell, presumably by reverse endocytosis, at the basal or basolateral surfaces of the cell for subsequent passage to the fetus. A mechanism of this type implies a saturable receptor system which under heavy load of IgG_R would be characterized by zero order transfer kinetics. Under conditions of light load first order kinetics would prevail whereby a constant fraction of the initial load would be transferred. This becomes apparent by analogy with the Michaelis-Menten equation that describes the kinetics of many enzyme reactions. If, as is reasonable to assume, some intracellular step, e.g., movement of IgG laden vesicles, is rate limiting in the overall passage of the IgG molecule through the absorptive cell, then the kinetics of this transfer should be described by the equation $V = \dfrac{C \cdot [IgG]}{1/K_A + [IgG]}$, where V is the rate of transcellular transfer; K_A, the IgG-receptor binding constant; [IgG], the concentration of IgG; and C, a constant which includes among other things the rate limiting factor. The reaction reduces to zero order when [IgG] is great relative to $1/K_A$, and becomes first order when [IgG] is small relative to $1/K_A$. Results of studies in which kinetics of in vivo transfer of IgG conform to zero order and first order will be described in the next section.

A knowledge of the K_A for these systems should enable one to predict the kinetics of transcellular transfer. Conversely, information on the

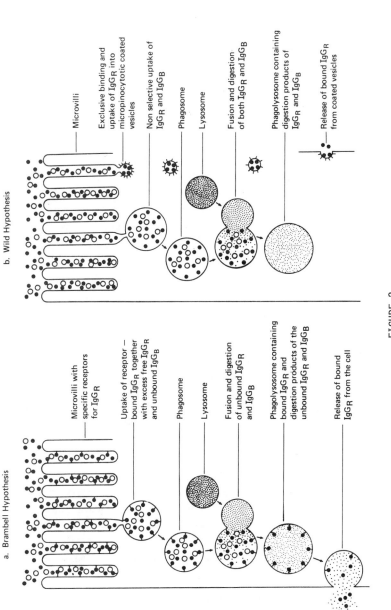

FIGURE 2

Mechanisms for selective transfer of immunoglobulins by the cell.
Adapted from Brambell (40) and Wild (20).

kinetics of transcellular passage of IgG should enable one to deduce values
for the K_A of the IgG-receptor system.

According to Brambell's hypothesis one would expect to detect the
presence of both IgG_R and IgG_B in the cell following pinocytotic uptake,
but only IgG_R, the transferable one, should be detected at or below the
basement membrane that underlies the endodermal cell layer. That this is
indeed the case was found by Wild (42) in studies of protein transfer across
the rabbit yolk sac membrane using mixtures of human IgG (IgG_H), a transferable protein, and IgG_B, a nontransferable one. Fluorescent antibodies to
these proteins showed both proteins were present in vesicles of the endodermal cells of the yolk sac splanchnopleur, whereas only the transferable
one, IgG_H, was detected in the basement membrane, the vascular mesenchyme,
and within vitelline vessels of the mesenchyme. Commenting on the results
of his studies, Wild states that they can be "interpreted as indicating
that all vesicles contained both proteins." No evidence has been reported
thus far to indicate otherwise.

The principal argument leveled against the Brambell hypothesis is that
no IgG containing vesicles fusing with the basal or basolateral surfaces of
the cell in the process of reverse endocytosis have ever been visualized.
Yet, (42, Fig. 4b; 43, Fig. 5) after five hours of intrauterine exposure
of rabbit yolk sac membrane to human gammaglobulin and subsequent staining
with fluorescein isothiocyanate conjugated rabbit antihumangammaglobulin
(FITC-Anti IgG_H) all vesicles in the apical and midregions of endodermal
cells fluoresced fully, whereas by the time the vesicles had reached the
cell's basal regions many of them fluoresce only on their peripheral surfaces, perhaps indicating that digestion and loss of excess unbound IgG had
occurred. Large vesicles, presumably the ones that have endocytosed proteins,
are seen principally in the apical region of the endodermal cell and only
small vesicles are seen in the basal region. Although, there is some gradation in vesicle size (42), the possibility should be considered that the
large vesicles seen in the apical region of the cell, shrink while passing
through the cell because of loss of water taken up in the endocytotic step.

Another criticism leveled at the Brambell hypothesis is that [125]I-
labeled rabbit immunoglobulin seen in the vesicles of rabbit yolk sac membrane appears as a bolus and is not restricted to the regions of the perimeter
of the vacuole as would be expected (20, discussion by Hemmings). This observation need not be considered a serious blow the Brambell hypothesis
when it is recognized (32) that the putative receptor containing material,

i.e., the glycocalyx, is not regidly fixed to and an integral part of the
bileaflet plasmalemma, but rather can be seen disengaged from it in many
regions, even in the intermicrovillar spaces. Sectioning through the polar
region of a glycocalyx lined vesicle could give rise to the appearance of
one that is almost filled with glycocalyx and glycocalyx-bound material.

The hypothesis of Wild (20), represented in FIGURE 2b, involves two
types of vesicles, large macropinocytotic vesicles and small, coated, micropinocytotic vesicles. The former are presumed to be involved in nonselective uptake of protein all of which is digested after fusion of the vesicles
with lysosomes. In this respect it is similar to Brambell's endocytosis and
fusion with lysosomes except that all uptake is presumed to be nonspecific.
The small coated vesicles are presumed to be exclusively engaged in taking
up transferable proteins, specifically binding them, and, after passing
through the cell, releasing these proteins by exocytosis at the basal or
basolateral cell surface. The mechanism does not spell out whether the
specific receptors for transferable proteins preexist on the cell surface
or are generated only during the formation of the coated micropinocytotic
vesicles. In the latter case they would not be detected in studies of the
nonpinocytotic cell. Careful inspection of the electron photomicrographs
(20, Fig. 4,5) of rabbit yolk sac membrane raises a number of interesting
questions. First, unlike the scheme where only small and large vesicles
are shown, the photomicrographs, in addition, depict vesicles of intermediate
sizes. Since the intermediate size vesicles are unlabeled in these figures,
it is not clear whether they are to be construed as small, coated, micropinocytotic vesicles with specific protein receptors or large macropinocytotic vesicles engaged in nonspecific uptake of protein. It is also difficult
to see morphological differences in the unit membranes, the glycocalyx coats
or the electron dense stained material coating their outer perimeters which
would give a clue to the basis for functional differences of these vesicles
of varying sizes. Further, the intermicrovillar invaginations that are
the prelude to vesiculation also range in size (20, Fig. 5a,b) and all,
including the one purporting to be a precursor of a coated micropinocytotic
vesicle, engulf extramembranous space and presumably the materials contained
therein. Therefore it is difficult to envision a mechanism through which
vesicles, whether small, intermediate, or large, are able to select for
inclusion one species of macromolecule, excluding all others. The void
area is retained even at the stage where the "coated micropinocytotic vesicles"
are close to and in confluence with the cell's basal membrane (20, Fig. 4).

The extent to which simple sectioning through different regions of spherical vacuoles may underlie the occurrence of small, medium or large vesicles is not discussed. This latter effect would be an artifact and not a real difference.

Similar gradation in size and inclusion of void space are seen in the apical intermicrovillar invaginations and in coated vesicles, either forming, formed or exocytosing of the proximal jejunum of mouse and rat gut, the segment purported to be engaged in specific uptake of protein in those species (44,45).

Wild (20) also states that the observed binding of macromolecules to the glycocalyx does not constitute specificity as required in the receptor hypothesis, in that the binding forces may reflect differences in charge, shape, size, hydrophobicity, etc. between the receptor and protein donor, as pointed out by Sonoda and Schlamowitz (46). At this point the problem becomes a semantic one. Specificity is a relative term and is operationally defined in terms of its thermodynamic parameter, the association constant, K_A. Values of K_A range from around 10^4 to 10^7 for relatively weakly interacting materials, such as haptens and substrates binding to antibodies and substrate binding sites of enzymes respectively, to around 10^{11} to 10^{13} for protein hormones binding to their cell receptor sites. The weaker binding systems, no less than the stronger binding ones, are "specific." The orientation and strength of binding forces rather than their nature confers the element of specificity. Only in retrospect and only when still consistent with the measured binding affinities are receptors described in terms of molecular composition and configuration.

The observation of Rodewald (45) that the capacities for selective and nonselective uptake of proteins in the rat gut are segregated in different regions of the small intestine cannot in and of itself be taken as proof of Wild's hypothesis inasmuch as combinations of identifiable proteins were not used in these studies. Taking Rodewald's observations into account, I. G. Morris (47) has set forth a transport mechanism which in essence combines the features of Brambell's and Wild's hypotheses.

A third mechanism, the one proposed by Hemmings (41), postulates that the major route of entry of proteins into the cell is by way of nonselective pinocytosis. It proposes that there are no specific receptors for homologous immunoglobulins on the cell's surface, but that the general stickiness of the glycocalyx attaches equally all protein presented to it. It then assumes that within the cell some of the vesiculated protein is released into the cytoplasm, by rupture of vesicles or by diffusion through them. It is this material, not that which remains vesiculated, that is transported through the cell by some diffusion related process for release to the fetus. The point is made that specificity for the overall transfer process is governed not by selective uptake, but by selective release by a "diffusion-carrier" process through the basolateral membrane of the cell. The basis for postulating this mechanism is the observation of a considerable cytoplasmic, non-vesiculated distribution of ferritin, and of ^{125}I-labeled IgG$_R$ and ^{125}I-labeled IgG$_B$ in the endodermal cells of the rabbit yolk sac exposed to these substances in utero. Similar observations were made on cells of the proximal and distal segments of rat gut, exposed to ferritin. Breakage of supranuclear vesicles have been seen in rat and rabbit yolk sac membranes by other investigators (42,48) but is regarded as an artifact of the processing of specimens for electron microscopy (41, discussion by Lloyd, Slade) and/or essentially as dead ends so far as protein transfer is concerned (42).

At its present stage of formulation this mechanism provides no insight into how specificity of transfer is achieved, i.e., how only pinocytosed protein, destined for transfer is distinguished from endogenous cytoplasmic protein and other exogenous, endocytosed, nontransferable protein for release from the cell. In addition, if movement of protein through the cell is a diffusion controlled process it would be an extremely inefficient one for transfer. Experimental results contradicting this hypothesis come from studies of the uptake of proteins from the isolated duodenal segment of the rat gut. Thus, Jones and Waldmann (49) found that 45 minutes after administering ^{125}I-IgG$_H$ the label was associated with the microvillus membrane component and not the cell sap. In addition, in experiments in which transferable IgG$_H$ and nontransferrable IgM$_H$ were administered intraduodenally, they found the percentage of the nontransported protein that remained in the duodenal lumen was considerably greater than that of the transported one. It is difficult to see how this could occur except through the agency of selective uptake.

In all fairness, it must be pointed out that the Hemmings hypothesis is chronologically the most recent of the three and that results of future experiments may provide it with the level of support that now appears lacking.

Experimental Support for IgG Receptors: In Vivo

Evidence supporting the presence of IgG receptors on absorptive cells of the fetal membrane and neonate gut have come from 2 types of studies:

(1) kinetics of transfer and (2) selectivity of transfer. The latter includes studies of direct selectivity as well as experiments in which selectivity is evidenced by competition.

(1) Kinetics of Transfer: Since both Brambell's and Wild's hypotheses
characterize saturable systems, it can be predicted that zero order or first
order kinetics of transfer should be demonstrable, depending on the level
of IgG presented to the absorptive cell.

Zero order kinetics has been demonstrated in vivo for the maternofetal
transfer of IgG_R and of rabbit serum albumin (RSA) via the rabbit yolk sac
membrane by Sonoda and Schlamowitz using high levels of IgG (50). They
also showed that RSA and IgG_R were taken up by the rabbit fetus independently
of each other, and suggested the existence of separate receptors for transfer
of the two substances.

First order kinetics of transfer of labeled mouse IgG was seen in the
data of Hemmings (51), where at low initial loads (10-1000 µg/ml) a constant
fraction of the load appeared as digestion products in the carcasses of young
mice. Several instances of first order transfer of antibodies across the
gut of stomach-fed young rats have also been recorded by I. G. Morris (47).
In both these experiments the results of transfer across the entire gut
were being measured.

It would be desirable to verify Hemmings' and Morris' results using
animals in which only the proximal segment of the gut was being used as the
transfer site. This is relevant in the light of the results of B. Morris
and R. Morris (52) and Waldmann and Jones (53) showing that only the
proximal segment is functional in transfer and that transfer across the ileal
segment is principally of degradation products (52).

(2) Selectivity of Transfer: The simplest and most convincing evidence that absorption by the fetus or neonate is selective, in a number of
animal species studied, including rabbit, rat, mouse, and man (49,54-59),
is the demonstration that numerous biological materials are not taken up by
the absorptive cells, nor transferred. Even other immunoglobulins (IgM,
IgA, IgD, and IgE) as well as proteins of nonimmunoglobulin nature are
transferred minimally or not at all.

In those species where there is cellular uptake and transfer of the
homologous IgG there is also uptake and transfer of heterologous IgGs, though
usually at lower rates as seen for example in the rabbit (60). This
probably reflects the degree to which their receptor-recognizable moieties
differ from those of the homologous material. Even within the homologous

IgG class, Hemmings et al. (60) noted a preferential uptake of a "slow," less anionic species of IgG, vis a vis a "fast" component, From this it would appear that the charge on the IgG is a factor involved in cellular uptake and transfer, although no correspondence was seen between isoelectric focussed fractions of IgG within each group and selectivity of uptake and transfer. Attempts to correlate isoelectric points of papain-produced F(ab) and Fc fragments of IgG with their transfer are complicated by the fact that fragments are metabolized differently from IgG in the absorptive cell and/or the fetus (61,62).

From studies of the competitive effect that unlabeled IgG has on the uptake of labeled IgG (49) it may be deduced that the number of receptor sites is finite.

The early work of Brambell (63) on the transfer of IgG and its Fc and F(ab) fragments indicated that the transfer of IgG across the rabbit yolk sac membrane was mediated through its Fc piece rather than through the F(ab) fragment. These conclusions were reinforced when it was found (64) that transmission of guinea pig antibodies across the gut of newborn mice was inhibited by rabbit IgG, even more so by its Fc fragment, but not at all by its F(ab) piece. The data also implies that transfer is mediated through receptor sites, finite in number, and specific for the Fc piece. The implied ability of the Fc piece to be transported more readily than the F(ab) fragment has also been verified in studies on absorption through the neonatal rat gut (49). The sum total of these data appears to be substantially correct, notwithstanding complications in quantitation because of the different catabolic half lives of the F(ab) and Fc fragments compared with intact IgG (61,62,65).

Experimental Support for IgG Receptors: In Vitro

Binding of IgG to receptors, one of the first, if not the first step in the multi-stage transfer process as postulated by either Brambell's or Wild's hypothesis can be studied from non-living systems.

The work of Schlamowitz et al. (66) demonstrated preferential localization of fluorescein isothiocyanate conjugated rabbit IgG (FITC-IgG$_R$) on the villus brush border of exposed rabbit yolk sac membrane prepared in frozen section. Little if any of the fluorescent probe bound to any of the intracellular components exposed in the cross section. In other experiments of that study FITC-IgG$_R$ and FITC-IgG$_B$ were mixed with vesicles derived from the endodermal cell's microvillus membrane; again surface localization was seen with the homologous IgG but not with the heterologous one, providing evidence of the species specific nature of this preferential binding. From these observations, seen by fluorescence light microscopy, and from observation of the distribution of glycocalyx on fixed preparations of these vesicles seen by electron microscopy (66) it would appear that the FITC-IgG$_R$ was preferentially associated with the glycocalyx coat substance. This confirms the previously cited evidence for localization of the antiferritin rabbit IgG to fixed membranes of the rabbit yolk sac (32).

As stated above there is almost an all versus none relationship of the binding of FITC-IgG$_R$ versus FITC-IgG$_B$ to microvillus membrane vesicles. Therefore, one would expect that the "receptors" would recognize IgGs from other species in relation to their chemical similarity to the homologous IgG in the receptor-binding region of the molecule. This would be reflected by a spectrum of relative selective binding for IgGs from different species. Recent measurements were carried out on the binding of IgGs from different species to formalin fixed rabbit fetal yolk sac membrane and are shown in descending order of their binding, as follows: rabbit > guinea pig > rat mouse > dog > sheep > pig > goat > human > cow > horse. On a scale of 100 for rabbit binding, horse was 3 and the others had values between these extremes. It was also shown by Sonoda and Schlamowitz (50) that bound ^{125}I-labeled IgG$_R$ is readily displaced from the formalin fixed yolk sac membranes by unlabeled IgG$_R$ but not by RSA. More recently Tsay and Schlamowitz showed that there are differences in relative binding (39) for fractions of IgG isolated by ion exchange chromatography (67). The binding strengths of the IgG fractions are in the order Fr-I > Fr-II > Fr-III, as indicated in FIGURE 1, and are inversely related to their electronegativity and sialic acid content. It remains to be established whether the rabbit IgG fractions examined are subpopulations of a single IgG class or represent populations enriched in one or another subclass of rabbit IgG. The quantitative results of this binding study, including K$_A$ values, are shown in FIGURE 1. It should be stressed that the data refer to binding of IgG to formalin fixed yolk sac membranes; however, Fr-I, the strongest binding of the IgG fractions, is the least anionic molecular species and corresponds to the "slow" component of rabbit IgG which was found, in the work of Hemmings et al. (60), most readily transferred in vivo.

We have recently demonstrated selective binding of ^{125}I-labeled IgG$_R$ vis a vis ^{125}I-labeled IgG$_B$ to a plasma membrane enriched fraction of cell free homogenates prepared from fresh rabbit yolk sac membrane, using sucrose gradient centrifugation. In much the same way, Jones and Waldmann (49) had earlier shown a preferential binding of IgG_H and of rabbit Fc piece but not of rabbit F(ab) fragment to homogenates and to membrane fractions derived from the neonatal rat gut. They also showed that complexing of labeled IgG could be inhibited in the presence of unlabeled IgG and that PVP did not bind to the same cell fractions. Additional evidence for the specificity of binding of IgG to cell homogenate components and plasma cell fractions, using sucrose density banding and column chromatography, has been provided by Balfour and Jones (68) working with homogenates of human placentae. The work of Jones and Waldmann and of Balfour and Jones, also demonstrated that while immunoglobulins of classes other than IgG, F(ab) fragments, and proteins unrelated to the IgG were not bound, the Fc piece of IgG bound. In contrast with the capability of homogenates from gut of untreated rats to bind IgG, homogenates of gut from rats treated with cortisone acetate had lost this capability (53), implying hormonal control of receptor formation and function. Shortly after birth the rat intestine loses its ability to absorb proteins (69-72). This is apparently due to cellular differentiation (43) and can be induced early in the life of the suckling rat by administration of adrenal cortical steroid hormones (73). Another indication that IgG receptors play an important role in transfer is the observation by Gitlin and Gitlin (56) that there is a qualitative correspondence between maternofetal transfer and placental binding.

Two other interesting reports that confirm the existence of cell surface IgG receptors, specifically Fc receptors, appear in the recent reports of Matre, et al. (74) and Jenkinson, et al. (75). The former authors showed strong adsorption of erythrocytes that had been sensitized with IgG antibodies of human, rabbit, and guinea pig origin to cryostat sections of human placental tissue, while no adsorption occurred using erythrocytes sensitized with $F(ab')_2$ fragments or with albumin. The reaction was strongly inhibited by intact IgG and by Fc fragments. In contrast with the results obtained with the human placental tissues, porcine placenta displayed no receptor activity, which is in line with the fact that there is no prenatal transfer of immunoglobulins from mother to fetus in the pig. The latter investigators, i.e., Jenkinson et al., have apparently pinpointed the location of the Fc receptor of placental tissue to the syncytiotrophoblast. This was demonstrated by rosette formation between antibody coated red blood cells and syncytiotrophoblast cells obtained from human term placenta.

Properties of IgG Receptors

The previous two sections cited evidence for the existence of IgG receptors and cited work which indicates that they are directed toward the Fc piece of the immunoglobulin. Much less is known of their biophysical or chemical nature.

Studies with rabbit yolk sacs show that IgG chemical receptors are associated with components of the glycocalyx coating on the microvillus surface of the endodermal cells. This was demonstrated by the deposition of ferritin, visualized by electron microscopy, on the glycocalyx of membrane previously incubated in utero with antiferritin rabbit IgG. Controls, in which normal rabbit IgG was used in place of antiferritin rabbit IgG, showed no deposition of ferritin (32).

The biochemical nature of the IgG receptor has been investigated by experiments using rabbit fetal yolk sac membranes, membranes of the gut of the neonatal rat, and human placental membranous tissues. In the case of the rabbit fetal yolk sac, treatment of formalin-fixed membranes with papain and trypsin results in a loss and/or solubilization of receptor activity to the extent of about 75% as measured by binding of purified rabbit IgG (76). These experiments testify to the protein nature of the IgG binding materials or receptors. On the other hand, treatment of fixed membranes with 0.025 M periodate for 6 hours at room temperature or with neuraminidase sufficient to eliminate about 70% of the total sialic acid has no influence on the binding of IgG, indicating that the sialic acid, in particular, and the carbohydrate component of the glycocalyx coat, in general, do not directly participate in the binding process (76).When binding of IgG to fixed membrane is carried out in the presence of increasing salt concentrations at pH 7.3 there is a marked drop in specific binding of 60-65% going from 0.01 M phosphate to 0.44 M NaCl, after which binding seems to level off even though NaCl concentrations as high as 1.1 M were used. The effect of the salt is reversible inasmuch as the binding characteristics of the membrane are restored to normal by removal of the salt, strongly implicating charge in the binding of IgG to receptors (76). Additional evidence for the involve ment of charge comes from experiments with membrane fractions of the gut of the neonatal rat (49) and from experiments with human placental membrane fractions (68). In both systems binding of human IgG to the receptors of the membranous tissues is strongly pH-dependent. Binding is low at values at or above physiological pH and stronger at pH values below 7. Waldmann and Jones

(18,53) have suggested that such a pH-dependent reversible binding of IgG to receptor may be the operative mechanism whereby IgG is bound by the cell surface receptors at the cell's apical surface, and then, following transport through the cell, ultimately released at the cell's basal surface for subsequent transfer to the fetus. They report that the luminal content of the proximal segment of the neonatal rat gut is about pH 6.5, which favors strong binding of IgG to gut membrane and is thereby conducive to uptake. Their presumption is that at the basal surface of the endodermal cell, where a pH in the neighborhood of 7.5 to 8 may prevail, the IgG would be released. In Waldmann and Jones work with the membrane fractions of the neonatal rat gut (53) and ours with fixed rabbit yolk sac membrane attempts to solubilize receptor with Triton X-100 have been unsuccessful. We have also been unable to solubilize IgG receptor materials from fixed yolk sac membranes using deoxycholate, 1.5%, Tween-20, 2-4%, or Triton X-100, 0.1-1%.

Clearly our understanding of the chemical nature of IgG receptors, whether from rabbit fetal yolk sac membrane, neonatal rat gut or human placenta, is still in its infancy.

Acknowledgement

This work was supported in part by Grant HD-7752 from the National Institutes of Health and by Grants PCM-73-02033 and PCM 76-09979 from the National Science Foundation.

References

1. Erlich, P., Z. Hyg. Infect. Kr., 12: 183, 1892.

2. Sterzl, J. and Silverstein, A. M., Adv. Immunol., 6: 337, 1967.

3. Martensson, L. and Fudenberg, H. H., J. Immunol., 94: 514, 1965.

4. Silverstein, A. M. and Lukes, R. J., Lab. Invest., 11: 918, 1962.

5. Fudenberg, H. H. and Fudenberg, B. R., Science, 145: 170, 1964.

6. Alford, C. A., Wu, L. Y. F., Blanco, A. and Lawton, A. R., in The Immune System and Infectious Diseases, edited by E. Neter and F. Milgrom, p. 42, S. Karger, New York, 1975.

7. Milford, C. A., Pediatr. Clin. North Am., 18: 99, 1971.

8. Deichmiller, M. P. and Dixon, F. J., J. Gen. Physiol., 43: 1047, 1960.

9. Patterson, R., Younger, J. S., Weigle, W. O. and Dixon, F. J., J. Gen. Physiol., 45: 501, 1962.

10. Grosser, O., Vergleichende Anatomie und Entwicklungs Geschichte der Eihaute und Placenta, Braumüller, Vienna, 1909.

11. Grosser, O., Frühentwicklung, Eihautbildung und Placentation der Menschen und der Saügetiere, Bergmann, München, 1927.

12. Needham, J., Chemical Embryology, Vol. 3, Biochemistry of the Placental Barrier, Cambridge University Press, Cambridge, 1931.

13. Brambell, F. W. R., The Transmission of Passive Immunity from Mother to Young. In Frontiers of Biology, vol. 18, North Holland Publ. Co., Amsterdam, 1970.

14. Waldmann, T. A. and Strober, W., Prog. Allergy, 13: 1, 1969.

15. Jeffcott, L. B., Biol. Rev., 47: 439, 1972.

16. Hemmings, W. A. in Proceedings of the First International Congress on Immunology in Obstetrics and Gynecology, edited by A. Centaro and N. Carretti, p. 252, Padua, 1974.

17. Gitlin, D. and Gitlin, J. D., in The Plasma Proteins, edited by F. W. Putnam, Second Edition, vol. II, p. 264, Academic Press, New York, 1975.

18. Waldmann, T. A. and Jones, E. A., in Protein Turnover, CIBA Foundation Symposium, 9 (new Series), p. 5, Elsevier, Amsterdam, 1973.

19. Wild, A. E. in Lysosomes, edited by J. T. Dingle, vol. 3, p. 169, North Holland Publishing Co., Amsterdam, 1973.

20. Wild, A. E., Phil. Trans. R. Soc. Lond. B., 271: 395, 1975.

21. W. A. Hemmings, editor, Maternofoetal Transmission of Immunoglobulins, Cambridge University Press, Cambridge, 1975.

22. Brunschwig, A. E., Anat. Rec., 34: 237, 1927.

23. Everett, J. W., J. Exp. Zool., 70: 234, 1935.

24. Mossman, H. W., Cont. Embryol., Carnegie Institute, Washington, 479: 129, 1937.

25. Brambell, F. W. R., Biol. Rev., 33: 488, 1958.

26. Barnes, J. M., J. Pathol. Bacteriol., 77: 371, 1959.

27. King, B. F. and Enders, A. C., Am. J. Anat., 130: 261, 1970.

28. Amoroso, E. C., in Marshall's Physiology of Reproduction, edited by A. S. Parkes, Third Edition, vol. II, Longmans, Green and Co., New York, 1952.

29. Amoroso, E. C., Brit. Med. Bull., 17: 81, 1961.

30. Tighe, J. R., Garrod, P. R. and Curran, C. R., J. Pathol. Bacteriol., 93: 559, 1967.

31. Boyd, J. D., Hamilton, W. J. and Boyd, C. A. R., J. Anat., 102: 553, 1968.

32. Sonoda, S., Shigematsu, T. and Schlamowitz, M., J. Immunol., 110: 1682, 1973.

33. Jeffcott, L. B., J. Reprod. Fertil. Suppl., 23: 727, 1975.

34. Jeffcott, L. B., J. Comp. Pathol. Ther., 84: 279, 1974.

35. Porter, P., Biochim. Biophys. Acta, 181: 381, 1969.

36. Smith, W. D., Wells, P. W., Burrells, C. and Dawson, A. MCL., Clin. Exp. Immunol., 23: 544, 1976.

37. Penhale, W. J., Logan, E. F., Selman, I. E., Fisher, E. W. and McEwan, A. D., Ann. Rech. Vet., 4: 223, 1973.

38. Porter, P., Immunology, 23: 225, 1972.

39. Tsay, D. D. and Schlamowitz, M., J. Immunol., 115: 939, 1975.

40. Brambell, F. W. R., Lancet ii, 1087, 1966.

41. Hemmings, W. A. in Maternofoetal Transmission of Immunoglobulins, edited by W. A. Hemmings, p. 91, Cambridge University Press, Cambridge, 1975.

42. Wild, A. E., J. Embryol. Exp. Morphol., 24: 313, 1970.

43. Wild, A. E., in Lysosomes, edited by J. T. Dingle, vol. 3, p. 511, North Holland Publishing Co., Amsterdam, 1973.

44. Clark, S. M., J. Biophys. Biochem., 5: 41, 1959.

45. Rodewald, R., J. Cell Biol., 58: 180, 1973.

46. Sonoda, S. and Schlamowitz, M., J. Immunol., 108: 1345, 1972.

47. Morris, I. G., in Maternofoetal Transmission of Immunoglobulins, edited by W. A. Hemmings, p. 341, Cambridge University Press, Cambridge, 1975.

48. Padykula, H. A., Deren, J. J. and Wilson, H. T., Dev. Biol., 13: 311, 1966.

49. Jones, E. A. and Waldmann, T. A., J. Clin. Invest., 51: 2916, 1972.

50. Sonoda, S. and Schlamowitz, M., J. Immunol., 108: 807, 1972.

51. Hemmings, W. A., IRCS Med. Sci., 3: 249, 1975.

52. Morris, B. and Morris, R., J. Physiol., 255: 619, 1976.

53. Waldmann, T. A. and Jones, E. A. in Maternal Transmission of Immunoglobulins, edited by W. A. Hemmings, p. 123, Cambridge University Press, Cambridge, 1975.

54. Gitlin, D. and Biasucci, A., J. Clin. Invest., 48: 1433, 1969.

55. Hemmings, W. A., Jones, R. E. and Williams, E. W., Immunology, 25: 645, 1973.

56. Gitlin, J. D. and Gitlin, D., J. Clin. Invest., 54: 1155, 1974.

57. Leslie, G. A. and Swate, T. E., J. Immunol., 109: 47, 1972.

58. Evans, H. E., Akpata, S. O. and Glass, L., J. Allergy Clin. Immunol., 50: 46, 1972.

59. Morris, I. G., Immunology, 13: 49, 1967.

60. Hemmings, W. A., Jones, R. E. and Page Faulk, W., Immunology, 28: 411, 1975.

61. Hemmings, W. A., IRCS Med. Sci., 2: 1453, 1974.

62. Hemmings, W. A., IRCS Med. Sci., 2: 1515, 1974.

63. Brambell, F. W. R., Hemmings, W. A., Oakley, C. L. and Porter, R. R., Proc. R. Soc. Lond. B., 151: 478, 1960.

64. Morris, I. G., Proc. R. Soc. Lond. B., 157: 160, 1963.

65. Gitlin, D., Kumate, J., Urrusti, J. and Morales, C., J. Clin. Invest., 43: 1938, 1964.

66. Schlamowitz, M., Hillman, K., Lichtiger, B. and Ahearn, M. J., J. Immunol., 115: 296, 1975.

67. Schlamowitz, M., Kaplan, M., Shaw, A. and Tsay, D. D., J. Immunol., 114: 1590, 1975.

68. Balfour, A. and Jones, E. A., in Maternofoetal Transmission of Immuno-globulins, edited by W. A. Hemmings, p. 61, Cambridge University Press, Cambridge, 1975.

69. Halliday, R., Proc. R. Soc. Lond. B., 145: 179, 1956.

70. Jordan, S. M. and Morgan, E. H., Aust. J. Exp. Biol. Med., 46: 465, 1968

71. Jones, R. E., Biol. Neonate, 24: 220, 1974.

72. Morris, B. and Morris, R., J. Physiol., 241: 1974, 1974.

73. Halliday, R., J. Endocrinol., 18: 56, 1959.

74. Matre, R., Tönder, O. and Endresen, C., Scand, J. Immunol., 4: 741, 1975

75. Jenkinson, E. J., Billingham, W. G. and Elson, J., Clin. Exp. Immunol., 23: 456, 1976.

76. Schlamowitz, M., Unpublished data.

Chapter 16

CELL SURFACE RECEPTORS AND LYMPHOCYTE MIGRATION

M. Schlesinger
Department of Experimental Medicine and Cancer Research
The Hebrew University-Hadassah Medical School
Jerusalem, Israel

Abstract

Lymphocyte subpopulations show a marked specificity in their distribution in various parts of the lymphoid system. It has been
suggested that specific localization of lymphocytes depends on the
presence of "homing" receptors on their cell surface. The present
paper summarizes some of the approaches used in the study of cell
surface receptors involved in the migration and localization of lymphocytes. The effect of exposure of lymphocytes to antibodies, enzymes or
lectins, on their capacity for localization is described. It is
concluded that the experimental procedures employed do not provide
conclusive evidence as to the mechanisms of lymphocyte localization,
and that the role of homing receptors is still far from clear.

Introduction

The mechanisms involved in the migration and localization of

lymphocytes in various lymphoid organs are still a mystery. Indeed, very

little is known on the mechanisms of cell dissemination via the vascular

route, and on the specificity of the localization of cells disseminated

by this route. It has been claimed that embryonic chick pigment cells

injected into the yolk-sac circulation of chick embryos localize

specifically in the feather germs (1), and that during avian development

germ cells migrate, via the vascular system, to the gonads (2). Curtis

(3) has raised, however, some criticism as to the evidence for

specificity of the localization of cells in these and other systems.

There seems little doubt, at present, that different lymphocyte
subpopulations localize in distinct, well-defined territor ies within
the lymphoid system. There is a clear delineation of the T- dependent
versus the T-independent areas of the lymphoid system (4). Procedures
which result in a depletion of T-lymphocytes, such as neonatal thymectomy
or the administration of anti-lymphocyte serum (5), also result in a
depletion of T-dependent areas (6). When T- or B- lymphocytes are
radioactively labeled prior to their injection into syngeneic hosts,
they can be detected in distinct T- or B- territories respectively (4).

An important approach to the analysis of the mechanisms involved
in the specific localization of lymphocyte subpopulations is a
morphological analysis of the migration routes of lymphocytes to
their destinated areas (cf. 7). Studies on the segregation of T- and
B-lymphocytes in the lymph-nodes indicate that T- and B-lymphocytes
enter the lymph-nodes through the same post-capillary venules which
run throughout the paracortex of the lymph-nodes. The distinctive
localization of B- and T- cells in their respective territories
becomes evident only after the lymphocytes have left the vascular
compartment. In the spleen, both B- and T-lymphocytes enter the
splenic pulp through the marginal zone. While T-cells remain confined
to the periarteriolar lymphocyte sheath, B-lymphocytes pass through
the peripheral zone of the periarteriolar sheath, and continue to
migrate to the germinal centers and to the lymphocyte corona surrounding the germinal centers.

The specific localization of lymphocytes in distinct areas of
the lymphoid system could, a priori, result from a number of
mechanisms, which are not mutually exclusive:

1. The migration of lymphocytes to various areas of the lymphoid
system may reflect their attraction by specific chemotactic factors,

distinct for each area. While considerable evidence is available for
the effect of chemotactic factors on the "directional locomotion" of
polymorphonuclear leukocytes (8,9) little is known on the role of a
chemotactic factor in the migration of lymphocytes.

2. The migration of lymphocytes may be random, but during their passage
through different areas, different lymphocytes may be retained selectively.
It is conceivable that such a selective localization of lymphocytes may
result from differences in the properties of the cell surface of the
migrating lymphocytes.

3. To some extent, at least, the migration routes of lymphocyte
populations may be affected by the physical resistance they encounter.
In some stages of their migration, T-lymphocytes seem to move towards
the area of least resistance (7).

The relative contribution of these and other factors to the migration
and localization pattern of lymphocyte subpopulations is at present far
from clear. It is the aim of the present paper to outline some of the
evidence that properties of the cell surface of lymphocytes, possibly
specific receptors, may contribute to the localization of cells in
various lymphoid areas. Some of the experimental approaches used in
the search for receptors for lymphocyte migration will be described,
and the difficulties encountered in the interpretation of the results
will be stressed.

Cell surface properties and migration patterns of lymphocytes. The
contention that the cell surface of lymphocytes may affect their
localization in different areas of the lymphoid system is supported by
the fact that lymphocyte-subpopulations which differ in their migration
pattern also differ markedly in their cell surface properties. The
antigenic difference between the cell surface of T- and B- cells are
now well established (10,11). Thus T-lymphocytes in the mouse are

charac terized by the presence of the θ- and Ly 1, 2, and 3 allo-

antigens, while B-lymphocytes are characterized by the PC and Ly-4

alloantigens. Antigenic differences between T- and B- lymphocytes

can also be demonstrated by the use of various heterologous anti-

sera (12,13). In addition, T- and B- lymphocytes differ by a number of

cell surface receptors, such as cell surface immunoglobulins, and

receptors for Fc (easily demonstrable on B-cells, but detectable on T-

cells only after activation), and receptors for C'3 on B-cells (cf. 14).

Differences can be detected between the antigens on the cell surface

of T-lymphocytes migrating to either the spleen or the thymus. The antigenic

differences between spleen seeking and lymph-node seeking thymocytes can

be demonstrated by analyzing the migration patterns of subpopulations of

thymocytes separated on BSA columns (15). The antigenic profile of spleen-

seeking versus lymph-node seeking thymocytes can also be ascertained by

studying the effect of various antisera on the localization of ^{51}Cr-labeled

thymus cells in syngeneic hosts (16). Thus, exposure of thymocytes to H-2

antiserum markedly reduces the proportion of cells localizing in the lymph-

nodes, but has a much lesser effect on the localization of thymus cells in

the spleen. The exposure of thymus cells to TL-antiserum has the opposite

effect - it causes a marked reduction of the proportion of cells localizing

in the spleen, but has only a minor effect on the lymph-node seeking

thymus cells.

The effect of antibodies to cell surface antigens on lymphocyte migration.

The finding that lymphocytes migrating to different parts of the

lymphoid system differ in their cell surface antigens, suggests the pos-

sibility that at least some of these antigens may be directly involved in

the specific localization of lymphocytes in predestined microenvironments.

If any of the lymphocyte antigens is associated with the cell-surface

receptors responsible for the specific migration
pattern, then it should be possible to "blindfold" the cells by blocking
the receptors with specific antibodies.

Analysis of this problem proved, however, to be a more complicated
task than seemed from the outset. Exposure of lymphocytes to antibodies
impedes their migration, but the overriding cause of this effect is the
opsonization of antibody coated lymphocytes by cells of the reticulo-endothelial
system, predominantly in the liver. It is difficult, therefore, to detect any
possible effect of antibodies on putative homing receptors.

The exposure of lymphocytes from the spleen, lymph-nodes and thymus to
various alloantisera inhibits their localization in the lymphoid organs of
syngeneic hosts (16,17). Antisera against the θ- and Ly- antigens could be
used, in the presence or absence of complement, to inhibit the localization
of T- cells in the spleen and lymph-nodes of mice (16,17). Similarly, the
incubation of bursa cells to anti-Ig serum inhibits their localization in
the spleen follicles of chicken (18). Most of the cells which do not reach
the lymphoid organs seem to be opsonized by cells of the RES. Of the
various alloantisera tested, H-2 antisera had by far the strongest inhibitory
effect on the localization of lymphocytes in the lymph-nodes. It seemed,
therefore, possible that H-2 antibodies may not only cause opsonization of
lymphocytes but may also block homing receptors. Exposure of lymphocytes
to Fab fragments of H-2 antibodies had only little, if any effect, on the
localization of lymphocytes (19). This may indicate that H-2 antigens on
the surface of lymphocytes are not directly involved in the migration of
lymphocytes. It should be borne in mind, however, that it is not clear
how effectively Fab fragments remain bound to the lymphocyte membrane
in vitro. It may be possible that Fab fragments become readily eluted
or shed with cell membrane fragments, so that their inhibitory effect
may be much weaker than that of intact H-2 antibodies.

An alternative approach to analysis of the role of lymphocyte antigens on their localization is to determine whether the migration of labeled lymphocytes is affected by antigenic differences between the lymphocytes and the recipients (20). Lymphocytes were found to localize normally in the lymph-nodes of recipients even when recipients and donors differ in their θ, Ly- or M- antigenicity. Diminished localization in the lymph-nodes of the recipients was, however, evident when the donors and recipients belonged to congenic strains differing in their H-2 antigenicity. Irradiation of the recipients prior to infusion of the lymphocytes prevented the depressed localization of the cells in the lymph-nodes of the allogeneic hosts. It seems, therefore, that incompatibility in the H-2 antigenicity of donor and recipient does not affect the process of localization of the lymphocytes. Rather, the depressed localization of lymphocytes in allogeneic hosts results from recognition by the host of the foreign antigens on the infused lymphocytes. Incompatibility at major transplantation antigens also did not seem to interfere with the migration of lymphocytes across the post capilary venules in isolated, perfused lymph-nodes of sheep (21).

The effect of enzymes on lymphocyte migration. Analysis of the function of lymphocytes exposed to various enzymes, under conditions which presumably do not impair their viability, may lead to the biochemical characterization of cell surface receptors. Indeed, the concept that the localization of lymphocytes in lymph-nodes requires specific "homing" receptors was originally proposed on the basis of results obtained with enzyme-treated cells (22,23).

The effect of trypsin. Woodruff and Gesner (24) were the first to note that the exposure of lymphocytes to trypsin inhibited their migration to lymph-nodes, while their localization in the spleen was affected to a lesser extent. The effect of trypsin on lymphocyte migration was transient. The deficiency of the localization in lymph-nodes was most marked 6 hours after the injection of enzyme-treated cells. At 24 hours after the injection of enzyme-treated lymphocytes they seemed to regain their capacity of migrating to the lymph-nodes of the recipients. Similar observations were made with trypsin-treated thymus cells (25). In a study by Jacobsson and Blomgren (26) the exposure of either normal or cortisone-resistant thymus cells to trypsin inhibited their migration to the lymph-nodes. The decrease in the localization on lymph-nodes was most noted 3 to 6 hours after the transfusion of trypsin-treated labeled lymphocytes, at which time the proportion of cells localizing in the spleen and in the liver was significantly increased. These changes in the migration pattern of enzyme-treated cells returned to normal 24 hours after the injection of the cells.

The diminished capacity of enzyme treated lymphocytes to migrate to the lymph-nodes lead to the development of the concept that the migration of lymphocytes to the lymph-nodes depends on specialized proteins on the surface of the lymphocytes, which function as lymph-node homing receptors (24). The specificity of these receptors seemed to be indicated by the observation of Gesner et al (27) that while trypsin treated lymphocytes were impaired in their capacity to home in T-dependant areas of the lymph-nodes, these cells migrated normally into the T-dependant zones of the spleen.

The notion that the migration of lymphocytes to the lymph-nodes depends on specialized cell surface proteins that function as lymph-node homing receptors is supported by a study of Woodruff (28) on the recovery of the capacity of trypsin-treated lymphocytes to home in lymph-nodes. Trypsin-treated lymphocytes were incubated in vitro for various time intervals prior to their injection into syngeneic hosts. After incubation for 6 hours the cells recovered about half of their capacity for migration to lymph-nodes, while after incubation for 12 hours the treated lymphocytes regained their full capacity for localization in lymph-nodes. The addition of puro= mycin to the medium prevented the regeneration of the migration capacity of the lymphocytes.

The membrane receptors required for homing in lymph-nodes seem to undergo rapid turnover. The in vitro exposure of lymphocytes to puromycin, without trypsin treatment, markedly impairs their migration to lymph-nodes (28). Similarly, the exposure of lymphocytes to mytomycin C impairs their localization in lymph-nodes (29).

While a number of laboratories have confirmed that trypsin treatment has a profound effect on the migration of lymphocytes to the lymph-nodes it now seems doubtful whether these findings indeed prove that lymphocytes possess specific protein receptors for homing. The proteolytic activity of trypsin can not only remove specific receptors but may have a profound effect on the lymphocyte surface. The biophysical properties of the membrane may become altered - its adhesiveness may decrease, and the distribution of cell surface receptors may become altered. This is illustrated by results of a freeze-etching study of the membrane of red blood cell ghosts (30). While the glycoproteins on the membrane of non-trypsinized cells are distributed uniformly, following trypsin treatment the glycoprotein fragments remaining on the cell membrane assume a reticular arrangements.

The effect of neuraminidase. The exposure of lymphocytes to neuraminidase, a treatment that removes much of their cell surface sialic acid, has a profound effect on many of the functions of lymphocytes. The treatment of lymphocytes with neuraminidase reduces their cell surface charge (31) and enables a more stable interaction between lymphocytes and various target cells (32). Neuraminidase treatment also increases the immunogenicity of various cell types (33,34). Antigenic analysis of neuraminidase-treated cells indicates that only a few antigen systems are "unmasked" by this treatment. Neuraminidase treatment of mouse lymphocytes does not seem to "unmask" H-2, Θ-, and TL-antigens but seems to expose antigenic determinants which interact with antibodies present in normal xenogeneic sera (35). In addition neuraminidase-treatment exposes antigens reacting with autoanti-bodies present in normal sera (35,37). The common feature of all the antigenic sites unmasked is that they possess carbohydrate determinants (35-37). This is in agreement with the fact that cell surface sialic acid is bound either to galactose or to N-acetyl-galactosamine residues, which become exposed upon neuraminidase treatment of the cells.

When lymphocytes are treated with vibrio cholerae neuraminidase prior to their infusion into syngeneic hosts they localize in increased amounts in the liver, while the number of cells reaching the lymph-nodes and spleen is reduced (39,40). Similar effects are obtained upon exposure of lymphocytes to Newcastle disease virus (41), which also possesses neuraminidase activity. The sequestration of neuraminidase treated lymphocytes in the liver is transient. It is highest during the first hours after the injection of the cells, and diminishes within 24 to 48 hours.

The role of the sialic acid coat of lymphocytes in their specific migration pattern is not clear. It has been postulated that sialic acid residues of glycoproteins found in the circulation enable these proteins to remain in the circulation. Glycoproteins from which the sialic acid residues are removed enzymatically are rapidly removed from the circulation and become concentrated in the liver (42). There are a number of mechanisms whereby neuraminidase treatment could affect lymphocyte migration. The treated cells may be unable to migrate normally as a result of their altered cell surface charge. Alternatively, enzyme-treated cells may become sequestered as a result of interaction with naturally occurring auto-antibodies reactive with neuraminidase treated cells. The fact that neuraminidase-treated cells, which are trapped in the liver shortly after

their infusion later regain their capacity of circulating, does not exclude

the possible effect of autoantibodies. It is conceivable that natural

19 s antibodies may trap the enzymatically altered cells without causing

their opsonization and that following the regeneration of their sialic

acid coat the cells become free again. An alternative interpretation of

the results obtained with neuraminidase is that the enzyme cleaves off

specific lymphocyte receptors for circulation. In view of the many known

effects of neuraminidase on the cell membrane this interpretation seems

unlikely at present.

The effect of lectins on the migration of lymphocytes. It has been suggested

that glycoproteins on the surface of lymphocytes are involved in the distinctiv

localization of lymphocytes in lymph-nodes. This is supported by the fact

that the exposure of lymphocytes to glycosidase (22), trypsin (24,25), neura-

minidase (39) or potassium periodate (43), inhibit the localization of

lymphocytes in lymph-nodes. Attempts were made to determine the role of

cells surface carbohydrates on the migration pattern of lymphocytes by the

use of lectins, known to interact with carbohydrate containing structures on

the surface of lymphocytes (44). In these studies radioactively labeled

lymphocytes were exposed in vitro to various concentrations of lectins,

and injected into syngeneic recipients. At various time intervals there-

after, the recipients were killed and the localization of lectin-treated

lymphocytes was compared with that of untreated cells (Table I).

Exposure of lymphocytes, derived from any lymphoid organ, to concana-

valin A (Con A) was regularly found to inhibit the localization of lymphocytes

in the lymph-nodes (19,40,45-49). Conflicting results were obtained,

however, by various investigators as to the effect of Con A on the spleen-

seeking lymphocyte subpopulation. In some studies, the exposure of

TABLE I

The effect of lectins on the localization of lymphocytes

Lectin	Cells injected	Effect on localization in:		Reference
		Lymph-nodes	Spleen	
Con A	LN	Decrease(+++)	Decrease(++)	19, 46-48
"	"	Decrease(+++)	Increase(++)	40, 45, 49
"	Spleen	Decrease(+++)	Decrease(+)	19, 47
"	Thymus	Decrease(++++)	Decrease(+++)	19, 47
PHA	LN	Decrease(+++)	Decrease(+)	19, 47
"	LN	Decrease(+++)	Increase(+)	49
"	Thymus	Decrease(++++)	Decrease(++++)	19, 47
"	Spleen	Decrease(+++)	Decrease(++)	19, 47
FBP	LN,Th,Spl	No effect	No effect	47
WGA	LN,Th,Spl	No effect	No effect	47
SBA	Thymus	No effect	Decrease(+++)	Unpublished
"	LN,Spl	No effect	No effect	Unpublished

Abbreviations: Con A= Concanavalin A, PHA= Phytohemagglutinin, FBP= Fucose

binding protein, WGA= Wheat germ agglutinin, SBA= Soybean agglutinin, LN= Lymph-

nodes, Th= thymus, Spl= Spleen.

lymphocytes to Con A was found to increase the proportion of lymphocytes

localizing in the spleen (40,45,49). In other studies, the migration

stream to the spleen was clearly inhibited by Con A (46-48), although the

inhibitory effect on the spleen-seeking population was always less marked

than on the lymph-node seeking cells.

Freitas and de Sousa, in whose study the exposure of lymphocytes to

Con A increased their localization in the spleen, studied the distribution

of Con A-treated cells in splenectomized hosts (40,49). In contrast to

the marked reduction of the localization of Con A treated cells in the

lymph-nodes of intact recipients, Con A treatment had no effect on the

migration of lymphocytes to the lymph-nodes of splenectomized mice.

This observation may indicate that the effect of Con A on lymphocyte migration is dependent on the presence of the spleen of the recipients. Preliminary observations in our laboratory (Schlesinger and Israel, unpublished data) indicate, however, that this may not be generally true. Table II illustrates our observation that Con A may indeed inhibit the localization of lymphocytes in the lymph-nodes of splenectomized BALB/c mice as effectively as in intact mice.

An additional indication that the lymphocytes found in the spleen of mice injected with Con A treated cells are not sequestered there, was obtained in secondary cell transfers (47). Con A treated lymphocytes localizing in the spleen of primary hosts, seemed to have a normal migration pattern when transferred to secondary hosts.

It is not clear, why in different studies Con A varied in its effect on the localization of lymphocytes in the spleen, and why splenectomy yielded different results. Differences in the technique of exposure of cells to lectins may be at fault and mice reared in different laboratories may vary in their response. It is possible, for instance, that mice kept

TABLE II

The effect of concanavalin A on the localization of lymph-node cells in splenectomized and intact BALB/c mice

Recipients	Lymphocytes suspended in:	LN	Localization in: Spleen	Liver
Intact	Saline	17.7 *	18.6	18.4
"	Con A (2 μg/ml)	2.4	13.5	38.1
Splenex	Saline	20.2	-	22.2
"	Con A (2μg/ml)	3.8	-	41.3

* Expressed as percent of the radioactivity injected

under different conditions may have different titers of natural antibodies
to either Con A or Con A modified cells, which, in turn, may affect the
migration of lymphocytes exposed to Con A.

Like exposure to Con A, the exposure of lymphocytes to phytohemagglutinin (PHA) inhibits their migration to lymph-nodes (19,47,49). In
our studies, PHA treatment inhibited the migration of lymphocytes to the
spleen and lymph-nodes to the same extent. Thymus cells were more
sensitive to the inhibitory effect of PHA than cells from the spleen or
lymph-nodes (19,47). In the study of Freitas and de Sousa (49) PHA
inhibited the migration of lymphocytes to the lymph-nodes, but, like in
their experiments with Con A, the proportion of PHA-treated lymphocytes
localizing in the spleen was increased.

Unlike Con A and PHA, wheat germ agglutinin and fucose-binding protein
failed to affect the migration of lymphocytes (47). Soybean agglutinin
(SBA), was without effect on migration of either spleen cells or lymph-node
cells. The migration of thymus cells to the spleen of syngeneic recipients
was greatly decreased, following exposure of the thymus cells to SBA,
although their migration to the lymph-nodes was unaffected (Schlesinger
and Israel, unpublished data).

What is the mode of action of lectins on lymphocyte migration?

The effect of lectins on lymphocyte migration does not seem to depend
on the selective killing of a cell subpopulation possessing specific migration properties. Thus, the effect of Con A can be reversed by exposure
to α-methyl-mannopyranoside (47), while that of SBA can be reversed by
exposure to N-acetylglucosamine (Schlesinger and Israel, unpublished data).
The effect of PHA on lymphocyte migration could not be abolished by N-
acetyl galactosamine (47) but fetal calf serum was reported to diminish
the effect of PHA (49). Another indication that the effect of lectins on

the localization of lymphocytes does not involve cell killing is that the
effect of lectins on the localization of lymphocytes is time-limited, and
is largely reversible by 72 hours after the administration of the treated
cells.

An alternative mechanism of action of lectins could be their mitotic
stimulation of lymphocytes which, in turn, could result in a decreased
capacity of cells to migrate. In support for this contention would be
the fact that only those lectins that have a strong mitogenic effect have a
profound effect on the migration of lymphocytes, while non-mitogenic
lectins (fucose-binding protein, wheat germ agglutinin, soybean agglutinin)
have little or no effect. The effect of the lectins on migration is evident
much earlier than their mitogenic effect, but possibly lectins may cause
changes in the cell membrane of lymphocytes triggered to mitosis long before
changes in DNA synthesis become evident in these cells. It should be noted,
however, that the magnitude of the effect of lectins on cell migration does
not correlate with their mitogenic effect. PHA has a relatively weak mitogenic effect on thymus cells, yet has a very strong inhibitory effect on the
migration of these cells. The lymph-node seeking subpopulation of thymus
cells and spleen cells are less responsive to Con A than the spleen-seeking
subpopulation (50), yet Con A has a stronger inhibitory effect on the
migration of the lymph-node seeking cells.

Many of the studies on the effect of lectins on lymphocyte migration
were motivated by an attempt to characterize glycoprotein receptors thought
to be involved in the homing of lymphocytes. Considerable doubt has been
raised, however, whether the effect of lectins on lymphocyte migration
results from blocking of such receptors. Some of the most striking arguments
against equating cell surface lectin binding receptors with homing receptors
are as follows:

1. The inhibitory effect of Con A on lymphocyte migration is maintained
for 24 hours after exposure to the.lectin, even under conditions in which
no Con A can be demonstrated on the cell surface (48). This suggests
that the effect of Con A is not due to blocking of cell surface receptors,
but rather reflects some changes induced in the membrane.

2. Exposure of lymphocytes to the monomeric fragment of Con A, rather than
to the whole molecule, largely abolishes the effect of Con A (49). The
slight activity maintained by the digested Con A preparation could be
attributed to contamination by the uncleaved molecule.

3. In some experiments, though not in all, splenectomy of the hosts
prevents the effect of Con A on the migration of lymphocytes to the lymph-
nodes (40,49). It must be concluded that, at least in these experiments,
Con A can inhibit the migration of lymphocytes to the lymph-nodes without
blocking any homing receptors.

It now seems that the effect of lectins does not result from blocking
of homing receptors, but rather from a non-specific alteration of the
biophysical properties of the cell surface. These changes either may
result directly from the attachment of the lectins to the cell surface, or
may be induced by the triggering of the cells.

Lymphocyte trapping.   Intensive antigenic stimulation of lymphoid organs
may alter their microenvironment in such a way that lymphocytes passing
through them may be retained to a greater extent than lymphocytes passing
through non-stimulated organs. Following the intra-venous injection of
antigens an increased number of cells become trapped in the spleen,
whereas following injection of antigens to the footpad the proportion of
cells localizing in the draining, ipsilateral popliteal lymph-node is
increased (51,52). Lymphocyte trapping by stimulated organs is under
the control of the thymus, since thymectomy of the recipients inhibits
trapping (53). Most of the cells trapped in this way are probably not

specifically committed to the stimulating antigen, and therefore this phenomenon, while important for the immune response, does not seem to be immunologically specific. Entrapment of specifically sensitized lymphocytes can, however, be demonstrated (54). Lymphocytes sensitized to an antigen become trapped in the presence of the sensitizing antigen, as compared with the localization of cells sensitized to a different antigen.

Some of the lymphocyte trapping may be caused by "inflammatory" changes elicited by the antigenic stimulation. It would be important to learn more on the changes in lymphoid organs that lead to increased retainment of lymphocytes passing through them. The fact that thymectomy inhibits lymphocyte trapping may mean that host T- cells or T-dependent areas may undergo changes which regulate the proportion of immigrating cells retained by these areas. It is not clear at present, what lymphocyte subpopulation is retained specifically, following sensitization with antigen. Once this becomes clear, it may be possible to correlate cell surface receptors involved in antigen recognition with the specific entrapment of sensitized lymphocytes.

CONCLUDING REMARKS

The hypothesis that the specific localization of lymphocytes in different parts of the lymphoid system is determined by specific cell surface receptors remains as attractive as ever. Experimental evidence has failed, however, to either prove or disprove the existence of such homing receptors. Many of the tools used so far for analysis of this problem have proved to be too clumsy. Each one of the approaches used might have affected not only the homing receptors but might have caused profound alterations of many properties of the cell surface. Some treatments may lead to the unmasking of autoantigenic determinants, others may lead to antigenic alterations of the cells. The cell surface charge may become altered, the fluidity of the membrane may change, the adhesiveness and stickiness of cells may be affected by the various treatments. All these changes must have a profound, non-specific effect on the capacity of the lymphocytes to migrate. Little wonder, then, that one remains puzzled whether any of the treatments had any effect on specific homing receptors. It is to be expected, however, that the development of more sophisticated methods will lead to a clarification of the mechanisms involved in the migration patterns of lymphocytes and in the trapping of uncommitted and sensitized lymphocytes. It has taken a great effort to unravel the nature of the "elusive" T-cell receptor. It might take the same kind of concerted effort to clarify the nature of the elusive homing receptor of lymphocytes, or even to determine whether such receptors are fact or fancy.

ACKNOWLEDGEMENTS

The original studies reported in this paper were supported by United States-Israel Binational Foundation Agreement No.714, and by a Lady Davis Endowment.

REFERENCES

1. Weiss, P. and Andres, G.M., J. Exp. Zool., 121:449, 1952.

2. Meyer, D.B., Dev. Biol. 10:154, 1964.

3. Curtis, A.S.G., The cell surface: Its molecular role in morphogenesis, Academic Press, London, 1967.

4. De Sousa, M.A.B., Clin. Exp. Immunol. 9:371, 1971.

5. Schlesinger, M. and Yron, I., Science 164:1412, 1969.

6. Turk, J.L. and Willoughby, Lancet 1:249, 1967.

7. Ford, W.L., Prog. Allergy, 19:1, 1975.

8. Boyden, S.V., J. Exp. Med., 115:435, 1962.

9. Sorkin, E., Stecher, V.J. and Borel, J.F., Ser. Haematol., 3/1:131, 1970.

10. Schlesinger, M., Prog. Allergy, 16:214, 1972.

11. Raff, M.C., Transpl. Rev., 6:52, 1971.

12. Raff, M.C., Nase, S. and Mitchison, N.A., Nature New Biol., 230:50, 1971.

13. Laskov, R., Rabinowitz, R. and Schlesinger, M., Immunology, 24:939, 1973.

14. Schlesinger, M., Ser. Haematol., 7/4:427, 1974

15. Schlesinger, M., Gottesfeld, S. and Korzash, Z., Cell. Immunol. 6:49, 1973.

16. Schlesinger, M., Schlomai-Korzash, Z. and Israel, E., Eur. J. Immunol.
    3:335, 1973.

17. Schlesinger, M. and Korzash, Z., Advances in Experimental Medicine and
    Biology, 29:71, 1973.

18. De Kruyff, R.H., Gilmour, D.G. and Thorbecke, G.J., J. Immunol. 114:1700,
    1975.

19. Schlesinger, M., Israel, E., Chaouat, M. and Gery, I., Ann. N.Y. Acad.
    Sci. 249:505, 1975.

20. Zatz, M.M., Gingrich, R. and Lance, E.M., Immunology, 23:665, 1972.

21. Frost, H., Cahill, R.N.P. and Trnka, Z., Eur. J. Immunol., 5:839, 1975.

22. Gesner, B.M. and Ginsburg, V., Proc. Nat. Acad. Sci. USA, 52:750, 1964.

23. Gesner, B.M., Ann. N.Y. Acad. Sci., 129:758, 1966.

24. Woodruff, J.J. and Gesner, B.M., Science, 161:176, 1968.

25. Berney, S.N. and Gesner, B.M., Immunology, 18:681, 1970.

26. Jacobsson, H. and Blomgren, H.H., Cell. Immunol. 5:107, 1972.

27. Gesner, B.M., Woodruff, J.J. and McClusky, R.T., Amer. J. Pathol.
    57:215, 1969.

28. Woodruff, J.J., Cell. Immunol., 13:378, 1974.

29. Romano, T.J., Ponzio, N.M. and Thorbecke, G.J., J. Immunol., 116:1618,
    1976.

30. Tillack, T.W., Scott, R.E. and Marchesi, V.T., J. Exp. Med., 135:1209,
    1972.

31. Nordling, S., Andersson, L.C. and Hayry, P., Science, 178:101, 1972.

32. Galili, U. and Schlesinger, M., J. Immunol., 112:1628, 1974.

33. Sanford, B.H. and Codington, J.F., Tissue Antigens, 1:153, 1971.

34. Bekesi, J.G., Roboz, J.P., Zimmerman, E. and Holland, J.F.,
    Cancer Res., 36:631, 1976.

35. Schlesinger, M. and Gottesfeld, S., Transpl. Proc., 3:1151, 1971.

36. Rosenberg, S.A. and Schwarz, S.J., Nat. Cancer Inst., 52:1152, 1974.

37. Schlesinger, M. and Chaouat, M., Behring Inst. Mitt, 55:216, 1974.

38. Schlesinger, M., Holland, J.F. and Bekesi, J.G., Proc. Third International
    Symposium on Detection and Prevention of Cancer, (In Press).

39. Woodruff, J.J. and Gesner, B.M., J. Exp. Med., 129:551, 1969.

40. Freitas, A.A. and de Sousa, M., Cell. Immunol., 22:345, 1976.

41. Woodruff, J.J. and Woodruff, J.F., Cell. Immunol., 10:78, 1974.

42. Morell, A.G., Gregoriadis, G., Schenberg, I.H., Hickman, J. and
    Ashwell, G., J. Biol. Chem., 246:1461, 1971.

43. Zatz, M.M., Goldstein, A.L., Blumenfeld, O.O. and White, A.,
    Nature New Biol., 240:253, 1972.

44. Sharon, N. and Lis, H., Science, 177:949, 1972.

45. Gilette, R.W., McKenzie, G.O. and Swanson, M.H., J. Immunol.,
    111:1902, 1973.

46. Taub, R.N., Cell. Immunol., 12:263, 1974.

47. Schlesinger, M. and Israel, E., Cell. Immunol.,14:66, 1974.

48. Rodriguez, B.A., Rich, R.R. and Rossen, R.D., J. Immunol., 115:771, 1975.

49. Freitas, A.A. and de Sousa, M., Eur. J. Immunol., 5 :831, 1975.

50. Lance, E. and Cooper, S., Transpl. Proc., 5:119, 1973.

51. Zatz, M.M. and Lance, E.M., J. Exp. Med., 134:224, 1971.

52. Zatz, M.M., Israel J. Med. Sci., 11:1368, 1975.

53. Zatz, M.M. and Gershon, R.K., J. Immunol., 112:101, 1974.

54. Emeson, E.E. and Thursh, D.R., J. Immunol., 113:1575, 1974.

Chapter 17

T-LYMPHOCYTE RECEPTORS FOR ALLOANTIGENS

H. Ramseier
Department of Experimental Microbiology,
Institute for Medical Microbiology, University of Zürich,
8028 Zürich, Switzerland

Abstract

Recent developments in research on T-cell receptors for alloantigens have
been reviewed. Recognition of these antigens is T-cell dependent and, when
measured in the PAR test, can be shown to take place with receptors present
on T cells, with those shed spontaneously from T cells and also with "recognition structures" of as yet unknown origin and structure present in posttransplantation alloantiserum but absent from B-cell induced alloantiserum.
All recognizing structures of parental T-cell origin mentioned above also
induce formation of anti-T cell receptor antisera when injected into appropriate F_1 hybrid animals. The highly specific inhibitory activity of the
sera can be demonstrated both in vitro and in vivo, and the most fascinating aspect constitutes cytotoxic elimination of T cells bearing receptors
for a given alloantigenic specificity without harming T cells with other
alloreceptors. The biochemical nature of T-cell receptors is not known and,
depending on the physical form in which it is investigated, high molecular
weight as well as considerably lower molecular weight structures have been
found. It has, on the other hand, become quite clear that the antigen-binding region of B- and T-cell receptors shows idiotypic similarity and it is,
therefore, likely that this part of the receptor molecule of both lymphocyte classes is similar if not identical (51).

Lymphocytes have long been known to be endowed with highly discriminative
cognitive potential for non-self antigens, but only during the last decade
has it become clear that recognition is mediated by receptors located in
lymphocyte membranes. Both thymus-independent (B) lymphocytes (1) and
thymus-dependent (T) lymphocytes (2) recognize alien antigens, such as
alloantigens, but the equal importance assigned to the receptors of the
two lymphocyte classes rests with functionality rather than with knowledge
on their precise immunochemical composition. There is no doubt that the receptor of B lymphocytes is an immunoglobulin (3,4) and there is also an

almost unanimous agreement that very little definitive is known on the nature of T-lymphocyte receptors (5).

This state of affairs will be reflected in this brief review on T-cell receptors for alloantigen: while a substantial body of evidence has been
accumulated on the function and discriminative power of T-lymphocyte receptors, pitifully little is known on their biochemistry.

Functional Demonstration of T-Lymphocyte Receptors for Alloantigens from Normal Animals and their Suppression by Specific Antisera

There are various technical means to measure alloantigen recognition, the
most widely used test being the MLC assay. MLC, however, has the important
shortcoming that recognition measurements take several days. A test system
of much shorter duration is the PAR assay. This test, technically described
in detail (6), has perhaps been given no application in laboratories other
than those of the reviewer. Nonetheless, it is employed because of its
surprising sensitivity. Basically the test is an MLC but, depending on the
reaction partners, PAR is formed within minutes or at the most within 4 to
10 h. The immunological basis of the assay is formation of the equivalent
of an antigen-antibody complex. Such complexes are known to have leuko-
tactic properties (7). In the case of PAR it is a granulotactic property
which is generated in at least two ways: T-cell receptor + alloantigen or
T-cell receptor + anti-T-cell receptor antibody. In the first reaction receptors appear to take the role of "antibody"; in the second that of
"antigen". Whatever the basis, the interaction product is responsible for
the accumulation of polymorphonuclear cells at sites of hamster skin (preferred because of its thinness) reactions developing within 24 h after

injection of concentrated cell interaction products. The recognition partner tested for (in most cases T-cell receptors for alloantigens) can be
estimated quantitatively by determining the magnitude of the granulotactic
response. For this skin lesions are excised, cut into fragments and are
subjected to a short course of trypsinization. This procedure liberates
PMN cells the nuclei of which are counted conventionally. The reason for
assigning PAR the nature of an antigen-antibody complex is the following:
for PAR as a complex of T cell receptor + alloantigen, the activity can be
destroyed by heating PAR for 30 min at 56° C. Heating inactivates the heat-
sensitive partner of the complex, namely alloantigen, for if heated PAR is

incubated with fresh alloantigen, activity is restored. The relative heat-stability of T-cell receptors can, on the other hand, be shown by failure to inactivate PAR formed as a complex of T-cell receptor + anti-T-cell receptor antibody. Only at temperatures known to denature proteins can this form of PAR (or the T-cell receptor alone) be inactivated. T cell receptor-alloantigen PAR can also be shown to be a complex in a positive way. Using appropriate columns, PAR can be retained either via its alloantigen component or via its receptor component.

In mixed lymphocyte interactions, recognition, by T cells of an inbred strain X, of alloantigen Y is being measured by cocultivating X T lymphoid cells (mostly from spleen) with $(X \times Y)F_1$ hybrid spleen cells for a period of time assuring maximal formation of PAR. As a population (not as single cells (8)) X T cells have receptors for a great variety of antigens, among which those for alloantigens are of interest here. A population of T lymphocytes of genotype X will have receptors or recognition structures (RS) for alloantigens A, B, C, D etc., but under normal conditions, not for its own alloantigen X, i.e. there will be no lymphoid cells with RS(X). The $(X \times Y)F_1$ hybrid target cells, on the other hand, will, of course, possess alloantigens X and Y, the latter being recognized by X T cells. As a population, F_1 cells also carry lymphocyte receptors for antigens A, B, C, D etc. Since, however, the F_1 animal has inherited alloantigens X and Y from its parents codominantly, F_1 lymphoid cells cannot possess RS(X) or RS(Y).

In order to recognize alloantigen the T-cell receptor does not have to be "bound" to viable lymphocytes. It has been shown that cultivated T cells from strain X spontaneously shed receptors for alloantigen A, B, C, D etc. (6). In accordance to what has been said above, culture supernates of X T cells cannot, of course, contain receptors for alloantigen X. If supernate of cultivated X T cells is confronted with $(X \times Y)F_1$ spleen target cells, PAR will be formed. Interactions of already shed T RS + alloantigen are very quick and maximal PAR formation is complete by 1 - 2 h at 37° C. PAR formation with shed T-cell receptors + alloantigen present on F_1 target cell is prevented when instead of untreated target cells, cells treated with formaldehyde are used. It can be concluded that formation of PAR as a complex of X T RS(Y) + antigen Y can only take place if alloantigen Y (or at least part of it) can be solubilized from the surface of F_1 spleen cells.

A seemingly completely different system of alloantigen recognition is that of employing alloantisera obtained after skin grafting. Together with F_1 target cells, posttransplantation sera form PAR. This interaction too is characterized by its extreme speed and, moreover, by the fact that when sera are diluted, PAR is formed exactly up to that dilution independently determined as the anti-H-2 titer of these sera. There is, therefore, a strict correlation between the capacity to recognize alloantigen and the ability to block alloantigen. This is in sharp contrast to alloantibodies raised by injecting isolated murine B lymphocytes into F_1 hosts. These alloantibodies (formed in the absence of T helper cells) are also capable of blocking alloantigen on F_1 target cells, but when confronted with alloantigen, are incapable of forming PAR.

The well established, rather simple genetic facts recalled above also form the basis of the induction of anti-T cell receptor (anti-RS) or anti-idiotypic antibodies (9,10). As discussed, an F_1 animal (no matter of which species) of genotype (X x Y) lacks, for genetic reasons, alloreceptors for X and Y. If, therefore, F_1 animals of this genotype are injected with T lymphocytes of genotype X which carry (among others) RS(Y), recipients will be capable of reacting against this foreign structure. (With respect to alloantigen, there will be no foreignness, because (X x Y)F_1 animals are naturally tolerant of antigens X and Y.) The reaction of F_1 animals to foreign receptor or idiotype is not limited to a cellular response (2), but results in formation of antibodies to T-cell receptors and it is likely that F_1 cell proliferation is the cellular expression of this antibody formation.

There is an additional remark to make. If the inoculum consists of unfractionated spleen cells, the F_1 serum contains in addition to anti-RS antibody also alloantibody activity. Responsible for the latter are B lymphocytes of the parental inoculum, for if only B cells are injected into F_1 animals, the F_1 serum displays alloantibody exclusively. Thus, in this standard protocol alloantibody activities of F_1 sera result from GVH reactions, anti-T cell receptor antibody activities from HVG reactions (10). This can also be shown by reversing the immunization protocol and, consequently, also the reaction type (11). Rather than injecting parental lymphocytes into F_1 hybrids (the standard protocol), F_1 spleen cells are injected in parental hosts. Whereas in the standard immunization parental donor cells are accepted, for genetic reasons, by their F_1 hosts, in the reverse protocol, there will be rejection of F_1 donor cells. Despite this, high titers of anti-RS antibody can be found in parental sera (in addition to alloantibody) provided bleeding is done before rejection of the anti-RS antibody-forming F_1 cell population. Even injection of F_1 spleen cells into specifically presensitized parents results in a brief burst of anti-RS antibody formation (again in addition to yet more alloantibody) before producer cells are rejected. Rejection by parental hosts of F_1 inocula can be prevented by lethal irradiation. This was found not to harm the antigen (parental T-cell receptors), thus permitting inoculated F_1 cells to produce increasing amounts of anti-receptor antibodies, plateauing, however, before animals died of irradiation damage (12). Studies like these in which the antibody-producing cell population could be manipulated revealed that anti-RS antibodies are formed by F_1 B lymphocytes and that this is a process dependent on syngeneic T cell help (12).

To induce anti-T cell receptor antibodies, presence of the T cell itself is not compulsory. This can be shown in two ways. First, parental T cells (or unfractionated parental spleen cells) can be heat-inactivated (56° C for 30 min), a process destroying cell viability and alloantigenicity, but not immunogenicity of alloreceptors; upon injection into F_1 hosts, there will be formation of normal amounts of anti-T cell receptor antibody, but not of alloantibody. Second, upon in vitro cultivation of parental strain lymphocytes, T cells spontaneously shed alloreceptors (6). The injection of shed T-cell receptors (so far only murine cells have been tested) into appropriate F_1 mice results in formation of high titers of anti-RS antibody. Injection of supernate of cultivated B cells, on the other hand, fails to elicit formation of these antibodies. Since the injected material is cell-free, such sera lack alloantibody (13).

Finally there is a last but extremely powerful way to induce anti-RS antibodies which does not seem to involve receptors bound to viable or to heat-killed T cells or receptors shed spontaneously from T lymphocytes. Sera obtained from animals grafted with allogeneic skin could be shown to be extremely well suited to provoke formation of anti-T-cell receptor antisera when injected into corresponding F_1 hosts (14). Thus, animals of strain X

grafted with just one unilateral full-thickness skin graft of strain Y
form, in the wake of graft rejection, posttransplantation alloantibodies X
anti-Y. These sera contain an immunogenic principle which, when injected
into $(X \times Y)F_1$ recipients, induces formation of $(X \times Y)F_1$ anti-(X anti-Y)
antibody, an activity which is functionally equal to anti-X RS(Y). Such
sera will contain no alloantibody. Remnants of the injected alloantibody
could never be detected in d 14 sera. The immunogenetic principle of posttransplantation sera is not known and it should be stressed that alloantibody raised by injecting parental B lymphocytes into F_1 recipients appear
to lack this property. These latter sera, however, share with posttransplantation sera certain properties: both are capable of blocking alloantigens present on F_1 hybrid target cells in PAR assays and both are able to
neutralize anti-T cell receptor antibodies (vide infra). Whether posttransplantation sera contain an immunogenically active population of molecules
absent from purely B-cell produced alloantisera, or whether the former
sera contain the equivalent of shed T-cell RS (perhaps due to an "over-
shooting" of T-cell receptors into the circulation during the (T-cell mediated) rejection process) or finally whether the difference between the
two forms of alloantisera is reducible to a physical state of existence
(aggregates in posttransplantation sera) or to differences in Ig classes,
are perhaps interesting but so far unproven hypotheses. The striking feature is that even diluted (e.g. 1:100) posttransplantation sera induce
formation of titers of anti-RS antibodies far exceeding those obtained with
any other immunization protocol (e.g. 1:260'000). In any case, it is clear
that recognition capacity as measured in the PAR assay and immunogenicity
of posttransplantation sera parallel each other. It should be mentioned
that all of these methods have so far always resulted in formation of anti-
RS antibodies. However, if in the standard protocol epidermal cells (lacking receptors) or native thymus lymphocytes (lacking at least functional receptors) are injected, the resulting F_1 serum will not contain anti-RS antibody.

The suppressive effect of anti-RS antisera obtained by any of the mentioned
procedures can be demonstrated in a number of ways both in vitro and in vivo.
So far two in vitro methods seem applicable. That adapted by McKearn et al.
(15) and also employed by Binz and Wigzell (16) measures anti-RS antibody

titers by a passive hemagglutination method in which sheep erythrocytes,
coated with specific alloantiserum IgG, serve as indicator cells for anti-
RS antibody. This method, incidentally, proved beyond doubt that anti-receptor antisera could not be complexes of alloantibody and alloantigen for if
they were, they would nonspecifically agglutinate coated SRBC. The method
thus confirmed earlier data excluding soluble complexes as the inhibitory
principle (10,17,18).

For the laboratory of the reviewer, however, the PAR test was a natural
choice to demonstrate anti-RS antibody titers, particularly because the
same test also revealed alloantibody titers (14). Thus, to test a given
antiserum for its anti-RS antibody titer, the cell population processing
the receptor in question, i.e. appropriate parental cells, is treated with
serum dilutions. Following this, aggressor cells are washed and cultivated
with untreated F_1 target cells for a period of time assuring maximal PAR
formation as determined for untreated partners. For the reaction of anti-
RS antibody with T-cell receptors it has become clear that within the short
period of treatment (15 min at 37° C), the antiserum solubilizes the T-cell
receptor. This is evidenced by the fact that anti-T receptor antibody reacting with T-cell receptors forms PAR within minutes. Thus, as mentioned
briefly, PAR in this case is a complex of T cell RS + anti-T cell RS antibody (whereas in allogeneic recognition it is a complex of T-cell RS +
alloantigen). In contrast to the interaction of T-cell RS with F_1 target
cell alloantigen, which takes only place at 37° C, that of T-cell RS with
anti-T cell RS antibody is temperature-independent. Studies with these so-
called direct PAR assays have thus revealed the mechanisms responsible in
titer determinations of anti-RS antisera: a serum can solubilize receptors
from parental T cells up to a certain dilution. Solubilization means formation of PAR (T-cell RS + anti-T-RS). This PAR, however, is removed from the
test system by washing treated parental cells. Parental cells (although
capable of resynthesizing receptors, vide infra) are for the duration of
the PAR test (4 to 6 h for H-2 differences) deprived of the particular receptor specificity and are thus incapable of recognizing alloantigens of
F_1 target cells with which they are now cocultivated. This lack of interaction means lack of PAR-formation. There will be a dilution of anti-RS
antiserum with which solubilization of T-cell receptors can no longer take

place; the receptors, therefore, remain on cell membranes and alloantigen of F_1 target cell will be recognized normally and PAR (T-cell RS + alloantigen) is formed.

The converse is done to test a serum for its alloantibody titer. For this, F_1 target cells carrying the appropriate antigen are treated, washed and cultivated together with untreated aggressor cells, again for a period of time assuring maximal PAR formation. For the demonstration of alloantibody titers, the mechanism responsible is not so clear. It seems that there is no solubilization of alloantigen of F_1 target cells by alloantibody (in contrast to T-cell receptors which are capable of solubilizing alloantigens). Rather, alloantibody seems to fix to alloantigens and thus appears to shield them. Again, a given serum can do that up to a certain dilution, which is its anti-H-2 titer. For both activities, the titers are always surprisingly sharp and, when determined accurately, ridiculously small dilution differences result in lack of PAR and in normal PAR formation, respectively. The reason for this precision is not clear but constitutes a wellcome feature of the test system.

Another in vitro demonstration of anti-RS antisera takes advantage of the spontaneous release of murine T-cell receptors (6). It has been shown that whereas untreated parental T cells upon cultivation in vitro almost immediately begin secreting alloantigen receptors, similar cells which had been treated with anti-RS antibody, washed and then cultivated, shed no receptors for the first 8 h. This delay can now easily be explained with the solubilization of T-cell RS by anti-RS antibody. The reaction is specific, for treatment of the same T-cell population with an anti-RS antibody of different specificity induces no delay in T-cell receptor shedding. A similar delay but of a nonspecific nature is induced when T-cell suspensions are treated with trypsin, washed and cultivated. The results indicate that resynthesis of T-cell RS for alloantigens is complete within 8 h. The criterium in these experiments is, however, based on T-cell RS matured to the stage of export. Other experiments employing binding of anti-RS antibody to T-cell receptors have shown that budding of receptors can be detected as early as 5 h after removal. There will be no resynthesis and, therefore, shedding of T-cell RS for at least 32 h if T lymphocytes are treated with anti-RS antibody and complement. Again this reaction is specific (anti-RS antibodies of different specificity + complement does not prevent shedding) and contrasts well with the nonspecific effect treatment with anti-θ and complement has, although the basic mechanism - cytotoxic elimination - might be similar (19).

Demonstration of the inhibitory activity of anti-RS antibody (or anti-idiotypic or anti-aliotypic antibody as they are also called) in vivo was important (20). It proved that even under biologically much more complicated conditions suppression of T RS was possible, thus confirming in vitro results. The first piece of evidence for in vivo T-cell receptor suppression was obtained by Joller (21) who showed that in a lethal GVH reaction, a significant proportion of newborn F_1 mice could be protected specifically from runting if parental donor cells were treated with anti-RS antibody and complement. These studies were refined and extended by Binz and collaborators (18,22) and by McKearn (23) to show that GVH-reactive parental rat T cells which normally induce severe local (popletial node) GVH reaction were prevented from doing so by anti-RS antibody and complement. As in Joller's experiments, the inhibitory effect was specific and suppression was shown with antisera raised by posttransplantation serum (18,22,23) or by T cells (24-26). Reactivity was observed in both rat and mouse systems. In these studies, inhibition of GVH-reactive T lymphocytes could be achieved either by actively immunizing adult F_1 hybrids with posttransplantation alloantisera (18,22) or with low doses of parental T lymphocytes (26) or by passive immunization of F_1 recipients with anti-RS antibody (26) before animals were challenged. Allogeneic immunization with parental B lymphocytes did not induce formation of the inhibitory serum principle (26). Another approach which, however, succeeded only with selected anti-RS sera was to kill reactive T cells. Thus, Binz and collaborators (16,18,25,26) demonstrated highly specific (leaving reactivity to third party alloantigens intact) cytotoxic removal of T cells reactive in GVH and in MLC by anti-RS sera when used together with complement. Specific inhibition of systemic GVH reactions in adult rats by anti-RS sera (anti-alloantisera) was reported by McKearn and collaborators. In concordance with in vitro results their study shows that irradiated F_1 rats fail to develop anti-RS antibody and thus remained susceptible to systemic GVH. Irradiated F_1 rats, however, could be protected by adoptively administered anti-RS antibody-forming spleen cells (15).

Visual Demonstration of T-Cell Receptors for Alloantigens and Idiotypic
Similarity with B-Cell Receptors of Identical Specificity

Direct visualization of T-cell receptors for alloantigens was only recently possible by means of autoradiography, immunoelectronmicroscopy and
immunofluorescence involving anti-RS sera raised by T cells or by alloantisera (8,24,27). These studies revealed that between 5 and 10% of T cells
appear to carry receptors for a given alloantigen. This surprisingly high
frequency is, however, matched by functional studies (28-38). Furthermore,
the techniques used together with quantitative intensity measurements
yielded results indicating a density of T-cell receptors for alloantigens
which is of the same order of magnitude as found for B cells (27).

These same methods showed that about 1% of B cells appear to carry the
same receptor specificity, a result constituting additional confirmation
of other experimental findings revealing that T- and B-lymphocyte receptors
must share idiotypic determinants. This means that an essential part (the
antibody-combining site) of the receptors of T and B lymphocytes is determined by an identical V gene. The fundamental similarity of T- and B-cell
receptors was first noticed when it was discovered that, on the one hand,
a B-cell product, posttransplantation serum, was capable of eliciting
antisera which interacted with T-cell receptors in a fashion indistinguishable from antisera raised by T cells (14) and that, on the other hand, such
anti-T cell receptor antisera could be neutralized by posttransplantation
sera and vice versa (39). However there is room for serious criticism.
As pointed out, such posttransplantation sera, capable of interacting with
alloantigen to form PAR, reveal that they contain an as yet unidentified
principle functionally similar to T-cell receptor. Taking the worst interpretation possible, i.e. presence of bona fide T-cell receptors in such
posttransplantation sera, both the anti-RS antibody induction and its neutralization would become trivial events. The experiments were, therefore,
repeated. With respect to anti-RS antibody inducibility, an important difference was discovered: While posttransplantation serum induces extremely
high titers of anti-RS antibody (anti-alloantibody), alloantibody raised
by injecting B lymphocytes into F_1 hybrids (and therefore induced in the
absence of T cells) failed to do so. At present, the reason for this discrepancy is not known, and the fact that alloantiserum as posttransplanta-

tion serum or as induced by injecting unfractionated F_1 spleen cells into parental hosts (8,25) is immunogenic, whereas B-cell induced alloantiserum is not, has for the moment simply to be accepted. However, with respect to neutralization, earlier results (39) were fully confirmed. In these processes, in which antigenicity, not immunogenicity, counts, antisera raised to spontaneously shed T-cell receptors were shown to be neutralized specifically by B-cell induced alloantisera. This was also true for the reverse: alloantisera were neutralized specifically with anti-T cell receptor antisera. Furthermore, similarly to findings of Binz and Wigzell for rat B cells (16), nude B lymphocytes fully absorbed anti-RS antibody activities. This, together with the finding that also normal mouse serum is endowed with the capacity to neutralize anti-T cell RS sera, raises the possibility that the neutralizing principle is the B-cell receptor itself. This is seemingly confirmed by the observation that in the PAR test normal murine serum in high concentrations blocks alloantigen and that culture supernates of B cells contain the same principle. It might thus be argued that normal serum, B-cell induced alloantiserum and possibly also posttransplantation serum harbor spontaneously shed B-cell receptors for alloantigens, just as supernates of cultivated B cells do. While it is too early to make definitive statements on the shedding of the B-cell receptors, one might, nevertheless, speculate whether this could not be the reason for the difference observed in alloantibody as posttransplantation serum as opposed to B-cell induced sera. It is known that T-cell independent alloantibody are of IgM nature, whereas T-cell dependent alloantisera contain IgG (40). B-cell induced (as well as normal sera) might be considered to be T-cell independent "alloantisera" - or, as speculated, shed B-cell RS, and it is possible that in this form, the B-cell receptor is antigenic but not immunogenic. An analogy to this is to be found for the T-cell receptor: thymus lymphocytes are perfectly capable of neutralizing anti-T receptor antisera but are themselves incapable of inducing them.

There is one disturbing fact. If the B-cell product in B-cell induced alloantiserum (X B cells injected into (X x Y)F_1 mice resulting in high titers (up to 1:512) of X anti-Y alloantiserum) is indeed shed B-cell receptor for alloantigen, an answer must be given to the obvious question of why this activity is present in host serum at all and is not absorbed by host tissue? Clearly it cannot be an excess for this would mean that F_1 host tissues are bathed in "alloantibody" and that the antigenic specificity derived from one parent is blocked permanently because parental B cells will, at least for genetic reasons, not be rejected by their F_1 hosts. There could be, however, a constant association with and dissociation from F_1 alloantigen of parental B-cell receptors. Furthermore, PAR might, in contrast to conventional tests, be particularly apt to measure low affinity alloantibody molecules (41). For normal mouse serum this problem does not apply, nor does it for T-cell dependent alloantibodies. For normal serum, one could imagine that a certain background level of B-cell alloantigen receptors is present and turns over constantly indicating, perhaps, some sort of preparedness. For T-dependent alloantisera, rejection of alloantigen disposes of this particular problem. There are various other lines of evidence proving similarity of T- and B-cell receptors for alloantigen worked out by Binz and Wigzell. They have very recently reviewed their work (20). In addition, work of Eichmann and Rajewsky (42) with an entirely different test system and anti-idiotypic antisera fits conclusions on similarity of T- and B-cell receptors for antigens perfectly well.

On the Possible Nature of T-Cell Receptors for Alloantigen

There have been numerous attempts to determine the immunochemical nature of the T-cell receptor, but so far the results must be considered inconclusive or controversial (4,43-48). The approach taken in this lab differs from others. Advantage was taken of the observation that upon cultivation in vitro, T cells spontaneously shed receptors for alloantigens (6). In a collaborative effort with Drs. Lindenmann and Aguet we attempted to analyze these shed receptors.

Two straight forward experiments ascertained specificity of reaction of shed T-cell receptors with alloantigen and binding of these receptors to the corresponding anti-receptor antiserum. A culture supernate of T lymphocytes from strain X contains receptors for a great variety of alloantigens. Incubation of supernate with alloantigen-carrying F_1 spleen cells of specificity $(A \times X)F_1$, $(B \times X)F_1$ etc. will all give PAR and, consequently, discrimination among receptor specificities is not possible. An easy way out of this is to treat F_1 spleen cells with formaldehyde, thus fixing allo-

antigens. T-cell receptor-containing supernate cultivated with formaldehyde-treated F_1 target cells fails to form PAR, because T-cell receptors
now cannot solubilize alloantigen but the receptor attaches to fixed alloantigen. This method can be used to selectively deplete a given supernate
of T-cell receptors. Absorption of a supernate of X T lymphocytes with
formaldehyde-treated $(A \times X)F_1$ spleen cells removes quantitatively all X
T-cell receptors for alloantigen A $[(X \ RS(A)]$ but leaves X RS(B), X RS(C)
etc. fully reactive in the supernate (19). This simple experiment settles
3 points, a) the supernate contains T-cell receptors capable of interacting with alloantigens to form PAR (a T cell-dependent product (49)), b)
the reaction is specific and c) in this type of PAR formation, T-cell receptors under normal conditions most likely solubilize alloantigen from the
surface of F_1 target cells. The other experiment was to show interaction of
shed T-cell receptors with anti-receptor antiserum. For this, mouse Ig was
covalently bound to Sepharose and was then saturated with rabbit anti-
mouse Ig and washed. Supernates of X T lymphocytes filtered through passed
this column without any loss of activity. After washings, the column was
now saturated with a murine anti-RS antiserum of specificity $(A \times X)F_1$
anti-X RS(A) with an anti-RS(A) titer of 1:6000. Being a mouse Ig, the
anti-RS antiserum was retained completely, for the 1:5 diluted eluate had
no titer. If supernates of X T lymphocytes were now filtered through this
column the following was observed: T-cell receptors for A alloantigens were
completely retained whereas T-cell receptors for B alloantigens were not.
This experiment also revealed three points, a) the T-cell receptor for
alloantigens is not retained by anti-mouse Ig (mostly an anti-Fc serum,
free of anti-κ activity, b) anti-RS antibody is an Ig and c) T-cell receptors bind specifically to anti-T receptor antibody.

With this basic information, separation on gel of T receptor-containing
supernates was attempted. Functionality of fractionated receptors was by
means of the direct PAR test: Fractions were cultivated for 2 h at 37° C
with appropriate F_1 target spleen cells carrying alloantigen (6,13,19).
Fractionations on Sephadex G-200 showed that the T-cell receptor eluted
with the void volume (marker blue dextran), indicating a high molecular
weight and a nature certainly different from IgG or from structures of
smaller molecular weights (markers BSA monomer, cytochrome C). Results on

G-200 columns were similar for supernates of spleen cells cultivated either
unfractionated or enriched for T cells, but no activity appeared when nude
spleen cell supernate was fractionated. A somewhat better characterization
of T cell receptors-containing material was obtained with Sepharose 4B
columns. The results, however, simply confirmed those obtained with G-200,
namely dissimilarity with IgG but elution of peak amounts of active material in fractions where 19S IgM would also be expected to elute. As before,
supernates from T lymphocytes resulted in active fractions, those from B
cells did not. Fractionation of T-cell supernates on Ultrogel AcA 22, so
far, revealed the most telling resolution. In this gel, 3 peaks were obtained of which that containing material of the lowest molecular weight
was slightly heavier than the marker thyroglobuline (MW 660,000) and the
peak with material of highest molecular weight slightly lighter than the
marker blue dextran (MW approximately 10^6). It is possible, but so far unproven that the two peaks with higher molecular weights are aggregates.
Clearly, no firm conclusions can be drawn from these attempts to determine
the molecular weight of shed T-cell receptors. Nevertheless, fractionation
of unmanipulated, spontaneously shed T-cell receptors with capacity to recognize alloantigen have revealed a molecular weight of around 660,000.
In this estimation, presence of cell membrane particles being shed together
with the receptor cannot be excluded. Such membrane fragments, however,
would contain lipids. These, presumably, are removed by treating ammonium
sulphate-precipitated T-cell receptors with acetone and washing preparations in acetone. As far as activity to recognize alloantigen is concerned,
this treatment resulted in no loss of activity. However, further fractionation has not been attempted so far.

Similar conclusions as to the molecular weight of spontaneously shed T-cell
receptors were obtained by fractionation on gradients and by filtrations.
Sucrose gradients used together with human serum markers showed peak activity close to IgM but certainly revealed no relation to IgG. Mannitol gradient used in conjunction with a rabbit intestinal sucrose marker (MW
290'000) showed activity peaking for far heavier material than the marker.
Filtrations likewise indicated that the shed T-cell receptor must be of
high molecular weight. Receptor-containing supernates failed to pass Amicon
filters with retention limits of 50,000, 100,000 and 300,000.

If the T-cell receptor were to be a molecule similar to IgM (as most attempts on its nature so far indicated), it would be retained strongly by conconavalin A. This was indeed the case, and elution from various concentrations of ConA-Sepharose slurry could be performed with mannoside. Binding of T-cell receptors to ConA proved to be very strong for in one extreme experiment, the T-cell receptor present in 320 ml of supernate was completely absorbed to 10 ml of ConA-Sepharose. Strong adsorption was also revealed by the observation that the T receptor appeared to have an affinity to ConA which was greater than that of mannoside: as measured by an about 50% retention, receptors could even displace mannoside from a ConA-column saturated with the sugar.

Other properties of spontaneously shed T-cell receptors can be summarized as follows. The receptor can be precipitated completely with 50% and 66% ammonium sulphate. The receptor is destroyed by treatment for 60 min at 37° C with 0.1 M mercaptoethanol. It is also susceptible to proteinases: treatment with trypsin and chymotrypsin at room temperature for 30 min is without effect, but after 2 h about 50% and after 14 h all activity is destroyed. Similar treatment with papain destroys the activity of shed T cell receptors to recognize alloantigen within only 30 min. With respect to pH, the receptor withstands 15 h buffer dialysis at 4° C (followed by 48 h redialysis) at pH values ranging from 2.0 to 8.3 without apparent loss of activity. At pH 8.6, about half and at pH 9.0 and 10.0 about 70% and 80% of activity, respectively, is lost. The murine T-cell receptor does not appear to have similarity to human β-2-microglobulin, for treatment of aggressor spleen cells with a rabbit anti-human β-2-microglobulin serum followed by washings revealed no impairment of alloantigen recognition. When used to treat F_1 target spleen cells, however, anti-β-2-microglobulin displayed a small titer (1:8). In addition, using the same preparation of anti-β-2-microglobulin bound to bromacetyl cellulose as an immunosorbent for T cell receptor-containing culture supernatant, there was no retention of T-cell receptors.

While most attempts to characterize the T-cell receptor lead to the conclusion that it might be a molecule similar to IgM, the following experiment employing an anti-IgM immunosorbens made this possibility unlikely. Being well aware of the critical importance the purity of the anti-IgM se-

rum has in such experiments (1), this serum was prepared by J. Lindenmann
by first injecting mice with RBC from rabbit No. 1. Mice were bled 4 days
later and the serum (presumably mostly 19S antibody) was passed over a
Sephadex G-200 column. The fractions corresponding to 19S IgM were absorbed onto RBC from rabbit No. 1. These RBC, now supposedly loaded with mouse
IgM, were inoculated back into the donor rabbit No. 1. Three further similar injections were done after 1, 2 and 4 weeks. The serum from this rabbit,
obtained 12 days after the last injection was tested by Ouchterlony and
found to have antibody mostly to IgM but also some to IgG. The serum was,
therefore, subjected to immunoelectrophoresis against mouse serum and IgM
precipitation arcs were cut from the gel with a razor blade. These mouse
IgM rabbit anti-IgM complexes were now injected together with Freund's complete adjuvant twice 2 weeks apart into rabbit No. 2 and the rabbit serum
obtained on day 35 was used in the same manner as the serum of rabbit No. 1
to provide a third injection (without adjuvant) for rabbit No. 2. The final
bleeding, performed 10 d later, yielded a serum which was absorbed with insoluble mouse IgG to remove antibodies reacting with L chains. The result
was an antiserum apparently specific for mouse IgM. T-cell receptors passed the immunosorbens prepared with this pure anti-IgM without loss of activity. To ensure that there was indeed no retention the columns were washed twice with Hanks BSS and then eluted with glycine buffer, pH 2.8. No
further activity could be recovered. In control experiments the capacity
of this column to retain mouse IgM was ascertained. It is, therefore, clear
that the T-cell receptor, despite an apparent similarity to IgM, particularly with respect to its molecular weight, differs antigenically from IgM.
It also seems to bear no similarity to other known Ig classes (anti-IgG
2A, anti-IgA, anti-IgM, anti-allotype b anti-a (kindly donated by Drs. H.
Jaquet, H. Wigzell and G. Mitchell), anti-IgG (anti-mouse Ig) being tested
so far) because treatments of recognizing (and also of recognized) cells
with these preparations either undiluted or diluted 1:2 failed to reveal
inhibition of alloantigen recognition in PAR assays. On the other hand,
similarity of T- and B-cell receptors of the antigen-combining region indicates a similar nature of the variable parts of the two receptor molecule classes. Since the B-cell receptor has been shown conclusively to be
a 7S IgM molecule (48), the decisive difference of the two receptor mole-

cule classes rests with the constant region, which, as pointed out, for T-cell receptors does not seem to conform with classical structures.

A different approach to analyze the nature of the T-cell receptor was taken by Binz and Wigzell (20,50). They found that normal serum and urine contain receptor-like material as evidenced by inhibition of the binding of anti-T cell receptor serum to T cells. Furthermore, early antisera to normal rat serum prepared xenogeneically (rabbit, chicken) proved to fix on T lymphocytes and these sera, when used with complement, blocked MLC. Upon analysis both for anti-RS antibody inhibition and for alloantigen binding activity on Sephadex G-200, receptor molecules present in normal serum appeared in two distinct groups of which that with the larger molecules appears to be slightly heavier than 7S IgG, while that with the smaller molecules is material with a molecular weight around 35,000. Normal rat serum contains both activities; urine only 35,000 molecular weight material. Further studies (50) have revealed that the 7S IgG material is of B-cell origin, whereas the low molecular weight molecules are T-cell products. There are obviously inconsistencies to several of the findings from this laboratory. The first concerns the molecular weight of the T-cell receptor. It is clear, as Binz and Wigzell (20) point out, that the material found in normal serum or in urine does not reflect the actual molecule as found on the T-cell membrane. Furthermore, its presence in urine indicates that it is almost certainly a degradation product of the native T-cell receptor. Thus, whereas Binz and Wigzell's results are likely to be underestimates, those obtained with spontaneously shed T-cell receptors are likely to be overestimates with respect to molecular weight. The second inconsistency concerns the immunogenicity of normal serum. As pointed out, allogeneic immunization with normal serum in this laboratory failed to result in formation of anti-T-cell receptor antiserum, as did also B-cell induced alloantiserum (in contrast to T-cell dependent alloantiserum), whereas the Uppsala group reported formation of anti-receptor antibodies. Normal serum, however, has been shown to be immunogenic in xenogeneic immunizations only, with the option of being immunogenic in syngeneic immunizations if polymerized (50). In this laboratory this has not yet been tested, but it is clear that just because of the similarity of T- and B-cell receptors this important problem is not easily resolved. Obviously, depending on the

physical state of the receptor molecule, on its treatment prior to injection, on the immunization procedure and on the test system employed, both
immunogenic and non-immunogenic B- and T-cell receptors might be detected.
Closer analyses of these conditions could perhaps reveal a unifying concept
on the nature of T- and B-cell receptors.

Conclusions

A survey has been attempted on the formation, inhibition and possible nature of T-cell receptors for alloantigen. Like all attempts, this brief review suffers from incompleteness. Furthermore a sin has been committed that
should not be excused nor belittled: much unpublished material has been
summarized here in a rather casual fashion. What then has been crystallized
to some extent? First, it has become clear what PAR is. For years the interest in its nature was almost absent, because a substitute was thought
to be around the corner. This was a mistake, for all attempts to find an
equally sensitive test failed. PAR, then, is nothing else but a granulo-
tactically active complex of antibody and antigen. In the allogeneic system, as used in this laboratory, it is formed whenever matching structures
meet. Second, the genetic facts which have been applied to alloantigenic
recognition studies and to induction of antisera to T-cell receptors for
alloantigen still hold. For the in vitro demonstrations, some more data
have been added, but most important, there has been confirmation in vivo.
Third, T-cell receptors, although notoriously difficult to define biochemically, have allowed one glimpse into their mystery: the antigen-binding region of the receptor appears to be determined by the same V genes as
the corresponding region of B cell-receptors. Fourth, thanks to the efforts
of Binz and Wigzell, we now also know the frequency of T and B cells bearing receptors for alloantigens. With about 6% T and 1% B cells this is a
high figure, but for T-cell receptors it is confirmed with independent
assays. Fifth, the nature of the T-cell receptor as it is present on or in
the cell membrane is still enigmatic. That certainly will change, but all
of those who work on it will, hopefully, realize the complexity introduced
by a given technique and the one-sided view that might result from it.

References

1) Vitetta, E.S. and Uhr, J.W., Science 189:964, 1975.

2) Elkins, W.L., Progr. Allergy 15:78, 1971.

3) Vitetta, E.S., Baur, S. and Uhr, J.W., J. Exp. Med. 134:242, 1971.

4) Marchalonis, J.J., Cone, R.E. and Atwell, J.L., J. Exp. Med. 135:956, 1972.

5) Rubin, B., in Regulation of Growth and Differentiated Function in Eukaryote Cells, edited by G.P. Talwar, p. 249, Raven Press, New York, 1975.

6) Ramseier, H., J. Exp. Med. 140:603, 1974.

7) Boyden, S.V., J. Exp. Med. 115:453, 1962.

8) Binz, H., Bächi, T., Wigzell, H., Ramseier, H. and Lindenmann, J., Proc. Nat. Acad. Sci. USA 72:3210, 1975.

9) Ramseier, H. and Lindenmann, J., Path. Microbiol. 34:379, 1969.

10) Ramseier, H. and Lindenmann, J., Transpl. Rev. 10:57, 1972.

11) Ramseier, H. and Lindenmann, J., Immunogenetics 2:551, 1975.

12) Ramseier, H. and Lindenmann, J., Immunogenetics 2:561, 1975.

13) Ramseier, H., Eur. J. Immunol. 5:23, 1975.

14) Ramseier, H. and Lindenmann, J., J. Exp. Med. 134:1083, 1971.

15) McKearn, T.J., Hamada, Y., Stuart, F.P. and Fitch, F.W., Nature (London) 251:648, 1974.

16) Binz, H. and Wigzell, H., J. Exp. Med. 142:197, 1975.

17) Binz, H., Ramseier, H. and Lindenmann, J., J. Immunol. 111:1108, 1973.

18) Binz, H., Lindenmann, J. and Wigzell, H., J. Exp. Med. 140:731, 1974.

19) Ramseier, H., Eur. J. Immunol. 5:589, 1975.

20) Binz, H., Kimura, A. and Wigzell, H., Scand. J. Immunol. 4:413, 1975.

21) Joller, P., Nature New Biol. 240:214, 1972.

22) Binz, H., Lindenmann, J. and Wigzell, H., Nature (London) 246:146, 1973.

23) McKearn, T.J., Science 183:94, 1974.

24) Binz, H. and Wigzell, H., J. Exp. Med. 142:1231, 1975.

25) Binz, H. and Askonas, B.A., Eur. J. Immunol. 5:618, 1975.

26) Binz, H., Scand. J. Immunol. 4:79, 1975.

27) Binz, H. and Wigzell, H., J. Exp. Med. 142:1218, 1975.

28) Wilson, D.B., J. Exp. Med. 122:143, 1965.

29) Wilson, D.B., Blyth, J.L. and Nowell, P.C., J. Exp. Med. 128:1157, 1968.

30) Simmons, M.J. and Fowler, R., Nature (London) 209:588, 1966.

31) Simonsen, M., Cold Spring Harbor Symp. Quant. Biol. 32:517, 1967.

32) Szenberg, A. and Warner, N.L., Brit. J. Exp. Pathol. 43:123, 1962.

33) Szenberg, A., Warner, N.L., Burnet, F.M. and Lind, P.E., Brit. J. Exp. Pathol. 43:129, 1962.

34) Bach, F.H., Graupner, K., Dan, E. and Klostermann, H., Proc. Nat. Acad. Sci. USA 62:374, 1969.

35) Ford, W.L. and Atkins, R.C., Adv. exp. Biol. Med. 29:255, 1973.

36) Wilson, D.B., Howard, J.C. and Nowell, P.C., Transpl. Rev. 12:3, 1972.

37) Howard, J.C. and Wilson, D.B., J. Exp. Med. 140:660, 1974.

38) Ford, W.L., Simmons, S.J. and Atkins, R.C., J. Exp. Med. 141:681, 1975.

39) Ramseier, H. and Lindenmann, J., Eur. J. Immunol. 2:109, 1972.

40) Klein, J., Livnat, S., Hauptfeld, V., Jerabek, L. and Weissmann, I., Eur. J. Immunol. 4:41, 1974.

41) Ramseier, H. and Lindenmann, J. in Immunopathology, edited by P.A. Miescher, p. 308, Schwabe, Basel, 1971.

42) Eichmann, K. and Rajewsky, K., Eur. J. Immunol. 5:661, 1975.

43) Unanue, E.R., Grey, H.M., Rabellino, E., Campbell, P. and Smiдtke, J., J. Exp. Med. 133:1188, 1971.

44) Lamelin, J.P., Lisowska-Bernstein, B., Matter, A., Ryser, J.E. and Vassalli, P., J. Exp. Med. 136:984, 1972.

45) Nossal, G.J.V., Warner, N.L., Lewis, H. and Sprent, J., J. Exp. Med. 135:405, 1972.

46) Cone, R.E., Sprent, J. and Marchalonis, J.J., Proc. Nat. Acad. Sci. USA 69:2556, 1972.

47) Lisowska-Bernstein, R., Rinuy, A. and Vassalli, P., Proc. Nat. Acad. Sci. USA 70:2879, 1973.

48) Vitetta, E.S. and Uhr, J.W., Transpl. Rev. 14:50, 1973.

49) Ramseier, H., Nature (London) 246:351, 1973.

50) Binz, H. and Wigzell, H., Scand. J. Immunol. 4:591, 1975.

51) Abbreviations:

GVH: graft versus host
HVG: host versus graft
MLC: mixed leucocyte culture
PAR: product of antigenic recognition
RBC: red blood cells
RS: recognition structure = receptor

Chapter 18

THE FUNCTION OF T CELLS CARRYING RECEPTORS

FOR COMPLEXES OF Ig AND ANTIGEN

F. Paraskevas, S.T. Lee and K.B. Orr
Departments of Medicine and Immunology
and the
Manitoba Institute of Cell Biology
University of Manitoba
Winnipeg, Manitoba, Canada

ABSTRACT

Thymocytes exposed briefly in vitro to a variety of particulate substances (such as mycobacteria, erythrocytes of allogeneic cells) or to substances known to act in vivo as adjuvants (LPS or poly A:U), generate supernates which are able to induce cytophilic Ig in normal mouse serum in the presence of a foreign protein (antigen). This cytophilic Ig is taken up by 20-25% of splenic T cells. Hydrocortisone resistant thymocytes show the same property, while bone marrow cells are inactive. This activity is similar to that reported previously as being present in the 4S fraction of mouse serum, collected 6 hours after injection of complete Freund's adjuvant. It is proposed that this factor is responsible for the formation of complexes of Ig and antigen which have been detected in the serum 6 hours after immunization. Thymocytes collected 6 hours after priming in vivo with SRBC (when a subpopulation among them carries easily demonstrable surface Ig) are able to amplify markedly the antibody response particularly the 7S. It is postulated that the factor by generating the cytophilic Ig (complexes ?) which is taken up by T cells, sets up a mechanism which markedly amplifies their helper cell function.

INTRODUCTION

A few years ago we reported that 6 hours after injection of soluble

protein antigen in Freund's complete adjuvant (FCA) or heterologous erythrocytes

without FCA, a subpopulation of T cells in the spleen acquires easily

demonstrable surface Ig (1,2). Similarly in a subpopulation of thymocytes

injected into irradiated hosts, surface Ig could easily be detected within 6

hours after immunogenic challenge (2). It is believed that the surface
Ig on T cells represents complexes of Ig and antigen since for soluble
protein antigens such complexes were detected in the serum of Balb/c mice
6 hours after injection of I^{125} labelled bovine serum albumin (BSA) in
FCA (3). The Ig which participated in the complex formation was shown to
be IgG2a. Complexes of BSA and IgG have also been demonstrated in the
serum of rabbits 5 hours after injection of radiolabelled BSA in FCA (4).

Since the cytophilic complexes induced by protein antigens were
detected in the serum only when FCA was used for immunization, its role
was investigated by examining the serum of mice collected 6 hours after
the injection of 0.2 ml of an emulsion of FCA. The 4S fraction of such
a serum was shown to contain a factor which when added to the 7S fraction
of normal mouse serum (NMS), generated cytophilic Ig for T cells (5). The
cytophilic Ig can be detected only when a heterologous protein antigen was
added to the mixture in vitro and the evidence presented was compatible
with the postulate that such a factor mediated the formation of Ig-antigen
complexes.

It is quite likely that the cytophilic complexes which are detected
in vivo 6 hours after immunization with an antigen given in FCA are formed
through the mediation of the 4S factor. In favor of such a mechanism for
complexing Ig and antigen is the fact that following injection of FCA alone,
the 4S factor appears in the serum at the same time as the cytophilic
complexes (6 hours) and furthermore this factor generates in vitro cytophilic
Ig (and most likely complexes) for T cells, only in the presence of a heterologous foreign protein. Such a factor was not induced by Freund's incomplete
adjuvant (IFA) (5). We would call the activity in the 4S fraction of serum
collected 6 hours after FCA injection as immunoglobulin-antigen complexing
factor (IACF).

The data reported from our laboratory indicated that only a subpopulation of about 25% of T cells in the spleen takes up the cytophilic complexes, whether such complexes are formed in vivo (serum 6 hours after immunization) (3) or in vitro through the mediation of the IACF(4S fraction of serum collected 6 hours after injection of FCA) (5). The same number of T cells in the spleen were shown to become Ig+ within 6 hours after immunization (1,2). Our data have also indicated that the Fc fragment is important for the uptake of the complexes formed in vitro with the IACF (5).

More recently Fc receptors have been reported on a subpopulation of normal non-activated T cells by a variety of different techniques (6-9). The Fc receptor positive T cells in the spleen of mice have been found to be 24% of all spleen T cells (8,9). This is identical to the number of T cells which we have found to take up the 6 hour cytophilic complexes (3,5). The data presented here extend our studies on the IACF and demonstrate that it is produced by thymocytes. Furthermore, the T cells which in vivo have been induced to acquire surface Ig (or the cytophilic complexes) at 6 hours after antigenic stimulation markedly amplify antibody production acting as highly efficient helper cells.

METHODS AND RESULTS

The in vitro induction of IACF

Our previous results indicated that IFA did not induce IACF in vivo. In order to study the cell which is responsible for the production of IACF we have devised a short term culture in vitro using thymocytes or bone marrow cells. Thymocytes ($1x10^7$ or $4x10^7$ /ml) or bone marrow cells ($3x10^7$) were suspended in Hanks' solution and were exposed for 30 minutes at 37°C to 100 µg of mycobacteria (BCG, Connaught Laboratories, Toronto). After washing (x3) they were cultured in fresh medium for 3 more hours and the supernates were collected by removing the cells with

centrifugation. As controls we have used supernates from cells treated in
a similar way without exposure to BCG.

For detection of IACF we have used the same assay as before (5). It
consists in the ability of IACF to induce cytophilic Ig when added to NMS
only in the presence of an antigen (such as BSA or human fibrinogen or
horse spleen ferritin or egg albumin) (5). Such cytophilic Ig is taken
up by T cells in the spleen when the mixture of IACF, NMS (or its 7S
fraction) and a foreign protein is incubated with normal spleen cells in
vitro. The uptake of Ig by T cells in vitro in such a system produces
an increase of the Ig+ cells which is detected by the hybrid antibody technique which was developed in our laboratory and is known as reverse immune
cytoadherence or RICA (10,11). In our experience RICA detects the same
number of Ig+ cells in the spleen of mice as other techniques. Furthermore,
it is highly reproducible and very sensitive which enables us to detect
consistently, small changes of Ig+ cell populations. The number of Ig+ cells
by RICA in the spleen is 300-320 rosette forming cells (RFC) per 1,000
spleen cells. We use the term RFC by RICA and Ig+ cell interchangeably.
The results are shown in Table I.

The results of experiments 1 and 2 show that thymocytes exposed to BCG
for thirty minutes give a supernate which can induce the production of
cytophilic Ig in NMS only in the presence of a foreign protein. This
is reflected by the significant increase of RFC as shown in experiments
1 (b), (c), (d) on the last column of Table I. No such increase is
detected by supernates of thymocytes cultured without BCG (exp. 1(a)).

The thymocyte cultures were carried out at two different cell concentrations with identical results (exp. 1 and 2).

In another series of experiments thymocytes were collected from mice
treated with hydrocortisone acetate (2.5 μg) given 2 days earlier. The

TABLE I

THE INDUCTION OF IACF IN VITRO

Exp.	Supernate	Control[1]	Sup.[2]	Sup+Ag[3]	Sup+NMS	Sup+NMS+Ag
				RFC/1000 spleen cells ± S.D.		
1.	Thymocytes (1 x 10^7)					
(a)	No BCG	307±13(5)	302±10(3)	303±10(3)	303±14(3)	311±8(3)
(b)	100 µg BCG	–	304±11(5)	302±4 (4)	304±10(5)	357±12(5) (p < 0.001)
(c)	100 µg BCG	296±10(10)	–	–	–	389±16 (7) (p < 0.001)
(d)	100 µg BCG	275	–	–	–	396
2.	Thymocytes (4 x 10^7)					
(a)	No BCG	301± 1(3)	302± 1(3)	299± 3(3)	–	305±5(3)
(b)	100 µg BCG	–	302± 2(3)	307± 5(3)	300± 1(3)	367±21(3)
3.	Hydrocortisone Thymocytes (1 x 10^7)					
(a)	No BCG	296± 7(3)	307± 8(3)	289± 5(2)	304± 9(3)	300±11(3)
(b)	100 µg BCG	–	304± 4(3)	300±10(3)	291± 7(3)	353±11(3)
4.	Bone Marrow Cells (3 x 10^7)					
(a)	No BCG	301± 1(3)	299± 7(3)	304± 4 (3)	–	299±1(3)
(b)	100 µg BCG	–	302± 5(3)	299± 4(3)	–	301±5(3)
5.	Thymocytes (1 x 10^7)					
(a)	5x10^8 SRBC	305±24(4)	320	–	326	400±15(4)
(b)	6x10^6 C_3H/He spleen cells	–	–	–	–	365± 8(3)
(c)	250 µg LPS	–	–	–	–	360± 7(3)
(d)	300 µg PolyA-U	–	–	–	–	385± 8(3)
(e)	500 µg BSA	–	303± 5(3)	–	300± 2(3)	303± 5(3)
(f)	SRBC-lysed[4]	292	–	–	310	294

1. The number of RFC in normal spleen cells.
2. Culture supernate.
3. In all experiments 250 µg of BSA was used except for 1(c) where 250 µg of human fibrinogen was used and 1(d) where a sonicated solubilized material from SRBC was used.
4. SRBC were lysed with distilled water. The culture supernate was tested with NMS and human fibrinogen.

hydrocortisone resistant thymocytes gave an active supernate as shown in
experiment 3.

Bone marrow cells (exp. 4) failed under similar conditions of culture
to provide active supernates.

From these experiments we may conclude that in terms of the biological
assay used here (i.e. induction of cytophilic Ig in the presence of a
heterologous protein) these thymocyte supernates are similar to the 4S
factor detected in the sera of mice six hours after the injection of FCA
and furthermore they establish that the cell of origin of IACF activity
is the thymocyte.

More work is required to investigate further the nature of this activity
both in vivo and in vitro, as well as the mechanism of generation of cytophilic Ig. The ability of other substances to induce IACF activity in vitro
from thymocytes was investigated. Some of the results are shown in experiment
5. Other particulate substances such as sheep erythrocytes (SRBC) (exp. 5a)
or allogeneic cell (exp. 5b) generate active supernates like the BCG.
However, active supernates were also obtained by soluble substances which
are known to act in vivo as adjuvants such as lipopolysaccharide (LPS)
from E coli (026:B6, Difco Laboratories, Detroit, Mich.) (exp. 5c) or the
mixture of polyadenylic-polyuridylic acid (Poly A:U) (Miles Laboratories,
Kankakee, Ill.) (exp. 5d).

Other soluble substances such as BSA failed to induce IACF (exp. 5e).
SRBC which were lysed with distilled water lost their ability to induce an
active supernate when tested with NMS and human fibrinogen (exp. 5f). The
results obtained earlier (5) indicated that the cytophilic Ig induced by
the 4S factor (FCA serum) was taken up by T cells.

As shown in Table II, following anti-θ treatment, the number of Ig+
cells in the spleen increases proportionally because of the elimination of

TABLE II

UPTAKE OF CYTOPHILIC Ig DEPENDS ON THE PRESENCE OF T CELLS

Cell treatment	RFC/1000 spleen cells	
	Normal	anti-θ treated
None	310	414, 430
Thymocyte supernate[2] (BCG)	374	415, 412, 444

1. Spleen cells were treated with an AKR anti-θC3H serum and guinea pig complement. The number of Ig+ cells proportionally increases because of the elimination of the T cells which are Ig negative.

2. The BCG stimulated thymocyte culture supernate was added to NMS and human fibrinogen was used for the detection of cytophilic Ig.

the Ig- T cell population. However, the anti-θ treated spleen cells did not show any increase of Ig+ cells upon exposure to an active thymocyte supernate (in the presence of NMS and antigen). This experiment indicates that the increase of Ig+ is due to the uptake of the cytophilic Ig (generated by the thymocyte supernate) by an Ig- cell and it is T cell dependent.

The helper cell function of 6 hour T cells

As we have shown before, thymocytes injected into irradiated mice and challenged with an antigen acquire easily demonstrable surface Ig (2). It is believed that such Ig represents cytophilic complexes (3). The ability of such thymocytes collected only 6 hours after antigenic challenge, to function as helper cells for antibody formation was investigated.

Thymocytes (5×10^7) were injected IV into lethally irradiated Balb/c mice and challenged with 5×10^8 SRBC. Spleen cells were collected 6 hours after antigenic challenge. Such "educated" thymocytes carry easily demonstrable surface Ig and will be referred to as 6 hour T cells. One spleen

equivalent (approximately 6×10^6 cells) was combined with 1.5×10^7 bone
marrow cells and injected into sublethally irradiated mice which were then
challenged with 5×10^8 SRBC or horse erythrocytes (HRBC). The number of
direct and indirect PFC were determined 7 days later. The results are shown
in Table III.

TABLE III

THE HELPER CELL FUNCTION OF 6 HOUR T CELLS[1]

| | | | | PFC/spleen x 10^{-3} ± S.E. | |
Exp.	Priming[2]	Challenge[3]	PFC	19S	7S
1	Nothing	SRBC	SRBC	0.92±0.12 (13)[4]	0.62±0.11 (14)
2	SRBC	SRBC	SRBC	2.35±0.25 (13) ($p < 0.001$)	3.50±0.40 (14) ($p < 0.001$)
3	Nothing	HRBC	SRBC	0.11±0.07 (9)	0.07±0.04 (9)
4	SRBC	HRBC	SRBC	0.06±0.01 (9)	0.03±0.02 (9)
5	Nothing	HRBC	HRBC	3.25±0.62 (8)	2.67±0.78 (8)
6	SRBC	HRBC	HRBC	3.94±0.72 (9)	2.42±0.50 (9)
7	Nothing– α MIg[5]	SRBC	SRBC	0.90±0.13 (5)	0.62±0.15 (5)
8	SRBC– α MIg[5]	SRBC	SRBC	0.61±0.17 (5)	0.55±0.21 (5)

1. All mice received approximately $6-7 \times 10^6$ T cells (primed or unprimed) and
   1.5×10^7 bone marrow cells and were challenged with 5×10^8 SRBC or HRBC.
   PFC were determined on the seventh day.

2. The antigen used for "education" of thymocytes in the primary hosts.

3. The antigen used in the secondary hosts.

4. The number of animals used.

5. Unprimed (exp. 7) or SRBC primed (exp. 8) T cells treated with a rabbit
   anti-mouse Ig serum and guinea pig complement.

The results indicate that within 6 hours after priming the thymocytes act as much more efficient helper cells than unprimed cells. (exp. 1 and 2, $p < 0.001$). The enhancement of antibody response is particularly pronounced for the 7S antibody production which is known to be thymus dependent. This enhancement is highly specific since SRBC primed 6 hour T cells challenged with HRBC, induced no enhancement of the anti-SRBC PFC. (exp. 3 and 4). The SRBC 6 hour "educated" T cells can only be triggered for enhancement of antibody formation by the homologous antigen used for priming, since challenge with HRBC exerts no enhancing effect on the anti-HRBC responses (exp. 5 and 6).

The marked amplification of antibody synthesis was due to the 6 hour T cells since no response was obtained using spleen cells from the primary hosts without injection of thymocytes whether antigen was injected or not. Similarly spleen cells collected from primary hosts 6 hours after challenge with SRBC from mice which were injected with bone marrow cells could not provide helper cell activity.

In another series of experiments, spleen cells were collected from thymocyte reconstituted lethally irradiated mice without priming or 6 hours after priming with SRBC. The cells were treated with a rabbit anti-serum to mouse Ig and guinea pig complement. The antiserum had been previously absorbed with normal mouse thymocytes to remove any non-specific cytotoxicity.

As shown in Table III, the helper cell function of unprimed T cells is not affected by such treatment. This is in agreement with the view of the majority of the investigators that normal T cells do not carry significant amounts of surface Ig. However, treatment of 6 hour primed T cells with rabbit anti-mouse Ig (exp. 8) completely abolished the marked amplifying effect which was exerted by such cells (Compare experiments

2 and 8). It is also important to note that the number of PFC by anti-
mouse Ig treated 6 hour T cells is not completely abolished. These results
indicate that some of the 6 hour primed T cells carry surface Ig and
furthermore that these Ig+ antigen activated T cells in some way
exert a marked amplifying effect on antibody formation. Although such treatment abolished the enhancement of helper cell function of 6 hour primed
T cells, it did not completely eliminate the ability of the remaining cells
to function as helper cells at a level equal to unprimed T cells.

DISCUSSION

Our work previously has established that within 6 hours after
immunogenic stimulation the number of Ig+ cells in the spleen of Balb/c
mice increases significantly (by about 20-25% above the pre-immunization
level (1,2). This phenomenon was shown to be due to the appearance of
easily demonstrable surface Ig on a subpopulation of about 25% of T cells (2).
At the same time the serum of animals contained complexes of Ig and antigen
which were shown to be cytophilic for T cells (3). It is considered likely
that the surface Ig detected in vivo on the T cells at 6 hours may actually
represent such complexes although this was never conclusively demonstrated.
Subsequent work from our laboratory established that a factor which was
detected in the 4S fraction of sera from mice injected 6 hours earlier with
FCA, when added in vitro to 7S fraction of NMS generated cytophilic Ig for
T cells (5). Such cytophilic Ig could be detected only when a foreign
protein was added to the mixture and some of the data suggested that complexes
of Ig and antigen were formed. Thus it seems quite likely that the Ig-antigen
complexes detected in vivo at 6 hours were formed by the mediation of such
a factor. Complexes of Ig and antigen 5 hours after immunization have also
been detected in rabbits (4).

We would call the activity of the 4S fraction (5) immunoglobulin antigen complexing factor (IACF). The data presented in this paper show that a similar activity can be induced in vitro in thymocyte cultures. Such supernates obtained from thymocytes exposed to mycobacteria, when added to NMS generate cytophilic Ig in the presence of a soluble heterologous protein (Table I, exp. 1 b,c and d). No activity was detected in unstimulated thymocytes (exp. 1 a). The cytophilic Ig could be detected with a variety of soluble foreign proteins and even with a solubilized preparation from SRBC obtained after sonication (exp. 1 d). The cortisone resistant thymocytes are able to give active supernates (exp. 3). Whether the production of such a factor is the property of a subpopulation of thymocytes remains to be seen. Bone marrow cells do not give active supernates.

The results obtained from these experiments are similar to those using the 4S factor from mouse serum 6 hours after FCA injection, and it is possible that the IACF may be the same whether isolated from mouse sera or induced in vitro.

Not only mycobacteria but other particulate antigens such as SRBC and allogeneic cells are able to induce IACF activity (exp. 5a and b). The ability of SRBC to induce such activity is lost upon hemolysis (exp. 5f) and soluble BSA is also incapable of inducing IACF (exp. 5e). Although it may appear that only large particles are able to produce active thymocyte supernates, other soluble substances such as LPS or the mixture of polyadenylic-polyuridylic acids are also potent inducers of IACF (exp. 5c and d). The last two substances are known to be potent adjuvants and it is possible that the ability of a substance to induce IACF activity may reflect its adjuvanticity. SRBC may be considered to have what Dresser calls intrinsic adjuvanticity (12) while mycobacteria, LPS and poly A-U would fall into the group of extrinsic adjuvants.

The ability of LPS to induce strong IACF activity is intriguing in view of the fact that LPS is considered to affect only B cells. It has been demonstrated, however, that in the presence of antigen LPS significantly enhances the helper T cell function (13). The authors could not decide whether such enhancement of helper T cell function was mediated by an effect of LPS on T or B cells or macrophages. It seems attractive to speculate that LPS may enhance the helper cell function in the presence of antigen by the release of IACF. Since LPS induced IACF is produced by thymocytes the data suggest that LPS does not act exclusively on B cells.

IACF was also induced by allogeneic cells and the question arises about the relationship of this factor to the allogeneic effect factor (AEF) described by Armerding and Katz (14). The answer to this question would require further studies of the characterization and functional significance of IACF.

Some similarities may superficially exist between the FCA 4S factor induced in vivo (5) and the active thymocyte supernates described in this paper on one hand and an immunoglobulin binding factor (IBF) described by Fridman and Goldstein (15).

Again, more work is necessary in order to decide whether both activities are identical. One notes however, that IBF is not recovered from normal thymocytes but only cortisone resistant thymocytes while IACF activity is obtained from normal as well as hydrocortisone resistant thymocytes. This difference however, may be only quantitative. The fact that IACF was never detected in supernates from unstimulated cultures while IBF seems to be spontaneously released appears to be another difference. IBF is produced in larger quantities by activated thymocytes (14) and the evidence suggests that IBF may represent released Fc receptors (16).

After elimination of the T cells by anti-θ serum and guinea pig complement no increase of Ig+ cells is detected upon exposure to active supernate generated cytophilic Ig (Table II). This result is similar again

to that obtained with the 4S fraction and indicates that T cells are

involved in the uptake of the cytophilic Ig.

The ability of the IACF to generate cytophilic Ig is not dependent

on any particular antigen, thus no specificity is shown by the 4S fraction (5)

or by the in vitro induced active supernates. The role of the complexes

detected in vivo at 6 hours in mice (3) or at 5 hours in rabbits (4) is at

the present time speculative. It is possible that such complexes taken up

by 20-25% of T cells in mice may function as a mechanism for concentrating

antigen as originally suggested by Michison (17). Such a non-specific

concentration could likely be mediated by the Fc fragment of the Ig in the

complexes reacting with an Fc receptor on T cells. Since cytophilic Ig

could not be generated with the 5S fragment the above mechanism seems quite

plausible. It is interesting that the number of T cells carrying Fc receptors

as determined by different techniques (8,9) is identical to the number of

T cells becoming Ig+(1,2) or the number of normal splenic T cells which take

up the cytophilic complexes when they are exposed in vitro to 6 hour sera (3).

As an alternative, the acquisition of such Ig-antigen complexes may

trigger the activation of T cells. In this case one would expect to find

the helper T cells among the Fc receptor positive cells. The evidence so far

is controversial. While Stout and Herzenberg (8) and Rubin and Hertel-Wulff

(18) have not been able to detect helper cells among Fc receptor carrying T

cells, Parish (19) presented evidence that the helper T cells were distributed

between the Fc receptor positive and negative cells.

The data presented in Table III is an attempt to provide an answer

towards the question of the role of the T cells 6 hours after activation by

antigen when it is known that they demonstrate surface Ig on their surface

presumably by the uptake of the complexes (2). Such "educated" thymocytes

are different to the ones used by other investigators in the sense that they are collected only 6 hours after priming instead of 5 or 7 days. It is

shown that the 6 hour "educated" thymocytes act as much more efficient helper

cells in comparison with the unprimed thymocytes (Table III. exp. 1 and 2).

The amplification of the antibody response is particularly marked for the

7S class (over 6.5 fold increase) which is known to be thymus dependent.

This amplification is highly specific since the 6 hour T cells primed to

SRBC could not be triggered by another antigen such as HRBC (exp. 3 and 4).

The background level of PFC obtained in these experiments indicate further-

more that no significant amount of antigen is transferred with the 6 hour

T cells to the secondary hosts since the primed 6 hour T cells (exp. 4)

challenged with HRBC gave fewer anti-SRBC PFC than unprimed cells (exp. 3).

The primed T cells are still able to exert the same helper cell function

for another antigen as the unprimed cells (compare exp. 5 and 6).

Finally the results of experiments 7 and 8 show that the amplification

of the antibody formation was abolished following treatment with an anti-

serum to mouse Ig and complement. There are two important points which

need to be stressed here. First, the treatment of unprimed T cells with

the rabbit anti-mouse Ig serum exerts no detrimental effect on antibody

synthesis. This supports the view of the majority of the investigators

that normal T cells do not carry easily demonstrable surface Ig and is in

agreement with the data of Aden and collaborators that the T cells are not

the targets for the suppression of antibody synthesis by heterologous

antisera (20).

The second point is that the amplification of the immune response

shown by 6 hour activated T cells is abolished after treatment with anti-

Ig serum. This implicates an Ig+ cell as being important for such

amplification. Only thymocyte but not bone marrow reconstituted mice

(unpublished data) were able to provide such an amplifying helper cell

function after priming. This suggests that a thymus derived cell becomes

Ig+ within 6 hours after priming and is responsible for the amplification. These data bring more evidence in support of our earlier observations that within 6 hours after antigenic stimulation, a subpopulation of T cells becomes Ig+, most likely by the acquisition of cytophilic complexes (1 - 3). Preliminary evidence in our laboratory shows that the amplification of anti-body formation by SRBC primed 6 hour T cells could be abolished by treatment with a mouse antiserum to SRBC (unpublished data).

The fact that anti-mouse Ig treatment does not abolish completely the helper cell function of 6 hour T cells suggests that some of the primed T cells are not sensitive to this treatment and are still able to function as helper cells for a level equal to that of unprimed cells. If the Ig+ primed T cells represent the subpopulation which carries Fc receptors while the Ig- subpopulation does not, the data are in agreement with those reported by Parish (19) that the helper cell function of T cells may be distributed between Fc receptor positive and negative subpopulations. Our earlier data (1-3) showed that only 25% of splenic T cells became Ig+ 6 hours after priming. Thus we may conclude that the Ig+ subpopulation,present among the T cells 6 hours after activation,is implicated in the enhancement of helper cell function of the 6 hour T cells.

It is proposed that the function of IACF is to generate cytophilic complexes very early in the immune response which by being taken up by T cells markedly amplify their helper cell function.

This work was supported by grants from the National Cancer Institute of Canada and the Medical Research Council of Canada.

The expert technical assistance of Mrs. P. Sheppard is gratefully acknowledged.

REFERENCES

1. Paraskevas, F., Orr, K.B. and Lee, S.T.   J. Immunol. 109:1254, 1972.

2. Lee, S.T. and Paraskevas, F.   J. Immunol. 109:1262, 1972.

3. Orr, K.B. and Paraskevas, F.   J. Immunol. 110:456, 1973.

4. Yuan, L., Harvey, J.S. and Campbell, D.H.   Immunochemistry 7:601, 1970.

5. Orr, K.B. and Paraskevas, F.   Can. J. Microbiol. 20:535, 1974.

6. Soteriades-Vlachos, C., Gyöngyössy, M.I.C. and Playfair, J.H.L.   Clin. Exp. Immunol. 18:187, 1974.

7. Anderson, C.L. and Grey, H.M.   J. Exp. Med. 139:1175, 1974.

8. Stout, R.D. and Herzenberg, L.A.   J. Exp. Med. 142:611, 1975.

9. Basten, A., Miller, J.F.A.P., Warner, N.L., Abraham, R., Chia, E. and Gamble, J.   J. Immunol. 115:1159, 1975.

10. Paraskevas, F., Lee, S.T. and Israels, L.G.   J. Immunol. 106:160, 1971.

11. Paraskevas, F., Lee, S.T., Orr, K.B. and Israels, L.G.,   J. Immunol. Methods 1:1, 1971.

12. Dresser, D.W. Nature 191:1169, 1961.

13. Armerding, D. and Katz, D.H.   J. Exp. Med. 139:24, 1974.

14. Armerding, D. and Katz, D.H.   J. Exp. Med. 140:19, 1974.

15. Fridman, W.H. and Goldstein, P.   Cell Immunol. 11:442, 1974.

16. Neauport-Sautes, C., Dupuis, D. and Fridman, W.H.   Eur. J. Immunol. 5:849, 1975.

17. Mitchison, N.A., in "Immunological Tolerance" edited by M. Landy and W. Braun,   p.115, Academic Press, New York, 1969.

18. Rubin, B. and Hertel-Wulff, B.   Scand. J. Immunol. 4:451, 1975.

19. Parish, C.R.   Transpl. Reviews 25:98, 1975.

20. Aden, D.P., Manning, D.D. and Reed, N.D.   Cellular Immunol. 14:307, 1974.

Chapter 19

EVOLUTIONARY AND DEVELOPMENTAL ASPECTS
OF T-CELL RECOGNITION

Gregory W. Warr, Janet M. Decker and John J. Marchalonis
Laboratory of Molecular Immunology, The Walter and Eliza
Hall Institute of Medical Research, P.O., Royal Melbourne
Hospital, Victoria 3050, Australia

ABSTRACT

Studies relating to the nature of the antigen-specific T-cell receptor are reviewed in the light of present knowledge of phylogenetic and ontogenetic development. It is suggested that this evidence supports the concept that immunoglobulin (Ig) is the T-cell receptor, and that the following conclusions may be tentatively drawn.

1) T cell membrane Ig differs in its physical and functional properties from that on the B cell membrane.

2) The divergence between the T and B cell surface Ig molecules was apparent at the time of emergence of the ancestors of modern fish.

3) Specific T cell recognition in the form of antigen binding appears early in ontogeny, and is blocked by antisera to Ig.

The nature of the antigen-specific surface receptor of T lymphocytes has generated much interest and debate. The obvious candidate for this receptor is immunoglobulin (Ig) which possesses variable region combining sites for antigen similar to those of serum Ig and B cell surface Igs (1, 2). Controversy has surrounded attempts to establish this hypothesis because, in a formal sense at least, positive attempts to demonstrate (1 - 5) or isolate (6 - 10) surface Ig-like molecules of T cells could be countered by negative reports of workers using "similar" (but not identical) experimental approaches (11 - 14). Studies

from a number of laboratories now document the conclusion that murine
T cells express a type of surface Ig termed IgM(T) or IgT, which is
distinct from the 7S IgM and IgD-like molecules isolated from the B
cell surface (2, 7 - 10, 15 - 20). Evidence showing that this T-cell
Ig is synthesized by T cells and not by contaminating B cells or plasma-
cytes was obtained using monoclonal T lymphoma cells (15, 21, 22)
characterized by strict T cell phenotypes. Moreover, it has now been
established in some cases that T cells and B cells possess surface receptors for antigen which bear the same variable (V) region idiotypes
as do circulating antibodies directed against the same antigen (23 - 25).
We believe that much of the difficulty reported in the isolation of T cell
Ig by some workers (11, 12) stems from the use of detergent extraction
procedures shown by other workers to be unsuitable for isolation of
T cell Ig (10, 26). This work has been reviewed elsewhere (2, 27, 28),
and will not be reiterated here except insomuch as it impinges upon the
phylogenetic and developmental results under consideration.

The criteria which are classically considered necessary for
the demonstration of a true immune phenomenon are specificity of the
evoked response, and an altered level of reactivity to subsequent antigenic exposure. Although apparently all vertebrate classes show
immune responsiveness of this type (29 - 31), the bulk of our knowledge
of the immune system comes from intensive investigations of a few
species, namely, man, the laboratory rodents, and to a lesser extent
chickens, and the comparative aspects of the generation and expression
of immune competence have been unduly neglected. This is, in our
opinion, unfortunate, since it is through an investigation of the proto-
typical and universal features of immunity that we may eventually hope
to achieve a comprehensive understanding of the subject, and be able to
place in perspective the overwhelming amount of often contradictory data
which abounds in current cellular immunology.

The present article represents an attempt in this spirit to review the current state of our knowledge regarding the nature, evolution and ontogenetic emergence of receptors on antigen-specific T-cells.

The salient conclusions arising from phylogenetic considerations are
that (a) all vertebrates possess cells which exhibit specific T cell functions, (b) antigen-specific T cells bearing Ig receptors appear early in
ontogeny of mammalian species, and (c) virtually all lymphocytes of
lower species, including thymus lymphocytes, express and synthesize
readily detectable surface Ig.   Furthermore, recent evidence supports
that hypothesis that surface Igs of T and B cells represent distinct
molecules which diverged from a common precursor early in vertebrate
evolution.

Phylogenetic Origins of T Cell Functions

Lymphocytes in certain birds, mice and other mammals have
been subdivided, on the basis of origin, function and surface markers,
into thymus-derived T cells, and bursa (or bursal equivalent) derived
B cells (32, 33).   The B cells are the precursors of antibody-secreting
cells and bear large amounts of readily detectable membrane immunoglobulin.   The T cells appear to be more heterogeneous in function than
B cells, being involved in the phenomena of cell-mediated immunity
(CMI) such as graft rejection, delayed-type hypersensitivity, graft-
versus-host reaction (GVHR), and the in vitro mixed lymphocyte reaction (MLC), and also being implicated in collaborating with B cells to
bring about efficient antibody production by the latter.   In considering
the nature of the T cell receptor, it is as well to bear in mind that the
heterogeneity of classes of T cells subserving various functions may be
reflected in a heterogeneity of receptors.   Since the true immunological
nature of GVHR and MLR phenomena has been questioned (31, 34), we
will confine our attention to the recognition of classical foreign antigens
and strong histocompatibility antigens.

It is clear (Table 1) that the thymus plays a central role in the
generation of immune competence in teleost fish (35), amphibians (36, 37),
birds and mammals (32, 33), and that lymphocytes of lower vertebrates
show responses typical of T cells, e.g. graft rejection (38), MIF pro-

TABLE I

STATUS OF IMMUNE RESPONSIVENESS IN VERTEBRATE CLASSES

Class	Graft Rejection	Presence of strong H-Antigens	Carrier effect in AB production	Depression of Immunity by Thymectomy*	Readily detectable Ig on Thymocytes	MLR or GVH Reactivity
AGNATHA	+ ve	- ve	?	?	?	+ ve ?
CHONDRICHTHYES	+ ve	- ve	?	?	+ ve	- ve
OSTEICHTHYES (super order Teleostii)	+ ve	+ ve	+ ve	+ ve	+ ve	- ve ?
AMPHIBIA (order Urodela)	+ ve	- ve	+ ve	+ ve	+ ve	+ ve
(order Anura)	+ ve	+ ve	+ ve ?	+ ve	+ ve(larval)	+ ve
REPTILIA	+ ve	- ve	?	?	?	+ ve
AVES	+ ve	+ ve	+ ve	+ ve	- ve	+ ve
MAMMALIA (order Rodentia)	+ ve	+ ve	+ ve	+ ve	- ve	+ ve
(order Artiodactyla)	+ ve	+ ve	?	- ve	- ve	+ ve

All extant vertebrates show classical antibody production. This table was compiled from data in refs. 32, 35, 36, 38, 40 - 48, 53, 54, 88 - 95. * Thymectomy at an early developmental stage.

duction, MLR (in anuran amphibians, 35), responsiveness to certain
mitogens (39), GVH reactions (40), and a carrier effect on secondary
challenge (41 - 44).   However, the inference that T and B cells constitute distinct populations of lymphocytes in all vertebrate species can
only be drawn indirectly from this evidence, as appropriate cell markers
and cell separation techniques are not available.   However, there
emerges a consistent phylogenetic picture suggesting that all vertebrates show responses typical of those mediated by T and B cells in
the higher vertebrates (Table 1) even in those animals such as the
agnathan hagfish which do not possess organized lymphoid tissue
(35, 38).   The possibility that T and B characteristics are shared by
a single class of lymphocytes is intriguing, but a number of observations suggest that there is a distinct and apparently universal division
of function.   The possibility that T and B cell functions are expressed
by one cell type in lower species might be supported by observations
which demonstrate that thymocytes in fish (45- 47) and larval amphibians (48) bear readily detectable, endogenously synthesised membrane
Ig.   This property is generally considered a B cell characteristic in
mammals and birds; however, demonstration of binding of anti-Ig to
thymus (4, 18, 49) and T lymphoma cells (22, 50) raises questions regarding the stringency of this criterion.   Antibody production can occur
in the thymus of certain fish (51), although all this may suggest is that
a small number of B cells occur in fish thymus, as they do in mouse
thymus.   The following evidence can be marshalled in support of the
idea that T and B cells are functionally distinct populations throughout
vertebrate phylogeny: (a) In a larval amphibian (37), thymus, but not
spleen, is essential to the reconstitution of graft rejection. (b)Lympho-
cyte differentiation into functional subgroups has occurred in fish, because subpopulations of lymphocytes reactive to certain mitogens are
demonstrable (39).   (c) In the goldfish, the membrane of thymocytes
differs in several physical respects from that of splenocytes (47).   In
particular, goldfish thymus lymphocytes resemble those of mice in that
use of the nonionic detergent Nonidet P-40 under conditions routinely

used for mouse B cells (11, 26), does not allow fully efficient extraction
of Ig. By contrast, "solubilization" of goldfish spleen lymphocytes
under conditions used for mouse B cells allows surface Ig to be precipitated in good yield. Since surface iodination and various solubilization
conditions will be used increasingly in attempts to characterize membrane receptors and markers of lymphocytes of lower species, we would
caution the reader that techniques which work satisfactorily for a subpopulation of mammalian lymphocytes cannot be expected to be directly
applicable to cells of other subpopulations within a species or to cells
of other species. As an illustration, the Nonidet P-40 approach was
not suitable for the isolation of surface Ig of chicken B cells (52).

Surface Immunoglobulins of Lymphocytes of Lower Species

Data collated in Table 1 show that surface Ig is demonstrable
by immunofluorescence on thymus lymphocytes of adult elasmobranchs
(45), teleosts (46, 47), urodele amphibians (53) and larval anuran amphibians (48). Moreover, studies involving shedding and reappearance
provided evidence that the thymus lymphocyte populations were capable
of synthesizing this surface receptor. This point might prove of
crucial importance because species such as rays and urodeles are thought
to lack strong histocompatibility antigens (38), but, nevertheless, carry
out antigen-specific T cell functions and bear surface Ig.

The presence of readily detectable surface Ig on thymus lymphocytes of lower species provides a sharp contrast with the result
usually obtained in studies of adult mice (1, 2, 54). It is worthwhile
considering possible explanations for this discrepancy. One obvious
possibility is that thymus lymphocytes of lower vertebrates and mammals
differ in the number of Ig molecules expressed on the membrane.
Another explanation might involve the use of antisera made in rabbits
or other mammals to investigate Igs of other mammals. Since
mammals evolved relatively recently and their Igs show extensive
crossreactions, it is possible that antibodies made in another vertebrate
class, e.g. chickens, might possess stronger reactivities for minor

determinants, particularly those of V-regions and the F_d fragment of
the heavy chain. Consequently, non-mammalian antibodies might
serve as better probes for the demonstration of cell surface Igs of
mammalian cells than do antibodies made in closely-related species.
Consistent with this notion are the observations that antisera made in
chickens to purified mouse or human Igs will bind to mouse thymus (18)
and human thymus lymphocytes (49), respectively. A third possible
factor is that surface Ig of mouse T cells might be "buried" with only
its V_H-V_L combining site exposed (55, 56), whereas most of the H-chain
constant region is exposed on B cells (57). These explanations are not
mutually exclusive and data to be considered below suggest that surface
Igs of T and B cells possess different H-chain constant regions (15 - 17,
47) and consequently manifest a number of functional (58 - 60) and
physical differences (16, 47, 61).

A meaningful role for surface Ig of lymphocytes of lower
species is indicated by studies which show that all specific antigen-
binding lymphocytes of anuran amphibians (62) including putative
"carrier-specific" helper cells (62) are inhibited by antisera directed
against light chains and μ chains. Antigen specific functions of T cells
of chickens (63) and mice (64, 65) have also been inhibited by antisera
to Igs, particularly those possessing antibodies to light chains.

At this time, surface Ig has been isolated from lymphocytes of
only one lower species, viz., the goldfish (Carassius auratus), a
teleost. This species, like other cyprinoids (66, 67), possesses only
one serum Ig, a tetrameric IgM molecule. Rabbit antibodies specific
for this serum Ig precipitated significant counts in ^{125}I-labelled surface
proteins from both goldfish thymus and spleen lymphocytes (Table 2).
Table 2 also illustrates that Ig of goldfish thymus lymphocytes, like
that of mouse T lymphoma cells and thymus lymphocytes, is not adequately extracted by use of Nonidet P-40 under mouse B cell conditions.

A distinction between surface Igs of goldfish thymus and spleen
lymphocytes is observed when polypeptide chains of these Igs are com-

TABLE II

EFFECT OF EXTRACTION CONDITIONS UPON ISOLATION
OF SURFACE IG

Lymphocyte Source	Extraction Procedure	% Macromolecular ^{125}I-Counts Precipitated		% Specifically Precipitated
		Anti-Ig PPT	Control PPT	
GOLDFISH				
Spleen	1% NP 40	5.3	2.2	3.1
Thymus	1% NP 40	0.8	0.3	0.5
Thymus	Metabolic release	4.5	1.5	3.0
MOUSE				
CBA Thymus	0.5% NP 40	0.3	0.2	0.1
nu/nu Spleen	0.5% NP 40	3.6	0.4	3.2
CBA Thymus	Acid Urea	1.7	0.3	1.4
nu/nu Spleen	Acid Urea	1.4	0.3	1.1
T LYMPHOMA				
WEHI 22	1% NP 40	0.1	0.1	0
WEHI 22	1% NP 40/ 6M Urea	1.2	0.1	1.1
WEHI 22	Acid Urea	6.1	1.7	4.4
WEHI 22	Metabolic release	2.9	0.8	2.1

In all experiments, lymphocytes were surface labelled with ^{125}I in a lactoperoxidase-catalysed reaction. Details of experimental techniques employed will be found in refs. 10, 26, 47, from which some of these data were taken.

pared by polyacrylamide gel electrophoresis in SDS-containing buffers under conditions which provide high resolution of proteins in range of apparent mass, 40,000 - 90,000 daltons. Most importantly, the Ig heavy chain shows a distinctly faster mobility in gel electrophoresis than that of splenocyte Ig heavy chains. This difference is illustrated

in Fig. 1, and it is significant that analogous differences in the mobility
of splenocyte and thymocyte Ig heavy chains have been described for the
mouse (15, 16).   Fig. 2 presents a composite diagram based upon
studies which have compared the mobilities on SDS-PAGE of surface μ
and δ -like chains of B cells with heavy chains of surface Igs of thymus
(16) and T lymphoma cells (17).   Surface Ig of in vitro grown murine
fetal thymus anlagen (68; Haustein, D. and Mandel, T.E. in preparation) also possesses a heavy chain which is slightly, but significantly,
faster than μ chain.

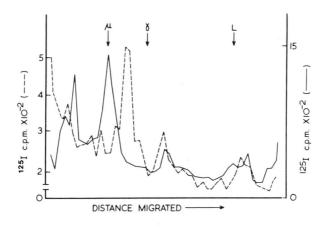

FIGURE 1.

A COMPARISON OF THE SURFACE IMMUNOGLOBULINS OF
SPLEEN AND THYMUS LYMPHOCYTES OF THE GOLDFISH

Polypeptide chains of Ig from ^{125}I-surface labelled spleen (——) and
thymus (-----) lymphocytes were analysed by SDS-polyacrylamide gel
electrophoresis after reduction with mercaptoethanol.   It can be seen
that the Ig from both cells dissociated into light (L) chains and heavy
chains.   Whereas the heavy chain of the splenocyte Ig migrated like a
true μ chain, the thymocyte Ig heavy chain showed a significantly faster
mobility.   Positions of standard murine and human μ, γ and L chain
markers are indicated by arrows (↓).   Both thymus and spleen lymphocytes show a polypeptide migrating faster than mammalian γ chain,
and which appears to be analogous to the molecule described in mammalian lymphocytes which has affinity for antigen-antibody complexes
(See also Fig. 2).

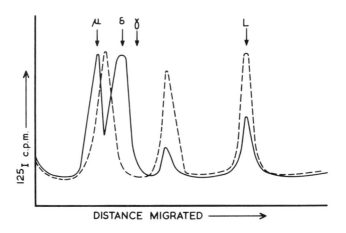

FIGURE 2.

A COMPARISON OF THE SURFACE IMMUNOGLOBULINS OF
MURINE T CELLS AND SPLENOCYTES

Polypeptide chains of Ig from ^{125}I-surface labelled spleen (———) and
T (-----) lymphocytes analysed after reduction with mercaptoethanol,
by SDS-polyacrylamide gel electrophoresis.   The positions of standard
murine and human μ, γ and light (L) chain markers are indicated by
arrows (↓).   Whereas the splenocytes show a light chain and two
heavy chains (one a true μ, the other designated ' δ '-like (96) and marked
(↓) as such in the figure, although it does not migrate like a human
δ chain, ref. 97), the T cell Ig resolves into a light (L) chain and a
heavy chain with a mobility distinctly faster than μ.   Both T and spleen
cell patterns show a polypeptide chain migrating in a position intermediate between γ and L chains.   This molecule has the property of
binding to antigen-antibody complexes.   The data shown in this figure
is a composite representation of that presented elsewhere (15, 16, 68).

Although mobility in SDS gels should not be taken as a sole

criterion for heavy chain homology, these results suggest that the

μ-like heavy chain of T cells is physically distinct from that of B cells

and that this distinction arose (unless we are dealing with convergent

evolution) in or prior to the divergence of the ancestors of modern

teleosts and mammals.   Quite clearly, further comparative studies

will be needed to confirm this hypothesis, and to determine whether or not the T cell Ig heavy chain differs from μ chain in some minor aspect (e. g. carbohydrate content) or whether the differences lie in the primary protein structure of the molecule.

Ontogeny of T Cell Recognition

The mammalian immune system is relatively immunoincompetent at birth. Stimulation with a foreign antigen will elicit at best only a weak antibody or cellular immune response, while the induction of tolerance in neonatal life is relatively easy (69 - 71). Nevertheless, the ability to recognize antigen, as measured by specific antigen binding to lymphocytes, appears rather early in the development of the lymphoid system during ontogeny. Antigen-binding lymphocytes have been detected in the thymus at 14 days gestation in the mouse and chicken, 20 days in the rat, and 12 weeks in the human (72 - 74). The frequency of binding cells in the thymus is high when first detected, often close to one percent of lymphocytes, and declines during gestation and childhood. This decrease in frequency might be explained by a change in the cell populations comprising the thymus (for example, a dilution of the binding population by a non-binding one), or by a decrease in the amount of receptor Ig per binding cell. In all cases antigen binding can be inhibited with antisera specific for immunoglobulin μ and kappa chains (72, 74, 75).

Although the newborn mammal is relatively immunoincompetent, there have been several reports of T cell function detectable at birth or even before. In the mouse, helper cells are present in the thymus at 48 hours after birth (76), and GVH reactivity has been measured in yolk sac cells at 9 days gestation (77). Silverstein has demonstrated helper activity in the fetal lamb, which is able to make some antibody responses in utero (78).

In general, conditions which demonstrate the presence of surface Ig on B cells by autoradiography will not detect Ig on T cells, although a high percentage of thymus cells from adult mice can be shown to have Ig when more sensitive conditions are used (3). It has been shown, however, that approximately 20% of thymocytes from young children do have surface Ig (49). The facts that only a subpopulation of cells show anti-Ig binding and that not all anti-Ig sera bind to thymocytes argue against passive acquisition of Ig by the cells. Moreover, Ig of the T cell type has been isolated from the surface of lymphocytes of in vitro grown murine fetal thymus anlagen (68; Haustein, D. and Mandel, T. E., in preparation).

Discussion

On the basis of the evidence available at present, we therefore incline to the view that lymphocytes exist, as low in the phylogenetic order as bony fish, which show distinct homologies, at both a functional and a molecular level, with the T cells of mammals. If this view is to be accepted, it implies that the thymocyte Ig, readily detectable by immunofluorescence in fish (45 - 47), has become, by the phylogenetic level of the birds and mammals, less readily detectable. The observations that larval amphibia have readily detectable Ig on their thymocytes, which becomes more difficult to detect during development (48), are readily compatible with this hypothesis. Possibly a parallel situation might obtain for fetal and neonatal mammalian thymus lymphocytes. If this interpretation of the comparative and developmental studies of thymocyte Ig is correct, several questions remain to be answered. Firstly, the teleological question: why should the display of T cell membrane Ig change? Secondly, do T cells possess receptors, other than Ig, that may, for example, be expressed in a reciprocal relationship?

In answer to the first question, the most attractive hypothesis is that the great bulk of membrane Ig on, for example, fish thymocytes and all B cells, is functionally redundant. Circulating human lymphocytes possess only about 2000 membrane bound receptors for insulin (79), so why should mammalian B cells require 80,000 or more membrane Ig receptor molecules (1, 80)? What distinguishes the Ig molecule from the insulin receptor is the capacity of the lymphocyte to syn-

thesize and secrete in response to stimulation large numbers of molecules closely related to its membrane Ig receptors.   This large quantity
of membrane Ig may merely represent a by-product of presecretory
activity of the protein synthetic apparatus of the lymphocyte, with very
few of the molecules being involved in the process of triggering.   As
far as we are aware, no one has yet investigated this important question.
It is clear that the antibody secretory apparatus becomes more complex
and diversified with movement up the phylogenetic scale, with appearance
of plasma cells (30), increasing complexity of lymphoid tissues (35) and
diversity of Ig classes (29, 30).   With this increasing complexity of the
B cell Ig secretory system, it is possible that T cells would not have
been subjected to selective pressure to retain superfluous membrane Ig
since the function of mammalian T cells is clearly not to secrete
large amounts of Ig.   The notion that some thymocytes are capable of
antibody secretion is however, supported by observations on the fish
thymus, and there appears to be an interesting intermediate stage in
the grass frog (62), where it has been reported that the thymus can
respond to immunization with an increasing number of antigen-binding
cells, but that these cells are non-secretors of antibody.

The dispute over the existence of Ig on mammalian T cells has
been argued for many years, but at the risk of boring the reader we
feel it is worth while summarizing what we consider to be the present
state of the field.   The evidence relating to the detection of T cell Ig
has recently been reviewed in detail (2), and although many investigators have reported positive results (4-10, 15 - 22), it is clear that some
workers have experienced difficulties.   In our opinion, these difficulties stem from the facts that (a) Ig in the membrane of T cells is obviously not readily accessible to anti-Ig reagents, (b) moreover, the
complication arises that readily detectable surface Ig of T cells is often
of passive nature, (c) Membrane Ig in T cells is not satisfactorily extracted by all methods used for B cells (10, 17, 26).   Some workers,
in addition, consider that Ig is present in the membrane of thymocytes
in relatively small amounts (1, 81), although no assumption-free methods

presently exist for the accurate quantitation of membrane molecules
which are neither completely accessible to binding reagents nor completely extractable under all "solubilization" conditions. The recent
phylogenetic evidence establishes that all lymphocytes of lower species
express surface Ig which is easily detected using antisera specific for
serum IgM.

The second question posed by the comparative studies, i.e.,
is it possible that phylogenetic development has seen the partial replacement of Ig by another receptor molecule? is worthy of serious consideration. It is clear that the plasma membrane of lymphocytes bears a
range of receptors (for hormones, antigens, Fc, heterologous erythrocytes, lectins), but it is difficult to see why the T cells should bear two
distinct types of antigen specific receptors both of which would bear the
V region of the Ig molecule (23 - 25). While a DNA translocation model
for the sharing of V regions by the various C regions, quite probably
linked to each other (82), has been proposed (82), it is less likely that
translocation between the unlinked V region genes and major histocompatibility complex (MHC) genes (82) is a feasible molecular mechanism.
The majority of published reports on the role of MHC specified cell
surface antigens suggest that they play a role in cell collaboration,
rather than mediating specific recognition (83 , but c.f. 84).

Many animals appear to have some powers of discriminating
self from non-self, a capacity which manifests itself in such phenomena
as incompatibility reactions in non-vertebrate groups like corals (85)
and tunicates (86). Such recognition does not appear, from our present
knowledge, to possess the characteristics of a true immune response,
and while such "quasi-immune" phenomena may have persisted in the
vertebrates, it is unclear to what extent they contribute to cell-mediated
immunity or are related to reactions against strong histocompatibility
antigens. Certainly all fish seem to lack MLR and elasmobranchs do
not show rapid first-set graft rejection as is also the case in urodele
amphibians (38). However, this apparent lack of major histocompatibility complex antigens does not prevent these animals from developing

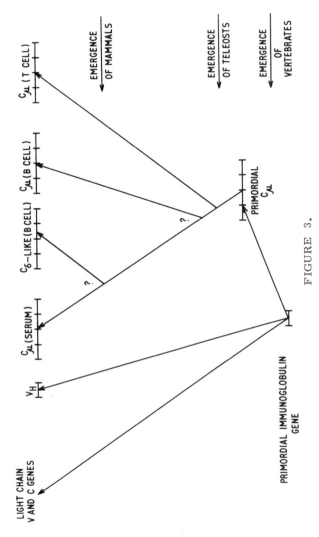

FIGURE 3.

HYPOTHETICAL MODEL FOR THE EVOLUTION OF
CISTRONS ENCODING SERUM IgM AND CELL-SURFACE
IMMUNOGLOBULINS

adequate immune responsiveness characteristic in other respects of T
and B cell function.   While it is unwise to draw conclusions from negative experiments, we do not feel that phylogenetic studies support the
view that antigen recognition by T cells is due in any great measure to
cell surface components specified by the major histocompatibility complex.

The recent phylogenetic and ontogenetic data reviewed here,
taken in conjunction with current investigations of the properties of surface Igs of murine T cells, indicate that (a) Ig of T cells is distinct from
the surface IgM and IgD-like molecules of B cells, and (b) the genes
specifying heavy chains of IgM(T) or IgT probably diverged early in vertebrate evolution from those encoding the μ and δ chains.   These concepts are depicted in Fig. 3 which gives a schematic representation of
the emergence of genes specifying μ chains and lymphocyte surface
H-chains.   This model proposes that the μ^T and μ^B genes diverged
prior to the divergence of the ancestors of mammals and teleosts,
whereas the divergence of μ^B and δ probably occurred much later
in evolutionary time.   The possibility is also included here that surface μ^B differs from serum μ in physical properties (87) and that this
might represent another gene duplication event.   This scheme must
be taken as tentative until sufficient primary sequence data accumulate
to allow a precise formulation.

ACKNOWLEDGMENTS

The authors' work was supported by grants from the
U.S. P. H. S. (AI-12565, AI-10886) and the American Heart Association
(AH 75-877).   JMD is a Postdoctoral Fellow of the National Multiple
Sclerosis Society, and GWW is in receipt of a Postdoctoral Fellowship
from the Science Research Council (U.K.).   We thank Sir F. M. Burnet
for helpful discussions.

This is publication number 2205 from The Walter and Eliza
Hall Institute.

REFERENCES

1. Warner, N. L., Adv. Immunol., 19 : 67, 1974.

2. Marchalonis, J. J., Science, 190 : 20, 1975.

3. Nossal, G. J. V., Warner, N. L., Lewis, H. and Sprent, J.,
    J. exp. Med., 135 : 405, 1972.

4. Santana, V., Wedderburn, N. and Turk, J. L., Immunology,
    27 : 65, 1974.

5. Roelants, G. E., Ryden, A., Hägg, L.-B., and Loor, F.,
    Nature (Lond.), 247 : 106, 1974.

6. Marchalonis, J. J., Cone, R. E. and Atwell, J. L., J. exp. Med.,
    135 : 956, 1972.

7. Moroz, C. and Hahn, J., Proc. Natl. Acad. Sci., U.S.A.,
    70 : 3716, 1973.

8. Rieber, E. P. and Riethmüller, G., Z. Immunitatsforsch.
    Exp. Ther., 147 : 262, 1974.

9. Boylston, A. W. and Mowbray, J. F., Immunology, 27 : 855, 1974.

10. Haustein, D., J. Immunol. Methods, 7 : 25, 1975.

11. Vitetta, E. S., Bianco, C., Nussenzweig, V. and Uhr, J. W.,
    J. exp. Med., 136 : 81, 1972.

12. Grey, H. M., Kubo, R. T. and Cerottini, J.-C., J. exp. Med.,
    136 : 1323, 1972.

13. Lisowska-Bernstein, B., Rinuy, A. and Vassalli, P., Proc. Natl.
    Acad. Sci., U.S.A., 70 : 2879, 1973.

14. Lamelin, J. P., Lisowska-Bernstein, B., Matter, A., Ryser, J. E.
    and Vassalli, P., J. exp. Med., 136 : 984, 1972.

15. Haustein, D., Marchalonis, J. J. and Harris, A. W., Biochemistry,
    14 : 1826, 1975.

16. Haustein, D. and Goding, J. W., Biochem. Biophys. Res. Commun.,
    65 : 483, 1975.

17. Cone, R. E., J. Immunol., 116 : 847, 1976.

18. Hämmerling, U., Mack, C. and Pickel, G., Immunochemistry,
    In press.

19. Hämmerling, U., Pickel, H.G., Mack, C. and Master, D.,
     Immunochemistry, In press.

20. Feldmann, M., Boylston, A. and Hogg, N.M., Eur. J.Immunol.,
     5 : 429, 1975.

21. Harris, A.W., Bankhurst, A.D., Mason, S. and Warner, N.L.,
     J. Immunol. 110 : 431, 1973.

22. Krammer, P.H., Citronbaum, R., Read, S.E., Forni, L. and
     Laing, R., Cell. Immunol. 21 : 97, 1976.

23. McKearn, T.J., Science, 183 : 94, 1974.

24. Black, S.J., Eichmann, K., Hämmerling, G.J. and Rajewsky,K.,
     in Membrane Receptors of Lymphocytes, edited by M.Seligmann,
     J.L. Preud'homme and F.M.Kourilsky, pp. 117-130,
     North-Holland, Amsterdam, 1975.

25. Binz, H., Bächi, T., Wigzell, H., Ramseier, H. and
     Lindenmann, J., Proc.Natl.Acad. Sci. U.S.A., 72 : 3210, 1975.

26. Cone, R.E. and Marchalonis, J.J., Biochem. J. 140 : 345, 1974.

27. Marchalonis, J.J., Contemp. Top. Mol. Immunol. In press.

28. Cone, R.E., Progr. Allergy. In press.

29. Grey, H.M., Adv. Immunol., 10 : 51, 1969.

30. Marchalonis, J.J., Immunity in Evolution, Harvard University
     Press, Cambridge, in press.

31. Hildemann, W.H., Nature (Lond.) 250 : 116, 1974.

32. Miller, J.F.A.P., Int. Rev. Cytol. 33 : 77, 1972.

33. Greaves, M.F., Owen, J.J.T. and Raff, M.C., T and B Lympho-
     cytes, Excerpta Medica, Amsterdam, 1973.

34. Dausset, J., LeBrun, A. and Sasportes, M., C.R.Hebd. Seances
     Acad. Sci. Ser. D. Nat. Sci., 275 : 2279, 1972.

35. Cohen, N., in The Lymphocyte : Structure and Function, edited
     by J.J.Marchalonis, Marcel Dekker, Inc., New York, in press.

36. Horton, J.D. and Manning, M.J., Transplantation, 14 : 141, 1972.

37. Brown, B.A. and Cooper, E.L., Immunology, 30 : 299, 1976.

38. Hildemann, W.H., in Transplantation Antigens, edited by B.D.
     Kahan and R.A.Reisfeld, pp. 3-73, Academic Press, New York,
     1972.

39   Lopez, D. M. , Sigel, M. M. and Lee, J. C. , Cell Immunol. , 10 :
        287, 1974.

40.   Borysenko, M. and Tulipan, P. , Transplantation, 16 : 496, 1973.

41.   Ruben, L. N. , van der Hoven, A. , and Dutton, R. W. , Cell Immunol. ,
        6 : 300, 1973.

42.   Yocum, D. , Cuchens, M. and Clem, L. W. , J. Immunol. , 114 :
        925, 1975.

43.   Ruben, L. N. , in Comparative Immunology, edited by J. J. Marchalonis,
        pp. 120-166, Blackwell, Oxford, 1976.

44.   Stolen, J. S. and Mäkela, O. , Nature (London.), 254 : 718, 1975.

45.   Ellis, A. E. and Parkhouse, R. M. E. , Eur. J. Immunol. , 5 : 726, 1975.

46.   Emmrich, F. , Richter, R. F. and Ambrosius, H. , Eur. J. Immunol. ,
        5 : 76, 1975.

47.   Warr, G. W. , DeLuca, D. and Marchalonis, J. J. , Proc. Nat. Acad.
        Sci. (U. S. A.) In press.

48.   Du Pasquier, L. , Weiss, N. and Loor, F. , Eur. J. Immunol. ,
        2 : 366, 1072.

49.   Jones, V. E. , Greaves, H. E. and Orlans, E. , Immunology, 30 :
        366, 1972.

50.   Boylston, A. W. , Immunology, 24 : 851, 1973.

51.   Ortiz-Muniz, G. and Sigel, M. M. , J. Reticuloendothel. Soc. ,
        9 : 42, 1971.

52.   Szenberg, A. , Cone, R. E. and Marchalonis, J. J. , Nature (London.)
        250 : 418, 1974.

53.   Charlemagne, J. and Tournefier, A. , in Immunologic Phylogeny,
        edited by W. H. Hildemann and A. A. Benedict, pp. 251-255,
        Plenum, New York, 1975.

54.   Raff, M. C. , Sternberg, M. and Taylor, R. B. , Nature (London.) 255 :
        553, 1972.

55.   Hogg, N. M. and Greaves, M. F. , Immunology, 22 : 967, 1972.

56.   Marchalonis, J. J. and Cone, R. E. , Transplant. Rev. , 14 : 3, 1973.

57.   Fu, S. M. and Kunkel, H. G. , J. exp. Med. , 140 : 895, 1974.

58.   Feldmann, M. , Cone, R. E. and Marchalonis, J. J. , Cell
        Immunol. 9 : 1, 1973.

59. Cone, R. E., Feldmann, M., Marchalonis, J. J. and Nossal,
     G. J. V., Immunology, 26 : 49, 1974.

60. Stocker, J. W., Marchalonis, J. J. and Harris, A. W., J. exp.
     Med., 139 : 785, 1974.

61. Cone, R. E., Immunochemistry, in Press.

62. Edwards, B. F., Ruben, L. N., Marchalonis, J. J. and Hylton,
     C., in Immunologic Phylogeny, edited by W. H. Hildemann and
     A. A. Benedict, pp. 397 - 407, Plenum, New York, 1975.

63. Theis, G. A. and Thorbecke, G. J., J. Immunol., 110 : 91, 1973.

64. Lesley, J. F., Kettman, J. R. and Dutton, R. W., J. exp. Med.,
     134 : 618, 1971.

65. Feldmann, M., J. exp. Med., 136 : 737, 1972.

66. Marchalonis, J. J., Immunology, 20 : 161, 1971.

67. Shelton, E. and Smith, N., J. Mol. Biol., 54 : 615, 1970.

68. Haustein, D., Marchalonis, J. J., Harris, A. W. and Mandel, T. E.,
     in Proceedings Tenth Leucocyte Culture Conference, in Press.

69. Dietrich, F. M. and Weigle, W. O., J. exp. Med., 117 : 621, 1963.

70. Dresser, D. W., Immunology, 4 : 13, 1961.

71. Sercarz, E. E. and Coons, A. H., J. Immunol., 90 : 478, 1963.

72. Dwyer, J. M., Warner, N. L. and Mackay, I. R., J. Immunol.,
     108 : 1439, 1972.

73. Dwyer, J. M. and Mackay, I. R., Lancet, 1 : 1199, 1970.

74. Hayward, A. R. and Soothill, J. F., in Ontogeny of Acquired
     Immunity, pp. 261 - 273, Associated Scientific Publishers,
     Amsterdam, 1972.

75. Dwyer, J. M. and Mackay, I. R., Immunology, 23 : 871, 1972.

76. Chiscon, M. O. and Golub, E. S., J. Immunol., 108 : 1379, 1972.

77. Hofman, F. and Globerson, A., Eur. J. Immunol., 3 : 179, 1973.

78. Silverstein, A. M. and Segal, S., J. exp. Med., 142 : 802, 1975.

79. Gavin, J. R., Gorden, P., Roth, J., Archer, J. A. and Buell,
     D. N., J. Biol. Chem., 248 : 2202, 1973.

80. Rabellino, E., Colon, S., Grey, H. M. and Unanue, E. R.,
     J. Exp. Med., 133 : 156

81. Grey, H. M., Colon, S., Campbell, P. and Rabellino, E.,
     J. Immunol., 109 : 776, 1972

82. Gally, J. A. and Edelman, G. M., Annu. Rev. Genet., 6 : 1, 1972

83. Katz, D. H. and Benacerraf, B., Transplant. Rev., 22 : 175, 1975

84. Munro, A. J., Taussig, M. J., Campbell, C., Williams, H. and
     Lawson, Y., J. Exp. Med., 140 : 1579, 1974

85. Hildemann, W. H., Linthicum, D. S. and Vann, D. C., in
     Immunologic Phylogeny, edited by W. H. Hildemann and A. A.
     Benedict, pp. 105-114, Plenum, New York, 1975.

86. Reddy, A. L., Bryan, B. and Hildemann, W. H., Immunogenetics,
     1 : 584, 1975

87. Melcher, U. and Uhr, J. W., J. Immunol., 116 : 409, 1976

88. Cooper, A. J., in Fourth Leucocyte Culture Conference,
     edited by O. R. McIntyre, pp. 137, Appleton-Century-Crofts,
     New York, 1971

89. Hildemann, W. H. and Thoenes, G. H., Transplantation, 7 : 506, 1969

90. Clark, J. C. and Newth, D. R., Experientia, 28 : 951, 1972

91. Warner, N. L. and Szenberg, A., Nature (London.), 196 : 784, 1962

92. Hudson, L. and Roitt, I. M., Eur. J. Immunol., 3 : 63, 1973

93. Weinbaum, F. I., Gilmour, D. G. and Thorbecke, G. J., J. Immunol.,
     110 : 1434, 1973

94. Morris, B., Contemp. Top. Immunobiol., 2 : 39, 1973

95. Silverstein, A. M. and Prendergast, R. A., Adv. Exp. Med. Biol.,
     29 : 383, 1973

96. Melcher, U., Vitetta, E. S., McWilliams, M., Lamm, M. E.,
     Phillips-Quagliata, J. M. and Uhr, J. W., J. Exp. Med., 140 :
     1427, 1974

97. Warr, G. W. and Marchalonis, J. J., J. Immunogenetics,
     in Press; Finkelman, F. D., van Boxel, J. A., Asofsky, R. and
     Paul, W. E. J. Immunol., 116 : 1173, 1976

Chapter 20

THE ROLE OF RECEPTORS FOR T CELL PRODUCTS IN ANTIBODY FORMATION

Tomio Tada, Masaru Taniguchi and Toshitada Takemori
Laboratories for Immunology, School of Medicine
Chiba University, Chiba, Japan

Abstract

Immunocompetent cell interactions are achieved via direct contact between functionally different cell types or via interactions between soluble
factors elaborated by regulatory T cells and specific receptors on responding cells for the T cell factors. In either case, there exist certain restrictions with respect to the effective interactions, which depend on the
state of differentiation and genetic background of the responding cell type.
Such restrictions are considered to be mainly determined by the development
and nature of the receptor site on responding cell types for different T cell
factors, which is now refered to the "acceptor" for the T cell factors. The
presence of such acceptor sites on different populations of both T and B cells
has been demonstrated in various experimental systems, and they are now considered to be the site by which responding cells receive appropriate signal
for destination of their further differentiation.

We have tried to review the nature and possible role of acceptor sites
on both B and T cells for different T cell factors with respect to the induction and regulation of immune responses. A special emphasis was put on
the genetic nature of the acceptor site. The observed genetic restrictions
in the acceptance of T cell factors by responding cells suggest that such
restrictions are needed for meaningful and unmistakable communications between funcionally different immunocompetent cells. Furthermore, the presence
or absence of acceptor sites for certain T cell factors is supposed to be a
very important factor for determination of the immune responsiveness of animals against certain antigens, and thus in some cases the Ir gene effect may
predominantly affect the expression of acceptor site. Possible implications
of acceptor site in the regulation of antibody response and in the network
of immunocompetent cell interactions are discussed.

In recent years the study of the "acceptor" has taken a remarkable start

according to discoveries of various T cell factors involved in interactions

of functionally distinct types of lymphoid cells. The term "acceptor" refers

to a receptor site on responding cells with which "factors" can specifically

interact. Thus, it is seemingly clear that the factor-acceptor model is one
of the central issues for understanding the molecular basis of various types
of cellular interactions.

The theoretical requirement for the presence of acceptor sites on B cells
for various T cell factors is based on observations that B cells only at certain maturation stages can accept the effects of enhancing and suppressive T
cell factors for their further differentiation into antibody-forming cells
(1-5). More definite evidence for the presence of the acceptor site on B
cells has been demonstrated by the fact that an efficient T-B cooperation
can take place only if these cells were derived from histocompatible donors
in the adoptive secondary antibody response (6-9).

This notion has been considerably strengthened by the recent identifi-
cations of some of the known T cell factors as the products of genes in the
I region of major histocompatibility complex (10-14). These include the
allogeneic effect factor (AEF), antigen-specific cooperative and suppressive
T cell factors, all of which show variable degrees of strain specificity in
their effectiveness. In some experimental models it has been shown that the
acceptor itself is, in fact, an I region gene product (15,16).

By generalizing above findings, it is tempting to assume that the factor-
acceptor link is an essential triggering event for the generation of immune
responses. If I region genes are rendered a role in coding for both the
factor and acceptor in a complementary fashion on the surface of functionally
different lymphoid cells, these molecules may in fact be useful devices by
which different cell types can make unmistakable and meaningful communications. Although little is known about the actual biochemical processes
following the factor-acceptor complementation, we would like to review some
properties of the acceptor site on B and T cells, and discuss possible biological significances of such molecules.

B Cell Acceptors for T Cell Factors

a) Acceptor site as cell interaction (CI) molecule.

An important contribution to the study of the nature of acceptor site
was made by Katz and his associates (7-9, 17) indicating that the gene products of major histocompatibility complex(MHC) control the interaction between T and B cells. This notion has been strengthened by the recent discoveries of certain T cell factors bearing determinants coded for by genes
mapped in the H-2 complex (10-14). It has been shown that under a physiologic
condition only H-2 histocompatible T and B cells can most effectively cooper-
ate to induce secondary antibody response (7-9). Their results led to a hypothesis that there are cell interaction (CI) genes located in MHC that code
for molecules responsible for the effective cell to cell interactions (reviewed in 17). Such CI molecules are indeed obvious candidates for the structural
entities possessed by both T cell factors and their acceptor sites.

By a sophisticated mixed cell transfer technique using primed T and B
cells sharing restricted subregions in the H-2 complex, Katz and Benacerraf
(18) showed that certain critical identities among genes in the MHC are
required for effective cell-cooperation. With respect to the responses
against hapten protein conjugates, they localized CI genes in I-A and I-B
subregions, where the majority of known immune response (Ir) genes have been
mapped. This raises an important question as to whether the apparently
distinct functions governed by Ir and CI genes reflect the activities of
identical genes or they are independently exerted by different products of
closely linked multiple genes. One interesting hypothesis presented by
Katz and Benacerraf (17) suggests that cell interactions are combined results
of Ir and CI gene effects, more precisely that Ir-CI bimolecular complex of
the T cell factor can interact by homology with comparable sites on other
lymphocytes. This predicts that Ir-CI molecular complex would be a focussing

device of antigen to other lymphoid cells by virtue of the complementation

of homologous CI molecules and antigen-specificity of Ir gene product.

However, the aforementioned close association of CI and Ir genes does

not necessarily imply that the acceptor site itself is an Ir gene product.

In fact, the problem is more complicated by the discovery that two comple-

mentary Ir genes control the response to a single defined antigen, e.g.,

GLØ. Katz et al.(19) obtained evidence that for effective cell interaction

between T and B cells the presence of both of the two complementary genes

in respective cell types is required. Since acceptor molecules are sup-

posed to be non-clonally expressed on B cells of syngeneic and semisyngeneic

haplotype strains, the failure of collaborative interaction between non-

responder B cells (lacking one of the complementary genes) and responder

T cells, and vice versa, suggests that the joint action of CI and Ir genes

is needed for the induction of T cell-dependent antibody response. No clear

conclusions can be drawn with respect to the identity or interrelationship

between CI and Ir gene products at the present moment.

b) B cell acceptor site as Ia molecule.

Another important discovery concerning the acceptor site on B cells has

emerged from the work of Taussig and Munro and their associates (11,12,15,20

- 22). They have demonstrated that an antigen-specific T cell factor can

be obtained in the culture supernatant of educated T cells, which can replace

the helper T cell function for the bone marrow B cell to induce in vivo primary

adoptive antibody response. The antigens they used are synthetic copolymers,

the responses to which are known to be under Ir-1 gene control. The molecular

characteristics of the T cell factor has been reviewed by Munro and Taussig

(15). Some important properties are 1) the factor has antigen-specificity

despite lacking any known Ig determinant, 2) the factor is an I-A subregion

gene product being adsorbable by anti-Ia antisera raised against the same

haplotype, and 3) some low responder strains can produce the factor.

An important observation concerning the acceptor site for this factor is that the acceptor site is blocked by incubation of B cells with anti-Ia antisera (15). Close analysis of the specificity of anti-Ia antisera indicated that alloantisera reactive with I-A subregion gene products, including cross-reactive anti-Ia, could effectively block the reactivity of B cells to the T cell factor. The results suggest that the acceptor site for the T cell factor is present on bone marrow B cells, and that this site is either identical to or closely related to Ia molecule on B cells.

This was further corroborated by the fact that the activity of the factor was absorbed by incubation with B cells of responder mice, thus the factor may directly interact with the acceptor of B cells (15). This interaction is not antigen-dependent, and normal bone marrow B cells can successfully remove the T cell factor. The results indicate that the acceptor sites are expressed on B cells non-clonally without strict antigen-specificity. On the other hand, there was found a heterogeneity in the acceptor sites for T cell factors of different specificities; for example, B cells of $H-2^k$ mice which are low responders to (T,G)-A--L cannot absorb the (T,G)-A--L factor while being able to absorb the (Phe,G)-A--L factor to which they are high responders. Nevertheless, both acceptor sites for (T,G)-A--L and (Phe,G)-A--L are blocked by the same anti-Ia antisera. From these results,they concluded that there exist separate "classes" of acceptor sites corresponding to the separate "classes" of T cell factors, having a very closely related structure to each other (15).

The evidence that the acceptor site is a product of different I region gene from that codes for the factor was demonstrated by the fact that certain low responder mice can produce the factor but their B cells cannot accept the factor (B cell low responder), and vice versa (T cell low responder). Thus, some low responders cannot produce antibody because they carry a defect in acceptor site on B cells. From these results they concluded that the factor

and acceptor are encoded by segregated genes both present in I-A subregion,
and that the specific Ir genes control the given antibody responses either
by governing the T cell responsiveness or B cell responsiveness via expression of the factor or acceptor. Thus, the results are most reasonably explained by assuming that two genes in I-A subregion code for distinct molecule on T and B cells, i.e., the factor and acceptor, and that the interaction
between these molecules would lead to the triggering of antigen-primed B cells.
Since the acceptor site is not clonally expressed on B cells, this triggering
mechanism may involve the Ig receptor antigen interaction as the first signal
and the acceptor-factor interactions as the second signal. The nature of
this second signal at the acceptor site is totally unknown. But preliminary
observations by Mozes (23) suggest that (T,G)-A--L primed spleen cells increase
intracytoplasmic cyclic AMP level upon exposure to (T,G)-A--L specific T cell
factor.

Although this two gene model is well entertained to explain the communication system between T and B cells through which signals are conveyed in
meaningful ways, there is a controversy in the above experimental system.
That is the fact that there is no strain specific restrictions for factor-
acceptor interactions with respect to allogeneic combinations, that have
often been observed in the cooperation of viable T and B cells. The factor
produced by one strain of mice is equally effective in activating B cells
of any other histoincompatible strains, provided the latter are genetically
capable of receiving T cells signal (B cell high responder). It was further
found that even a xenogenic combination, i.e., mouse T cell factor and human
peripheral B cells, can induce in vitro antibody response to sheep erythrocytes (24). It is hard to explain that B cells of histoincompatible mice as
well as those of humans do have a functionally similar acceptor site for the
T cell factors with I region product in nature, which probably have a considerable heterogeneity, inasmuch as the acceptor site itself is also an I region
product.

This apparent controversy is not solely explained. It is possible that
the primary IgM antibody response using unprimed B cells and T cell factor
requires much less strain and species specificity than usual adoptive secondary IgG antibody response. Although structural entity of the acceptor
site may be closely preserved throughout species, recent observations by
Taussig suggest that a preferential acceptance of known haplotype mouse T
cell factor by certain B cells with certain HLA phenotype (Taussig, personal
communication). Therefore, it is not yet completely excluded that there
exist certain genetic restrictions in factor-acceptor interactions in their
system. The latter observation would in turn make it possible to study the
Ir gene expression of human B cells. Alternatively, the acceptor site may
not be an Ir gene product, and the blocking of the site by anti-Ia is merely
due to steric hindrance or to inactivation of certain activity of B cells,
since even a cross-reactive anti-Ia can block the activity of the acceptor,
while the factor is removed only with anti-Ia raised against the same haplotype strain. In the (T,G)-pro--L system, such a possibility that the
acceptor is not linked to MHC was presented by Mozes (23), showing that the
(T,G)-pro--L specific factor is I region product, but the B cell responsiveness is not linked to H-2 complex. Therefore, the exact nature of the acceptor
site on B cells in general is still open to further investigation.

T Cell Acceptor Site for T Cell Factors

a) Acceptor site on T cells for suppressive T cell factors.

Many experimental models dealing with T cell factors are measuring the
net effect of T cell factors on the magnitude of antibody responses, which is
a consequence of B cell differentiation. However, in view of the complexity
of the immunoregulatory system, this does not imply that the target of T cell

factors is a B cell. Contrary, it has been shown that there are several
examples showing interactions between different subsets of T cells, which
lead to the enhancement and suppression of immune responses (25-28). A
parallel situation appears to exist in T cell-macrophage interactions.

Our previous studies (13,14) demonstrated that an antigen-specific T
cell factor can be extracted from disrupted thymocytes and splenic T cells
of mice that were immunized with a relatively high dose of protein antigen.
When the factor was injected intravenously into syngeneic but not allogeneic
animals, it suppressed primary antibody response against a hapten coupled
to the carrier protein by which the donor of the suppressive factor was
immunized. The suppressive activity was also demonstrable in an in vitro
secondary antibody response, to which the factor was added at the time of
the start of cultivation.

The properties of the suppressive T cell factor have been extensively
reviewed previously (29,30). The important point concerning the acceptor
site for this suppressive T cell factor is that the acceptor is expressed
on certain T cells but not on B cells. This was demonstrated by direct
absorption of the factor by thymic and splenic T cells but not by B cells,
bone marrow cells and macrophages (16). It was also shown that the factor
could not suppress the antibody response unless T cells having the same
specificity as that of the factor coexist (14). Therefore, the observed
suppression of antibody response is the consequence of the factor-T cell
interaction rather than of the direct action of the factor on B cells.

It was further found that the factor could not suppress the responses
of mice having different H-2 histocompatibility (13,16). In keeping with this
observation, it was demonstrated that the factor is in fact a product of MHC
gene which maps in a newly defined I subregion designated as I-J (31). I-J
subregion has recently shown to be intercalated between I-B and I-E, the locus
designated as Ia-4 (32).

The failure to suppress the antibody response across the major histocompatibility barrier suggested to us that the acceptor site for the suppressive T cell factor may also be a product of MHC gene. To test this possibility, we have obtained T cell factors from various haplotype strains and tested their effect on the responses of other strains having restricted identities in MHC subregions. A summary of the results is shown in Table I. In general, all syngeneic, semisyngeneic (parents and F_1) and H-2 histocompatible combinations were effective in inducing suppression, whereas no suppression was observed in any of the allogeneic combinations. Furthermore, the identities of genes in the left side half (K, I-A, I-B) of H-2 complex between the donor and recipient were found to be both necessary and sufficient for the induction of suppression. Thus, CBA factor could suppress the response of A/J mice which share the same K, I-A and I-B subregions as CBA. The results indicate that the acceptor site for the T cell factor is determined by gene(s) in the left side half of H-2 complex.

However, there were found two types of defects in the expression of the factor or acceptor. As shown in Table I, A/J mice could not produce the factor which can suppress the antibody responses of syngeneic and H-2 compatible strains. On the other hand, all the congenic strains with B10 background and D2GD mice could not accept the suppressive T cell factor produced by non-B10 histocompatible partners. Thus, B10.A could suppress the responses of A/J and C3H, but not those of B10.A and B10.BR. If A/J (non-producer) and B10.A (non-acceptor) were crossed to make F_1 hybrids, these F_1 could both produce and accept the F_1 factor indicating that expressions of both factor and acceptor genes are dominant traits. It is also shown from the data with (BALB/c x CBA)F_1, both H-2^d and H-2^k acceptor sites are codominantly expressed on F_1 target T cells.

The presence of acceptor sites only on T cells of the acceptor strain was further confirmed by absorption experiment using spleen cells of acceptor

TABLE I

Complementary interaction of the factor and
acceptor both determined by $\underline{I}$ region

Donor of T extract	Responding spleen cells	Identities of $\underline{H-2}$ subregion	Suppression[*]
CBA	CBA	K,I,S,D	93 %
	A/J	K,I	74
	BALB/c	none	0
	(BALB/c x CBA)F$_1$	K,I,S,D	97
(BALB/c x CBA)F$_1$	(BALB/c x CBA)F$_1$	K,I,S,D	91
	BALB/c	K,I,S,D	88
	CBA	K,I,S,D	73
	A/J	K,I	76
	SJL	none	0
A/J$^{\underline{a}}$	A/J$^{\underline{a}}$	K,I,S,D	0
	CBA	K,I	0
	BALB/c	S,D	0
B10.A$^{\underline{b}}$	B10.A$^{\underline{b}}$	K,I,S,D	0
	A/J	K,I,S,D	53
	B10.BR$^{\underline{b}}$	K,I	0
	C3H	K,I	69
	(A/J x B10.A)F$_1$	K,I,S,D	73
(A/J$^{\underline{a}}$ x B10.A$^{\underline{b}}$)F$_1$	(A/J$^{\underline{a}}$ x B10.A$^{\underline{b}}$)F$_1$	K,I,S,D	72
	A/J	K,I,S,D	75
	B10.A	K,I,S,D	0
SJL	SJL	K,I,S,D	63
	B10.S$^{\underline{b}}$	K,I,S,D	0
B10.S$^{\underline{b}}$	B10.S$^{\underline{b}}$	K,I,S,D	0$^{\underline{c}}$
	SJL	K,I,S,D	72

[*]Percent suppression = $\dfrac{\text{Control PFC - PFC with factor}}{\text{Control PFC}}$ x 100

$\underline{a}$ non-producer $\underline{b}$ non-acceptor $\underline{c}$ enhancement (more than two times increase)

and non-acceptor strains. The cells from B10 congenic lines and D2GD mice failed to remove the suppressive T cell factors from corresponding haplotype strains of non-B10 mice.

The cell type which expresses acceptor site for the T cell factor was found to be nylon adherent T cells (33): Only the nylon adherent T cells could remove the suppressive T cell factor, and this absorbing capacity of adherent spleen cells was completely abrogated by treatment of the cells with anti-θ and complement. Since under the condition used for separation of adherent T cells with nylon wool, most helper T cells were recovered in the fraction passed through the column, the results indicate that the cells expressing the acceptor site are not actual helper T cells themselves, but may be a third cell type which transmits the suppressor activity. This raises a new problem for considerations of the mechanism of suppressive cell inter-actions, which will be discussed later.

In order to learn what identity of genes in the I subregions is required for the effective suppression, we have obtained T cell factors from various recombinant mice, and tested in the responses of various haplotype strains which share restricted subregions with donors of the T cell factors. The results shown in Table II indicate that the identity of I-J subregion between the donors and recipients is a prerequisite for the induction of effective suppression. Although the results do not exactly mean that the acceptor site itself is an I-J subregion product, it is presumable that paired genes closely linked to each other and present in the same subregion (cluster of loci) selectively code for the factor and its acceptor in a complementary fashion on different subsets of T cells.

The evidence that different I region gene products are selectively express-ed on different subsets of T cells was recently presented by Okumura et al. (34). In chronic allotype suppression reported by Herzenberg et al. (25), their sup-pressor T cell inhibits the given allotype-bearing memory B cell via removing

TABLE II

Requirement for identity of I-J subregion between the donor
and acceptor of the factor for effective suppression

T extract (A,B,J,E,C)	Responding spleen cells	Identities of I subregion	Suppression[*]
B10 (b b b̲ b b)	C57BL/6J	A B J̲ E C	68 %
	A/J	none	0
B10.A (k k k̲ k d)	A/J	A B J E C	53
	C3H	A B J̲ E	69
	BALB/c	C	0
B10.A(4R) (k b b̲ b b)	CBA	A	0[**]
	C57BL/6J	B J̲ E C	46
B10.A(5R) (b b k̲ k d)	C57BL/6J	A B	0[**]
	CBA	J̲ E	82
	BALB/c	C	0
B10.A(3R) (b b b̲ k d)	C57BL/6J	A B J	71
	CBA	E	0
	BALB/c	C	0
B10.S(9R) (s s k̲ k d)	SJL	A B	0
	C3H	J̲ E	94
	BALB/c	C	0
B10.HTT (s s s̲ k k)	SJL	A B J̲	57
	C3H	E C	0
	BALB/c	none	0

[*] Percent suppression = $\dfrac{\text{Control PFC - PFC with factor}}{\text{Control PFC}} \times 100$

[**] Definite enhancement.

the helper activity required for this allotype production. Therefore, the observed allotype suppression involves interaction between suppressor and helper T cells both specialized in the regulation of production of the given allotype. This effect is also mediated by a soluble factor released from the suppressor T cell (Herzenberg, L.A., personal communication). The allotype suppressor T cell was recently found to carry I-J determinant (32), while the helper T cell expresses a different cell surface molecule which is also coded for by genes in the I region of H-2 complex (34). Although the actual site at which above suppressive interaction between suppressor and helper T cells takes place is still unknown, it is tempting to assume that unique I region determinants expressed on suppressor and helper T cells are the devices involved in such cell to cell communications. One can hypothesize that the I region determinants detected on helper T cells could represent the acceptor site for suppressive factors, which are also I region gene products, and that the signal given at the acceptor site on helper T cells leads to the inactivation of the helper function. Alternatively, the surface bound effector molecules on such acceptor site are used for interaction site with allotype-bearing B cells. These possibilities should be examined in the future in order to constitute a conceptual framework of interactions between different subsets of T cells, as well as helper T and B cells, carrying different I region determinants.

In contrast to the above observations on the antigen-specific and allotype-specific suppressive T cell factors in the antibody response, Rich and Rich (35) have demonstrated that in the suppression of mixed lymphocyte reaction (MLR) the factor elaborated from alloantigen-activated T cells can suppress the MLR reaction only of responder cells of those who share I-C, S and D regions of H-2 complex with the cell producing the suppressor factor. Since this suppression is not due to the cytotoxicity of the factor for the responding cells, it is presumable that the factor exerts the inhibitory effect on the

responding cells by the homology relationship in the right side half of H-2
complex. This suggests that in the cell-mediated immune response too, the
acceptance of the regulatory factor by responding T cells has certain genetic
restrictions. It is probable that an histocompatibility-linked acceptor site
for the factor is present on MLR responding T cells, and that such molecule
is coded for by genes in the right side half of H-2 complex.

b) T cell acceptor site for enhancing T cell factors.

As mentioned above, various T cell functions are regulated by other T
cells and their products. We have recently observed that the thymocyte and
spleen cell extracts of KLH-primed mice, which were used for studying the
antigen-specific suppression of the in vitro antibody response against DNP-
KLH, were capable of enhancing the same secondary IgG antibody response if
the extracts were added to the cultured primed spleen cells 2 to 3 days after
the start of cultivation (33). This was clearly demonstrated especially when
the responding cells were derived from B10 congenic lines, the non-acceptor
strains for the suppressive T cell factor. The enhancement was usually in
the range of two to three time over the control response.

The physicochemical and immunochemical properties of the enhancing
factor were comparable to those of suppressive T cell factor in that the
active molecule is not of immunoglobulin nature and molecular weight being
less than 50,000. Nevertheless, the enhancing factor is a clearly distinct
molecule from the suppressive factor, since the non-producer strain (A/J) for
the suppressive factor could produce the enhancing factor, and non-acceptor
strains (B10 congenic lines and D2GD) for the suppressive factor can accept
the enhancing factor of H-2 histocompatible strains. It was also found that
the cell types producing the enhancing and suppressive factors are different:
The former is present in the nylon column-passed fraction, while the latter
is adherent to the low-yield nylon wool column. Absorption studies using
various anti-Ia antisera indicated that the factor is an I-A subregion gene

product unlike the suppressive T cell factor which was shown to be I-J sub-region gene product (see above).

The most noteworthy difference in the properties of the enhancing T cell factor from those of Taussig and Munro factor is that it is not a T cell-replacing factor. This was shown by the fact that the factor cannot reconstitute the response of T cell-depleted spleen cells. The KLH-specific enhancing T cell factor could augment the anti-DNP antibody response only if the responding spleen cells were primed with DNP-KLH but not with DNP-heterologous carrier. Thus, it appears that the factor can enhance the antibody response only if helper T cells having the same specificity as that of the factor are already present in the cultured spleen cells.

The cell type which expresses the acceptor site for the enhancing T cell factor was tested by absorption with cells from different sources as well as with those of nylon-purified fractions. It was found that the cells capable of removing the enhancing activity from the T cell extract are present in the spleen and thymus but not in the bone marrow. They are anti-thy 1 sensitive but adherent to a low yield nylon wool column.

These results indicate that there are two distinct regulatory factors derived from antigen-primed T cells; one is inhibitory and coded for by genes in I-J subregion, and the other is enhancing and I-A subregion gene product. This dichotomy may be of importance for explaining the differences of the magnitude of antibody responses in different strains of mice to various complex antigens as heterologous erythrocytes and proteins, since certain strains do not produce or accept one of these two distinct factors in response to certain antigens. The target of both these two factors was found to be the nylon adherent T cell, which may have a key role for the transmission of both suppressive and enhancing effects to B cells. However, this does not necessarily mean that the same nylon adherent cells express acceptor site for both factors, since B10 congenic lines cannot accept the suppressive T cell factor but can well accept the enhancing T cell factor.

In order to determine the genetic requirement for the acceptor site to receive the enhancing T cell's signal, we have tested the effect of the factors obtained from various recombinant mice on in vitro antibody responses of various strains of mice sharing restricted subregions of H-2 complex with the donor strain. As shown in Table III, the identities of genes in I-A subregion between the donor and recipient of the enhancing T cell factor were both necessary and sufficient for the effective enhancement. This was clearly demonstrated by the fact that B10.A(4R) and D2GD factors could enhance only $H-2^k$ and $H-2^d$ responses, respectively, in which only K and I-A subregion is identical between the donor and responding cells. Since the factor itself is an I-A subregion gene product, it was suggested that I-A subregion genes code for both the factor and corresponding acceptor site.

This was further confirmed by the fact that B10.BR($H-2^k$) factor can enhance the responses of strains sharing the same $I-A^k$ subregion (e.g., B10.A and B10.A(4R)), clearly indicating that the acceptor site for the enhancing T cell factor is encoded by a gene in I-A subregion. In fact, the enhancing T cell factor of $H-2^k$ was successfully removed by absorption with spleen cells of B10.BR, B10.A and B10.A(4R) but not with those of B10, B10.A(5R) and B10.D2.

These results now indicate that the factor-acceptor-links in the suppression and enhancement of antibody response are achieved via complementary interaction between cell surface molecules on different subsets of T cells, both of which are coded for by genes closely located in the same loci. It is probable that each locus in different subregions possesses complementary paired genes, i.e., genes for the factor and its acceptor, which are expressed on functionally different subsets of T cells.

The expression of acceptor site on adherent T cells may be dependent on the time after antigenic stimulation, since the same T cell extracts con-

TABLE   III

Requirement of identity of I-A subregion for effective enhancement

T extract (A B J E C)	Responding spleen cells	Identities of I subregion	Enhancement[*]
B10.A	B10.A	A B J E C	136 %
(k k k k d)	B10.BR	A B J E.	135
	B10.D2	C	θ[**]
B10.A(5R)	B10	A B	144
(b b k k d)	B10.BR	J E	θ
	B10.D2	C	θ
B10.A(4R)	B10.BR	A	141
(k b b b b)	B10	B J E C	θ
D2GD	B10.D2	A	77
(d b b b b)	B10	B J E C	θ
B10.BR	B10.BR	A B J E C	136
(k k k k k)	B10.A	A B J E	141
	B10.A(4R)	A	166
	B10.A(5R)	J E	θ

$$\ast \text{ Percent enhancement} = \frac{\text{PFC with factor} - \text{Control PFC}}{\text{Control PFC}} \times 100$$

** Less than 15 % enhancement (not significant).

taining both suppressive and enhancing T cell factors exert differential
effects depending on the time when they were added to the antigen-stimulated
responding cells. The enhancing effect predominates over the suppressive
effect when the mixture of factors was given 48 to 72 hours after the start
of cultivation. This is in keeping with observations of others (36-42), who
showed that antigen-nonspecific factors are most effective in augmenting in

368

TADA, TANIGUCHI, AND TAKEMORI

vitro antibody responses when they were given to the cultured spleen cells
after certain period of incubation with antigen. Therefore, it may be a
general fact for both T and B cells that the acceptor sites for enhancing
T cell factors would develop after contact with antigen. However, the absorption of the T cell factor by normal spleen cells indicates that such acceptor
sites are expressed in non-clonal fashion.

Discussion

We have heretofore tried to review the known general properties of
acceptor sites on B and T cells for various T cell factors. It is apparent
that our knownledge of the acceptor site is still very limitted as compared
to that about factors. Especially, almost nothing has been known about
the biochemical events occurring at the acceptor site in response to the
delivery of T cell factors, that subsequently trigger the intracellular
processes for activation and differentiation of target cells. This is, in
fact, an important issue for understanding the initiation and regulation
of immune responses, to which only unreasonably little attention has been
focussed.

However, several important clues are in our hands for studying above
problems. The acceptor sites are selectively expressed on certain T and B
cells, though not clonally, at a critical time after antigenic stimulation.
The presence of I region determinants on a few known T cell factors and
corresponding acceptor sites provides a new framework for immunocompetent
cell interactions. The location of the loci coding for both the factor and
its acceptor in close proximity to Ir loci suggests that the regulation of
immune responses is achieved via factor-acceptor complementation which in
many cases reflects the effects of Ir genes. Furthermore, it has been shown
that T-B and T-T interactions are, in general, successfully and unmistakably

achieved through interactions between the factor and its acceptor, which suggest that the network of various subsets of immunocompetent cells is maintained by genetic restriction in factor-acceptor complementations. Although the evidence is too fragmentary to draw any conclusions, it is tempting to consider that factor-acceptor link is a common general language by which functionally different subsets of immunocompetent cells make meaningful and unmistakable communications.

It is interesting to note that in the antigen-specific suppression and enhancement of antibody response a special type of T cells, which is adherent to nylon wool, possesses the acceptor for both suppressive and enhancing T cell factors, without which both effects can be exerted on B cells. This suggests that the cells which express the acceptor sites plays a key role in the regulation of antibody responses. If this is the case, this cell type may well be one of the sites of expression of Ir genes, with which the responsiveness of animals is critically determined. The Ly phenotype of this cell type has not been defined by the present time.

References

1. Gorczynski, R.M., Miller, R.G. and Phillips, R.A., J. Immun., 111:900,1973.
2. Rubin, A.S., MacDonald, A.B. and Coons, A.H., J. Immun., 111:1314,1973.
3. Schimpl, A. and Wecker, E., J. Exp. Med., 137:547,1973.
4. Kishimoto, T. and Ishizaka, K., J. Immun., 112:1685,1974.
5. Dutton, R.W., Transplant. Rev., 23:66,1975.
6. Kindred, B. and Shreffler, D.C., J. Immun., 109:940,1972.
7. Katz, D.H., Hamaoka, T. and Benacerraf, B., J. Exp. Med., 137:1405,1973.
8. Katz, D.H., Hamaoka, T., Dorf, M.E. and Benacerraf, B., Proc. Nat. Acad. Sci. U.S.A., 70:2624,1973.
9. Katz, D.H., Hamaoka, T., Dorf, M.E., Mauer, P.H. and Benacerraf, B., J. Exp. Med., 138:734,1973.
10. Armerding, D., Sachs, D.H. and Katz, D.H., J. Exp. Med., 140:1717,1974.
11. Munro, A.J., Taussig, M.J., Campbell, R., Williams, H. and Lawson, Y., J. Exp. Med., 140:1579,1974.
12. Taussig, M.J., Munro, A.J., Campbell, R., David, C.S. and Staines, N.A., J. Exp. Med., 142:694,1975.

13. Takemori, T. and Tada, T., J. Exp. Med., 142:1241,1975.

14. Taniguchi, M., Hayakawa, K. and Tada, T., J. Immun., 116:542,1976.

15. Munro, A.J. and Taussig, M.J., Nature, 256:103,1975.

16. Taniguchi, M., Tada, T. and Tokuhisa, T., J. Exp. Med., 144:20,1976.

17. Katz, D.H. and Benacerraf, B., Transplant. Rev., 22:200,1975.

18. Katz, D.H. and Benacerraf, B., in The Role of Products of the Histocompatibility Gene Complex in Immune Responses, edited by D.H. Katz and B. Benacerraf. p.355, Academic Press, New York, 1976.

19. Katz, D.H., Dorf, M.E. and Benacerraf, B., J. Exp. Med., 143:906,1976.

20. Taussig, M.J., Nature, 248:234,1974.

21. Taussig, M.J. and Munro, A.J., Nature, 251:63,1974.

22. Mozes, E., Isac, R. and Taussig, M.J., J. Exp. Med., 141:703,1975.

23. Mozes, E., in The Role of Products of the Histocompatibility Gene Complex in Immune Responses, edited by D.H. Katz and B. Benacerraf, p.485, Academic Press, New York, 1976.

24. Taussig, M.J., Munro, A.J. and Luzzati, A.L., in The Role of Products of the Histocompatibility Gene Complex in Immune Responses, edited by D.H. Katz and B. Benacerraf, p.553, Academic Press, New York, 1976.

25. Herzenberg, L.A., Okumura, K. and Metzler, M., Transplant. Rev., 27:57,1975.

26. Eardley, D.D., Staskawicz, M.O. and Gershon, R.K., J. Exp. Med., 143:1211,1976.

27. Jandinski, J., Cantor, H., Tadakuma, T., Peavy, D.L. and Pierce, C.W., J. Exp. Med., 143:1382,1976.

28. Cantor, H., Shen, F.W. and Boyse, E.A., J. Exp. Med., 143:1391,1976.

29. Tada, T., Taniguchi, M. and Takemori, T., Transplant. Rev., 26:106,1975.

30. Tada, T. and Taniguchi, M., in The Role of Products of the Histocompatibility Gene Complex in Immune Responses, edited by D.H. Katz and B. Benacerraf, p.513, Academic Press, New York, 1976.

31. Tada, T., Taniguchi, M. and David, C.S., J. Exp. Med., in press, 1976.

32. Murphy, D.B., Herzenberg, L.A., Okumura, K., Herzenberg, L.A. and McDevitt, H.O., J. Exp. Med., in press, 1976.

33. Tada, T., Taniguchi, M and David, C.S., in Cold Spring Harbor Symposium on Quantitative Biology, volume XLI, in press, 1976.

34. Okumura, K., Herzenberg, L.A., Murphy, D.B., McDevitt, H.O. and Herzenberg, L.A., J. Exp. Med., in press, 1976.

35. Rich, S.S. and Rich, R.R., J. Exp. Med., 142:1391,1975.

36. Gorczyski, R.M., Miller, T.G. and Phillips, R.A., J. Immun., 108:547,1972.

37. Rubin, A.S. and Coons, A.H., J. Immun., 108:1597,1972.

38. Watson, J., J. Immun., 111:1301,1973.

39. Kishimoto, T. and Ishizaka, K., J. Immun., 111:1194,1973.

40. Armerding, D. and Katz, D.H., J. Exp. Med., 140:19,1974.

41. Kishimoto, T. and Ishizaka, K., J. Immun., 114:1177,1975.

42. Schimpl, A. and Wecker, E., Transplant. Rev., 23:176,1975.

Chapter 21

IgD RECEPTORS OF LYMPHOID CELLS

Benvenuto Pernis and Luciana Forni
Basel Institute for Immunology
Basel, Switzerland

Abstract

While present in normal human serum in very low amounts and un-detectable in sera of non-human primates as well as of mice, IgD is found on the surface of the majority of B lymphocytes in all the above mentioned species.

Lymphocytes which carry IgD on their membrane also have IgM. The two molecules are present in relative amounts that can be very different in different cells.

Both IgM and IgD of a single cell are the actual product of the cell itself. They have the same light chain and, more importantly, the same combining site and idiotype.

IgD/IgM bearing lymphocytes are the majority of all B lymphocytes in spleen, lymph nodes and Peyer's patches, whereas in bone marrow they account for half of the immunoglobulin positive cells. Although the percentage of double IgD/IgM cells is very similar in different tissues, the total amount of IgD, as well as the relative amounts of IgD and IgM as detected by biochemical methods varies. In fact, lymph nodes, and even more Peyer's patches are much richer than spleen in cells having levels of IgD higher than those of IgM; conversely, in the bone marrow, all the positive cells have very low levels of IgD.

In ontogeny, as in evolution, IgD appears after IgM: in human foetuses IgD bearing cells are not detectable before 13 weeks of gestation, and in the mouse they appear only after birth.

IgD receptors seem to disappear from B cells which undergo matur-ation to secretion, as indicated by the fact that only a proportion of IgM secreting plasma cells show membrane IgD.

IgD is never found on the membrane of IgG-containing cells, and also lymphocytes bearing simultaneously IgD and IgG are very rare, and it might well be that for these cells, the double expression for short periods of time does not actually correspond to simultaneous synthesis.

These cells could be elements that have undergone a switch from the
synthesis of IgD to that of IgG, with no changes in the synthesis of
IgM, that is from a IgM/IgD cell to a IgM/IgG cell.

A hypothesis is proposed concerning the relationships between the
different steps of immunoglobulin expression and different developmental
stages of B lymphocytes.

IgD was identified as a separate immunoglobulin class by Rowe and

Fahey (1) on the basis of an unusual human myeloma protein. This

immunoglobulin, which lacks various biological activities such as comple-

ment fixation and reactivity with skin, mast cells or neutrophils (2),

is present in the serum of the majority of normal individuals at concen-

trations of 20-50 µg/ml. Normal humans with high levels (up to 400

µg/ml) of serum IgD, as well as individuals whose serum IgD levels are

below detectability have also been reported (3). The number of IgD-

containing plasma cells in normal human tissues is compatible with the

low level of molecules in the serum (4).

The light chain distribution of IgD myeloma proteins is unusual in

that IgD myelomas with λ light chains are much more frequent than those

with κ light chains. This reflects the predominance of IgDL over IgDK

containing plasma cells in normal human spleen (5).

Searches for antibody activity in IgD have not given clear-cut

results. Due to the minute amounts of IgD in normal sera, no studies

have been performed on purified IgD preparations. The only data avail-

able concern autoantibodies, such as anti-nuclear antibodies in lupus

erythematosus and rheumatoid arthritis, anti-thyroid antibodies, and

anti-insulin antibodies (see review by Spiegelberg, 6), always detected

at low titers by using highly sensitive indirect methods. Spiegelberg

(6) also reports negative results in his attempts to induce an IgD

antibody response to keyhole limpet hemocyanin by active immunization in

rhesus monkeys.

Recent results indicate that man is exceptional as far as secreted

IgD is concerned. The serum levels of IgD in two species of non-human

primates (Macaca mulatta and Macaca speciosa) are consistently much

lower than those found in humans, being on the average of the order of

100 ng/ml (7). IgD containing plasma cells could not be found in spleen,

bone marrow and tonsils from these animals, and no labelled IgD could be

precipitated from the supernatant of monkey spleen cell cultures incub-

ated with radio-active amino acids (7). In the mouse, another laboratory

animal in which IgD has been detected and studied in the last few years,

serum IgD is below the level of detectability (8), and IgD plasma cells

are absent from lymphoid tissues (9).

It seems therefore that the ability to secrete IgD is a late event

in human evolution. This may be due to mutation in the region of the

genome where the genes controlling the constant part of immunoglobulin

heavy chains are located. The correlation of the level of serum IgD in

humans with the Gm genotype (10) suggests it. It is therefore conceiv-

able that IgD plasma cells and secreted IgD may not play an important

role in the immune system.

IgD as a lymphocyte receptor

Interest in IgD is presently focused on its possible role as a

lymphocyte receptor, since this class of immunoglobulin has been found

to be present on a high proportion of peripheral blood lymphocytes of

adult humans (11) and newborns (12). Immunoglobulin molecules reacting
with antisera made against human IgD, and considered to be homologous to
it, have been detected on the membrane of a proportion of lymphocytes in
different lymphoid tissues of rhesus and stump-tailed monkeys (7,13).
The size of the heavy chain of these molecules, as assessed by their
mobility in SDS-polyacrylamide gel electrophoresis, is the same as that
of human δ chains; i.e., only slightly faster than that of μ chains.

Melcher et al. (14) and Abney and Parkhouse (15) were the first to
show by lactoperoxidase-catalized radioiodination of mouse lymphocytes
followed by SDS-polyacrylamide gel electrophoresis of the reduced and
alkylated molecules precipitated with a polyvalent anti-mouse immunoglobulin antiserum, the presence on mouse B lymphocytes of immunoglobulin molecules whose heavy chain had a mobility intermediate between
those of the μ and of the γ chains. They hypothesized that this "non-
IgM" and "non-IgG" membrane immunoglobulin was the murine counterpart of
human IgD, and their concept has been accepted by other investigators
(16). Recently on the basis of the fact that the mobility in SDS-PAGE
of the mouse "non-IgM" membrane immunoglobulin is not the same as that
of the human δ chain, Finkelman et al. (17) have raised some doubts that
these molecules are, in fact, IgD. However, even more recent data
(Sitia, personal communication) seem to have solved the problem in the
sense that the faster mobility of mouse δ chain may be attributed to
modifications of the original chain, which actually has a mobility
identical with that of the human counterpart. Furthermore a series of

parallels in the cellular expression of human IgD and mouse "non-IgM"
membrane immunoglobulins, that will be discussed later, leave no doubts,
in our opinion, that this mouse molecule is the physiological counterpart
of human IgD; in view of the problems involved in the preparation of
sufficient amounts of mouse IgD, a definitive demonstration of structural
homology (in addition to the known high susceptibility to proteolysis)
may have to wait some time.

An analogue of IgD is likely to exist in all species that have a
multiclass immunoglobulin system. We have detected the presence of an
unidentified immunoglobulin (that is non-IgM, non-IgG and non-IgA) on
the surface of lymphocytes in rabbits and chickens (unpublished results).

We have already said that IgD is poorly represented among serum
immunoglobulins in humans, and undetectable in monkeys and mice, in
striking contrast with the high frequency of IgD-bearing lymphocytes in
all the above mentioned species. It must be pointed out however that
such a dissociation between the serum level of an immunoglobulin class
and its presence on the lymphocyte membrane is not a characteristic
feature of the IgD, but applies also to IgM and IgG. It is accepted in
fact that IgM is the dominant lymphocyte receptor and IgG is present on
a minority of B lymphocytes only (17), whereas, in the serum, IgG is by
far the major immunoglobulin class.

Simultaneous presence of IgM and IgD on single cells

In the first reports on the presence of IgD on human lymphocytes
(11,12), it was clear that the total number of lymphocytes detected with

antisera each one specific for one immunoglobulin class exceeded the number of lymphocytes stained by a polyvalent anti immunoglobulin antiserum, a finding that was unexpected on the basis of previous reports supporting restriction to the production of one immunoglobulin class only by a single lymphocyte (18), a rule accepted for immunoglobulin-secreting plasma cells (reviewed in 19). In fact, further investigations have shown that the vast majority (80-90%) of IgD-bearing lymphocytes also carry membrane IgM (20,21), and that the two immunoglobulins can be redistributed independently of each other, thus excluding any cross-reactivity at the level of the antisera. In addition, the possibility that one of the two immunoglobulins could have been passively absorbed and not be the actual product of the cell carrying it, was ruled out by the finding that a single cell carried only one type of light chains, K or L, and by the observation that both IgM and IgD were resynthesized in vitro after removal of the surface molecules by "capping" with anti-IgM or anti-IgD antisera, as well as with antisera directed against light chains (20). More recently (22), reappearance of both IgM and IgD has been observed after removal from the membrane with the proteolytic enzyme pronase, although IgD may not reappear when the cells are treated with an excessive concentration of the enzyme. This observation is explained on the basis of the very high susceptibility to proteolysis of IgD, so that even minimal amounts of enzyme left in the culture medium are able to digest the newly synthesized membrane IgD.

The simultaneous presence of IgM and IgD on single cells has been observed also in monkeys (7) and mice (16,23). In the latter case, in

fact, for which no specific antisera to IgD were available, the
simultaneous presence of IgM and of a "non-IgM" molecule on the same
cells was taken as a support, in analogy with the human situation, of the
IgD-like nature of the "non-IgM" molecule.

The coexistence of IgD and IgM on the same cell raised the
problem of the specificity of the combining sites of the molecules of the
two classes. Since there was evidence that all the membrane immunoglobulin molecules on a single lymphocyte react with one antigen, and are
likely to have a uniform combining site (24), it was predictable that
IgM and IgD present on the membrane of the same cell, and synthesized by
the same cell, would have the same combining site. This prediction was
confirmed by all the investigations that have been aimed at this purpose
either directly, through the demonstration that both classes of receptors
reacted with the same antigen (25,26,27), or indirectly through the
demonstration of a common idiotype (28,29). This identity of the
combining site of two different immunoglobulin classes present on the
same cell, is a crucial element of the double expression maturation
scheme that will be discussed later (see also 30).

Another problem that is raised by the simultaneous presence of IgM
and IgD on the membrane of a single cell, is that of the quantitative
ratio between the two classes of receptors. In normal human peripheral
blood lymphocytes, the relative expression of the two immunoglobulins
varies all the way from an excess of IgM to an excess of IgD. As it will
be discussed later, the same variability of the IgM/IgD ratios from cell
to cell has been observed also in monkeys and mice, with some variations

of distribution in different tissues. It is not known and remains to be investigated whether this ratio is random or is correlated with a process of maturation or with other physiological parameters of the cell.

The simultaneous expression of IgM and IgD in human immunocytes can also be shown in plasma cells. Pernis et al. (25) have shown that a proportion, variable from one case to the other, of IgM-containing plasma cells in the bone marrow of patients with Waldenström macroglobulinemia have IgD on the membrane. Ferrarini et al. (31) made similar observations for IgM-secreting plasma cells in human tonsils, 10-20% of which carry on the membrane IgD molecules together with IgM. These observations are useful to try to construct a pathway of B lymphocyte development as related to immunoglobulin class expression, as it will be attempted at the end.

Once established the simultaneous expression of IgM and IgD by a single cell, the problem arises of the existence of cells bearing IgM without IgD, or IgD without IgM. Certainly a population of cells that carry IgM without IgD exists, and usually ranges in human peripheral blood around 18% of all immunoglobulin positive lymphocytes (20). Also cells with apparently IgD without IgM have been detected in human blood, and cases of chronic lymphocytic leukemia have been described (32) in which practically all the blood lymphocytes showed only IgD. However, Seligmann (personal communication) on a large series of cases of chronic lymphocytic leukemia, has observed that even when IgD is the dominant membrane immunoglobulin, traces of IgM can always be demonstrated by increasing the sensitivity of immunofluorescence with the use of an

indirect "sandwich" technique. Also in the mouse, the existence of

cells having only IgM as a surface immunoglobulin is clearly demonstrated

(16, 23), whereas some controversial results have been obtained for

cells having IgD without IgM. In fact, Abney et al. (33) reported a

high proportion of these cells in different tissues of mouse, whereas in

our laboratory (9) "pure" IgD cells in mouse tissues proved to be very

rare, and one cannot rule out the possibility that these cells, as the

leukemia cells observed by Seligmann, have levels of IgM which are below

the threshold of detectability of direct immunofluorescence.

Distribution of IgD-bearing lymphocytes in lymphoid tissues

Two sets of results are available on the distribution of membrane

IgD lymphocytes in different lymphoid tissues, one obtained by biochem-

ical methods, and the other by direct observation in double immunofluor-

escence.

With lactoperoxidase-catalized surface radioiodination, followed by

SDS-polyacrylamide gel electrophoresis analysis of the precipitate

obtained with a polyvalent anti-immunoglobulin, Vitetta and co-workers

(14,34,35) found no detectable IgD in thymus and bone marrow, roughly the

same amount of IgM and IgD in spleen, from two to five times more IgD

than IgM in lymph nodes, whereas in Peyer's patches nearly all (more

than 95%) the immunoglobulins precipitable with a polyvalent anti-immuno-

globulin antiserum were IgD. Similar results, limited to the tissues

studied, are reported by Abney and Parkhouse (15) for mouse, and by Corte et al. (7) for monkey. Recent results by Sitia (personal communication) indicates on the other hand presence of IgD, albeit in low amounts, in bone marrow of mice.

With double immunofluorescence, by staining cells first with an anti-IgM reagent in "capping" conditions, and then with either an anti-IgD or a polyvalent anti-immunoglobulin conjugated with a different fluorochrome, Pernis et al. (16) and Knight et al. (9) in mouse, and Corte et al. (7) in monkey have found that the majority (65-80%) of the cells in spleen, lymph nodes and Peyer's patches carry both IgM and IgD, very few (1-2% in spleen and lymph nodes, and 6% in Peyer's patches) have IgD without detectable IgM, and the remaining cells have IgM only. In bone marrow, about half of the B cells are "doubles" IgM/IgD, the rest showing IgM without IgD.

The results reported by Abney et al. (33) are in partial disagreement with the above reported values, in the sense that they found a high percentage of cells with IgD only in all the tissues examined, that is spleen, lymph nodes and Peyer's patches. This discrepancy remains to be explained, but recently M.C. Cooper (personal communication) has found that the IgM/IgD distribution of mouse lymphocytes is not very different from that of primates; that is, his results are comparable with those we reported above.

On the other hand, the apparent discrepancy between the biochemical data and the observations with immunofluorescence can be easily explained on the basis of the variability of IgM/IgD ratios that have already been discussed. In fact we (9) have observed that lymph nodes are richer

than spleen in high-IgD lymphocytes and that practically all cells in

Peyer's patches have very low levels of IgM. The reverse happens in the

bone marrow, where all the IgD-bearing cells are low-IgD.

The significance of a preferential localization of high-IgD cells

in some lymphoid tissues is unknown and must be investigated in view of

the existence of different B cell subpopulations and of their distribut-

ion in the lymphoid system.

Ontogeny of IgD lymphocytes

The first observations made on IgD on human lymphocytes (11,12),

reporting a higher percentage of IgD positive cells in cord blood than

in adult peripheral blood, seemed to indicate that IgD could preceed IgM

in ontogeny. However, Vitetta et al. (34) first have shown that newborn

mice lack IgD, and that these molecules reach amounts detectable by

biochemical methods only at two weeks of age, a finding that was

confirmed by other investigators (3), who also reported lack of IgD in

mouse foetal liver.

With immunofluorescence, that results to be more sensitive than the

biochemical methods used by the above investigators, rare IgD bearing

lymphocytes are present in 4-days old mice (9) and even at birth, with

some differences according to the strain (unpublished observations), and

their percentage increases rapidly after birth reaching adult levels, as

percentage of IgM/IgD bearing cells, at three weeks of age.

In human foetuses only pure IgM cells are found up to 13 weeks of

gestation, a moment when the first IgD/IgM bearing lymphocytes appear

in foetal spleen, to become the dominant population of splenic B
lymphocytes already at 16 weeks of gestation. At this time also the
first IgD bearing lymphocytes appear in foetal bone marrow, whereas no
IgD lymphocytes are found in foetal liver until 23 weeks of gestation.
(37). There is on the whole no discrepancy between the data in man and
mouse, if one takes into account that mouse at birth is equivalent, from
the point of view of immunological maturity, to a 3-month old human
foetus (38).

It therefore becomes clear that IgM preceeds IgD in ontogeny, and
that IgD can be considered as a marker of a more mature population of B
lymphocytes. We shall discuss this problem again, also in view of the
fact that some recent observations (30) make us feel that the cells
bearing IgM in the adult may actually be constituted of two different
cell populations, one less mature and the other more mature than the one
bearing simultaneously IgM and IgD.

It is interesting to remember that the limited sequence data
available, indicate that human IgD molecules show higher homology with
IgG and IgE than with IgM (39), and this can be taken as an indication
that IgD appeared in evolution, as it does in ontogeny, after IgM.

IgD-receptors and cell maturation

In contrast to the high percentage of lymphocytes synthesizing both
IgM and IgD, double synthesis of these two immunoglobulin molecules by
secreting plasma cells is a much rarer event.IgD-secreting plasma cells,

which although they are rare can be seen in human tonsils, have only IgD as a surface immunoglobulin. Only a proportion of IgM plasma cells in the same tissue show IgD, together with IgM on their membrane (31). In bone marrow of macroglobulinemia, where nearly all the cells can be considered as members of one clone at different stages of differentiation, the majority of lymphocytes express both IgM and IgD, whereas only a proportion of IgM-secreting cells have IgD on the membrane (25). Preliminary results from our laboratory on mouse spleen cells stimulated in vitro seem to confirm these observations, that is IgD is progressively lost by cells that undergo maturation to IgM secretion.

We can conclude from these observations that the simultaneous synthesis for long period of time of IgM and IgD is a feature of B lymphocytes at a given stage of differentiation, and that the maturation to IgM secretion coincide in time with a turning off of the synthesis of IgD, with progressive disappearance of IgD molecules from the cell membrane. Whether the loss of IgD is the cause or the consequence of the process of cell maturation, has to be clarified. More generally we can say that synthesis of two immunoglobulins by a single cell, IgM and IgD, IgM and IgG, and possibly others, is a normal and stable condition of lymphocytes (30), whereas plasma cells, which represent the end stage of differentiation of immunocompetent cells, tend, through a short period of double expression, which does not necessarily mean double synthesis, to restriction of the production of one immunoglobulin class only.

The problem of simultaneous synthesis of IgD and IgG

Lymphocytes carrying both IgD and IgG in human peripheral blood are extremely rare in our experience (unpublished results).In human tonsils IgD molecules are never found on the membrane of IgG secreting cells(31). Preliminary work of our laboratory indicates that this also applies to mouse, although rare double IgD-IgG lymphocytes and lymphoblasts seem to be present in mitogen-stimulated cultures.

Our opinion is that, whereas normal immunocytes can simultaneously synthesize IgM and IgD or IgM and IgG, they cannot simultaneously synthesize IgD and IgG. There may be, however, a sharp switch from the synthesis of IgD to that of IgG, as hypothesized by Vitetta and Uhr(40), with a very short period of time during which the "switched" cell carries on its membrane previously synthesized IgD molecules, together with the actually produced IgG. An indirect support for the assumption that IgG secreting cells may be derived from IgD bearing precursors is the fact that the treatment in vivo with anti-IgD antisera (and not with normal sera) results in a remarkable increase of the serum IgG levels (41). However, a direct proof of a δ to γ switch is still missing, as well as any indication in favour or against other possible switches such as δ to α and δ to ε, that, on the same basis, may well occur. Anyhow, if a δ to γ switch does in fact exist, it would be from IgD to one or another of the IgG subclasses, since we never found (30) whether in unstimulated or in stimulated spleen cells, double expression by a single lymphocyte or plasma cell of different IgG subclasses.

A hypothesis on the relationship between immunoglobulin

expression and the differentiation of B lymphocytes

From the evidence so far available, we can try to hypothesize the different steps in membrane immunoglobulin expression in the course of development of B lymphocytes and to correlate these developmental steps with the sequence of functional stages of B immunocytes as investigated by many authors (reviewed in 42). The following sequence can be outlined:

$$\text{IgM only} \rightarrow \text{IgM+IgD} \rightarrow \text{IgM+IgG} \rightarrow \text{IgG only}$$
$$\quad 1 \qquad\qquad 2 \qquad\qquad 3 \qquad\qquad 4$$

Cells of group 1 are relatively immature elements found early in ontogeny or in the adult bone marrow. The membrane IgM is easily modulated by anti-IgM antibodies (43,44). When stimulated with a B mitogen such as lipopolysaccharide these cells can generate IgM as well as IgG secreting plasma cells (44). The fact that plasma cells can differentiate from these cells is a puzzling phenomenon, and one may wonder whether this line of development might not go through a brief period of synthesis of small amounts of IgD and give rise to cells of group 2 and then, through a δ and γ switch, to cells of group 3. From the functional point of view, these cells could tentatively represent the tolerizable virgin cells.

Cells of group 2 are generated from cells of group 1 by an antigen-independent process of maturation (34) and express progressively increasing amounts of membrane IgD together with IgM. If stimulated by antigen or by a mitogen, these cells can either (a) proliferate, (b)

mature to IgM-secreting plasma cell with progressive loss of membrane
IgD and eventually also of membrane IgM, and (c) become cells of group 3,
through a switch from the synthesis of δ to that of γ chains. Functionally these cells could be the non-tolerizable virgin cells.

Cells of group 3 arise from cells of group 2 by a δ to γ switch as
a consequence of stimulation with antigens or mitogens. When further
stimulated by the same, they can either (a) proliferate, (b) mature to
IgG secreting plasma cells with progressive loss of surface IgM, and
(c) lose surface IgM and become cells or group 4. This last group of
cells, when stimulated by antigen or mitogens, can proliferate, or,
more easily, mature to IgG secreting plasma cells with loss of membrane
IgG. Memory cells could belong to these two groups.

What differentiates this hypothesis from others that have been
proposed (40) is the emphasis on the simultaneous synthesis of two
immunoglobulin classes by single cells at a given stage of differentiation. So, the μ to δ, and the μ to γ switches go through a more or
less prolonged period of double synthesis, and perhaps they should not
be called switches at all. The only real switch that remains is that
from δ to γ.

However, it should be emphasized that the existence of this switch
has not yet been definitely proved, and it might still be that the IgM/
IgD cells can only mature to IgM-secreting plasma cells. In this case,
the IgM/IgG cells that will mature to IgG secreting plasma cells should
derive from cells having IgM only as a consequence of antigenic or
mitogenic stimulation, and more precisely from a proportion of the pure

IgM cells present in adult tissues, like the ones that we have observed to express membrane IgG after only 6 hours of stimulation with mitogen, and which are probably already synthesizing undetectable amounts of IgG (30).

On the whole, the problem of the signal or signals delivered to a B lymphocyte by the IgD receptor is still completely open. Maybe it has merely a function in operating antigen-dependent switches to any of the immunoglobulin classes or subclasses (except IgM). But it might also have a function in lymphocyte stimulation, and in this case there is some evidence (45) that it stimulates lymphocyte proliferation rather than maturation to immunoglobulin secretion.

References

1. Rowe, D.S. and Fahey, J.L., J. Exp. Med., 121,171, 1965.

2. Henney, C.S., Welscher, H.D., Terry, W.D. and Rowe, D.S., Immunochemistry, 6,445, 1969.

3. Rowe, D.S. and Fahey, J.L., J. Exp. Med., 121,185, 1965.

4. Pernis, B., Chiappino, G. and Rowe, D.S., Nature, 211,424, 1966.

5. Pernis, B., Governa, M. and Rowe, D.S., Immunology, 16,685, 1969.

6. Spiegelberg, H.J., in Contemporary Topics in Immunochemistry, edited by F.P. Inman, vol. 1, page 165, Plenum Press, N.Y., 1972.

7. Corte, G., Ferrarini, M., Tonda, P., Bargellesi, A. and Pernis, B., Eur. J. Immunol., in press.

8. Melcher, U., Vitetta, E.S., McWilliams, M,., Lamm, M.E., Phillips-Quagliata, J.M. and Uhr, J.W., J. Immunol., 113,1326, 1974.

9. Knight, K.L., Schweizer, M., Forni, L. and Pernis, B., submitted
    for publication.

10. Walzer, P.D. and Kunkel, H.G., J. Immunol., 113,274, 1974.

11. vanBoxel, J.A., Paul, W.E., Terry, W.D. and Green, I., J. Immunol.,
    109,648, 1972.

12. Rowe, D.S., Hug, K., Faulk, W.P., McCormick, J.N. and Gerber, G.,
    Nature New Biol., 242, 155, 1973.

13. Martin, N.L., Leslie, G.H. and Hindes, R., Intern. Arch. Allergy.
    51,320, 1976.

14. Abney, E.R. and Parkhouse, R.M.E., Nature, 252,600, 1974.

15. Parkhouse, R.M.E. and Abney, E.R. In Membrane Receptors of Lympho-
    cytes, edited by M. Seligmann, J.L. Preud'homme and F. Kourilsky,
    page 51, North Holland-Elsevier, Amsterdam, 1975.

16. Pernis, B., Forni, L. and Knight, K.L. Ibidem, page 57.

17. Finkelman, F.D. and van Boxel, J.A., J. Immunol., 116,1173, 1976.

18. Pernis, B., Forni, L. and Amante, L., J. Exp. Med., 132,1001,1970.

19. Fröland, J.S, and Natvig, J.B., Scand. J. Immunol., 1,1, 1972.

20. Pernis, B., Cold Soring Harbor Symp. Quant. Biol., 32,330, 1967.

21. Rowe, D.S., Hug, K., Forni, L. and Pernis, B., J. Exp. Med., 138,
    965, 1973.

22. Knapp, W., Bolhuis, R.L.H., Radl, J. and Hijmans, W., J. Immunol.,
    111,1295, 1973.

23. Ferrarini, M., Corte, G., Viale, G., Durante, M.L. and Bargellesi,
    A., Eur. J. Immunol., 6,372, 1976.

24. Raff, M.C., Feldmann, M., and dePetris, S., J. Exp. Med., 137,1024, 1973.

25. Pernis, B., Brouet, J.C. and Seligmann, M., Eur. J. Immunol., 4, 776, 1974.

26. Stern, C., and McConnell, I., Eur. J. Immunol., 6,225, 1976.

27. Goding, J.W. and Layton, J.E., J. Exp. Med., 114,852, 1976.

28. Fu, S.M., Winchester, R.H., Feizi, T., Walzer, P.D. and Kunkel, H.G., Proc. Natl. Acad. Sci., 71,4487, 1974.

29. Salsano, F., Fröland, S.S., Natvig, J-B. and Michaelsen, T.E., Scand. J. Immunol., 3,841, 1974.

30. Pernis, B., Forni, L. and Luzzati, A.L., Cold Spring Harbor Symp. Quant Biol., 1976, in press.

31. Ferrarini, M., Viale, G. Risso, A. and Pernis, B., Eur. J. Immunol., 6:562, 1976.

32. Kubo, R.T., Grey, H.M. and Pirofsky, B., J. Immunol., 112, 1952, 1974.

33. Abney, E.R., Hunter, I.R. and Parkhouse, R.M.E., Nature, 259,404, 1976.

34. Vitetta, E.S., Melcher, U., McWilliams, M., Lamm, M.E., Phillips-Quagliata, J.M. and Uhr, J.W., J. Exp. Med., 141,206, 1975.

35. Vitetta, E.S., McWilliams, M., Lamm, M.E., Phillips-Quagliata, J.M. and Uhr, J.W., J. Immunol., 115,603, 1975.

36. Vossen, J.M. and Hijmans, W., Ann. N.Y. Acad. Sci., 254,262, 1975.

37. Soloman, J.B. Foetal and neonatal immunology, chapter 11, North-
     Holland, Amsterdam/London, 1971.

38. Spiegelberg, H.L., Nature, 254,723, 1975.

39. Vitetta, E.S. and Uhr, J.W., Science, 189,964, 1975.

40. Pernis, B., in Membrane Receptors of Lymphocytes, edited by M.
     Seligmann, J.L. Preud'homme and K. Kourilsky, page 27, North-
     Holland/Elsevier, Amsterdam, 1975.

41   Strober, S., Transplant. Rev., 24,84, 1975.

42. Raff, M.C., Owen, J.J.T., Cooper, M.D., Lawton, A.R.,Megson,M.and
     Gathings, W., J. Exp. Med., 142, 1052, 1975.

43. Kearney, J.F. and Lawton, A.R., J. Immunol., 115,677, 1975.

44. Vitetta, E.S., Forman, J. and Kettman, J.R., J. Exp. Med.,143,1055,
     1976.

Chapter 22

TRIGGERRING OF LYMPHOCYTES BY ANTIBODIES AGAINST β_2 MICROGLOBULIN

O. Ringdén, U. Persson and E. Möller

Transplantation Immunology Laboratory, Huddinge Hospital and

Division of Immunobiology, Karolinska Institutet Medical School,

Stockholm, Sweden

Abstract

$\beta_2$microglobulin (β_2m), structurally related to "domains" of immunoglobulin molecules, is associated with products of the major histocompatibility system on cell surfaces. Heteroantibodies against β_2m are mitogenic to a specific subpopulation of human and mouse B lymphocytes. This subpopulation is present in human blood which makes the antibody convenient and clinically useful as a fuctional marker for peripheral B lymphocytes. Absorption and elution experiments, as well as tests showing mitogenic activity of Fab monomers of anti-β_2m indicate that the interaction of the binding site of the antibody and relevant cell surface structures, probably β_2m itself, is directly responsible for lymphocyte activation. The relevance of these findings for cell receptors involved in lymphocyte activation is discussed.

Surface markers for lymphocyte subpopulations

Surface differences have made it possible to distinguish between human T and B lymphocytes (for references see 1). A fraction of human lymphocytes forming spontaneous rosettes with sheep erythrocytes (SRBC) (2) were demonstrated to be T cells (3,4) and this property is often used for their identification (5).

Mouse T lymphocytes express the theta (θ) isoantigen on the cell surface (6). By heteroimmunization and absorption, specific anti-T-cell serum can also be obtained for human cells and can be used to identify T cells (7,8).

Some lymphocytes bear immunoglobulin (Ig) bound to the cell membrane, first demonstrated by Möller (9). Later it was established that cells with surface Ig were B lymphocytes (10,11). B cells also possess receptors for several complement components, among these C3, detected by binding SRBC coated with IgM antibody and complement (2). B cells also possess receptors for the Fc portion of IgG, which can be detected by uptake of fluoresceinated heat aggregated IgG or binding of IgG coated SRBC (13,14). In humans, not all B cells that have surface Ig have detectable receptors for Fc and C3 (1,15) and it is possible that Ig-bearing lymphocytes express Fc and/or C3 only during part of their lifespan (16).

Heteroantibodies against B lymphocyte antigen in mice (17) and humans (18) can also be used for detection of B lymphocytes.

Functional markers of lymphocyte subpopulations

Substances that selectively activate T or B lymphocyte subpopulations can be used to study cellular functions. Mitogenic activation can also serve as a model for the specific activation of these cells by antigen (19,20).

The first described mitogen was phytohemagglutinin (PHA) (21) later shown to selectively activate T lymphocytes (22). Concanavalin A (Con A), another T cell mitogen (23,24), probably activates a lymphocyte subset distinct from that responding to PHA (23). The phytomitogen of pokeweed (PWM) has been shown to stimulate both T and B cells (25) but later was demonstrated to contain multiple mitogens (26).

B cells are postulated to be activated by "one non-specific signal", delivered to the cell by surface structures which are not the Ig receptors (20). Ig receptors serve a passive focusing function by binding antigen to the specific cell. Activation of B cells by thymus-dependent antigens is thought to depend upon an activating signal delivered by T cells (20).

Upon activation, B cells proliferate, synthesize and secrete antibodies. Polyclonal antibody secretion induced by mitogens can be measured in vitro using the hemolytic plaque assay, or by the demonstration of intracellular Ig using fluorescent techniques.
B cell mitogens or polyclonal B cell activators (PBA) for mouse lymphocytes include lipopolysaccharide of E.coli (LPS) (24), purified protein derivative of tubercle bacilli (PPD) (27) and other thymus independent antigens (28).

Different PBA stimulate different subsets of B cells in mice, which are believed to vary in their stage of maturation and homing patterns (29,30). Evidence has been presented indicating that the different subsets belong to one differentiation line of B cells (30).

The use of mitogens for functional characterization of lymphocyte subpopulations should have several clinical applications. However, none of the above mentioned selective PBA activate human blood B cells. Since we have shown that anti- β_2m has this property, it may be clinically useful for the functional characterization of B cells.

$\beta_2$microglobulin

$\beta_2$microglobulin (β_2m), a protein with a molecular weight of 11,800 daltons, was first isolated from the urine of patients with renal tubular damage due to Wilson's disease or chronic cadmium poisoning (31). This protein is also present in normal serum and other biological fluids (32).

Amino acid sequence analysis of β_2m indicated that this protein was related to the immunoglobulins (33) and was strikigly homologous to the Ig constant domains (34). β_2m wes demonstrated to be present on the surface of lymphocytes (34,35), polymorphonuclear cells and platelets (35) and was synthesized by lymphoid (36) and non-lymphoid cells (36,37).

Purified, papain solubilized HLA antigens were reported to be composed of two noncovalently linked polypeptide chains. The large chain was glucoprotein with a molecular weight of 33,000 daltons and carried the serological activity, and the small chain was a protein with a molecular weight of 10-12,000 daltons (38,39). The small chain was shown to be β_2m (40,41,42). In comparison, H-2 antigens were also shown to be composed of two polypeptide chains, one of which was similar to β_2m (43,44). The suggested association of the two subunit structures of HLA was not due to the solubilization procedures, because the subunits were also shown to be associated within the cell membrane. This conclusion was based on the finding that antibodies against β_2m, and also Fab-monomers of such antibodies, could block the cytotoxic effects of HLA antibodies on lymphoid cells (45,46). Thus, capping of the β_2m with antibodies specific for this molecule gave total co-capping of all HLA antigens (47). There appears to be more β_2m on the cell surface than is associated with HLA antigens since Fab-fragments of HLA antibodies gave incomplete inhibition of the cytotoxic effect of anti-β_2m. Furthermore, capping of all HLA antigens, with anti-HLA antibodies, did not give total co-capping of the β_2m structures present on the cell. Nilsson et al. (48), estimated the number of β_2m molecules present on peripheral blood lymphocytes (PBL) to be about 2×10^5 molecules/cell and Sanderson and Welsh (49) estimated the number of HLA molecules to be about $3-6 \times 10^4$ molecules/cell (for more references see 50). However, recent chemical analyses indicate that all $\beta_2$microglobulin on mouse spleen lymphocytes is present in association with H-2 molecular structures (51).

The two subunit structures of the HLA antigens are under separate genetic control. β_2m is coded for by a gene on chromosome 15 in human cells (52) while the genes for HLA antigens are located on chromosome 6 (53). Furthermore, there exists a human cell line, the "Daudi" Burkitt lymphoma cell line, which cannot produce β_2m (48) but has HLA like structures on the cell surface (54).

Anti-β_2m is a polyclonal B cell activator

During attempts to clarify the role of cell surface β_2m, we observed that addition of IgG rabbit anti-human β_2m to cultures of mouse spleen cells led to an increased DNA synthesis compared to cultures treated with control sera, such as rabbit anti-human lambda or kappa light chains, anti-Ig or normal rabbit sera (55). Further studies showed that rabbit anti-human β_2m reacted with almost all mouse lymphocytes to an equal extent. Thus, the rabbit antibody presumably cross-reacted with mouse β_2m. The mitogenic effect could be absorbed by soluble β_2m as well as by mouse thymocytes and bone marrow cells (55). Since all hitherto described pure mitogenic substances activate either T or B cells, we studied the target cell of anti-β_2m mitogenicity. Anti-β_2m was mitogenic for spleen cells, lymph node and bone marrow cells but not for thymocytes or cortison-treated thymocytes. In addition, it was highly mitogenic for anti-Thy.1 treated spleen cells and to cells from Balb/c Nude athymic mice but not for anti-MBLA pretreated spleen cells. We concluded that the target cells were of B and not of T origin. This was further supported by the findings that anti-β_2m, like other polyclonal B cell activating substances (PBA), induced non-specific antibody synthesis in spleen cells. The mitogenic effect seemed to be caused by a direct effect of the antibody on the B cells, since pretreatment of spleen cells to remove phagocytic cells did not interfere with mitogenicity (55).

As with all other PBA's, optimal effects of anti-β_2m in vitro was observed on days 2-3 of microcultures under standard conditions. In conclusion, anti-β_2m was found to be mitogenic for mouse B lymphocytes. No other sera of all those tested, including rabbit anti-mouse B cell antigens (MBLA) was found to have similar properties.

Mitogenic effects of anti-β_2m for human lymphocytes

Functional markers for lymphocytes are important tools for the characterization of immune reactivity. T cell mitogens have been widely used for studies of human lympocyte function. Hence, it was of importance to study whether rabbit anti-human β_2m would activate human B cells, since such substances had not been earlier described. We found that anti-β_2m was strongly mitogenic for human blood lymphocytes (56). Optimal in vitro activation occurred on days 3-4 of cells cultured at a density of $1\text{-}4\text{x}10^5$ cells/0.2 ml medium in microcultures. Other sera, such as various anti-Ig or light chain sera were not active. Anti-β_2m increased DNA synthesis from $1\text{-}5\text{x}10^3$ cpm/cultures in controls to $30\text{-}80\text{x}10^3$ cpm/culture in blood lympho-cytes, and to $100\text{-}250\text{x}10^3$ cpm in cultures of human spleen cells. For further technical details see (56,57).

Furthermore, mitogenicity of anti-β_2m for cells from different lymphoid organs was related to the proportion of surface Ig positive cells in such organs, indicating that the target cell, also in the human system, was a B cell. Proof for this was obtained when it was found that anti-β_2m-induced blast cells synthesized immunoglobulin (56). Only a certain proportion of blasts in the spleen were positive for intracellular Ig (approx, 30%) which is similar to the proportion of Ig positive blasts that develop in mouse spleen cells treated with LPS (58) as well as in human spleen cells treated with LPS (56), whereas no blasts, or less than 1-2%, of those induced in blood cell cultures were Ig positive. We believe these data to indicate that different subpopulations of B cells reside in the spleen and blood, respectively, especially since it is well established that different mouse PBA act of distinct subsets of B cells, reflecting their different degrees of differentia-tion (for ref. and extensive discussion see 59).

The binding site of anti-β_2m is the mitogenic principle

The mitogenic activity of rabbit anti-β_2m sera for both mouse and human cells was absorbed by highly purified soluble β_2m (56,57) as well as by β_2m coupled to Sepharose beads (60). Not only the mitogenicity was abolished but also the capacity to induce polyclonal antibody synthesis. Furthermore, some mitogenic activity could be eluted from β_2m Sepharose columns at low pH. These results strongly support the notion that specific antibodies against

$\beta_2$microglobulin are responsible for mitogenicity of IgG fractions of rabbit anti-β_2m sera.

Not only the IgG portion of anti-β_2m sera can activate cells, but also F(ab)$_2$ as well as Fab fragments of IgG antibodies. It was reported by Östberg et al. (61) that F(ab)$_2$ fragments of IgG anti-β_2m were equally mitogenic for human blood lymphocytes as the IgG antibodies, whereas in their experiments Fab monomers were inactive. However, recent experiments in our laboratory (62) have shown that also Fab monomers of such antibodies are mitogenic both for human spleen cells and for peripheral blood lymphocytes.

At equal protein concentrations, highly purified preparations of F(ab)$_2$ and Fab fractions of anti-β_2m induced equally high cell stimulation. Therefore, it was concluded that mitogenicity of anti-β_2m for human lymphocytes can be ascribed to the antigen-binding site of the antibody and not to the Fc portion of the molecule. This suggests that it is a direct interaction between B cell surface bound β_2m with the appropriate antibody that initiates lymphocyte activation. Furthermore, this result shows that bivalency of a ligand is not essential for lymphocyte activation, which contradicts the cross-linking hypothesis for cell triggerring. Recently, Sela et al. (64) reported that Fab fragments of carbohydrate binding chicken immunoglobulin were as mitogenic as the divalent antibody for mouse spleen cells.

The target B cell population for anti-$\beta_2$microglobulin.

Most probably mitogens interact only with distinct subsets among T and B cells. Thus, Con A and PHA activate partly distinct T cell subsets among B cells (30). There is good reason to believe that the sensitivity of B cell subsets reflects their stage of maturity. Thus, DxS which mainly gives rise to increased DNA synthesis and little antibody secretion upon stimulation activates relatively immature cells, which might yet lack the capacity to be stimulated to antibody synthesis; LPS gives rise to both antibody secretion and increased DNA synthesis and PPD, finally, gives very low increase of DNA synthesis but a marked antibody production, and probably interacts with mature plasma cell precursors (for ref. see 59).

G. Möller first described (64) that a quantitative principle applies for the activation of lymphocytes by mitogens and by antigens, in so far as two different mitogens or a mitogen and an antigen in optimal doses result in inhibition of activation, whereas two suboptimal concentrations could give a synergistic response. If completely different subsets of lymphocytes are activated by two mitogens, such as is the case when a T and a B cell mitogen are mixed in culture, additive effects are observed. We performed experiments with mouse spleen cells and found that DNA synthesis was completely additive when anti-β_2m was mixed in optimal concentrations either with DxS or with LPS (60). This indicates that the B cell subset that responds to DxS, LPS and anti-β_2m are distinct. However, a slightly different conclusion was reached when induction of polyclonal antibody synthesis was studied instead of DNA synthesis. In such experiments mixing of optimal concentrations of anti-β_2m with either LPS or PPD resulted in slight inhibition, i.e. less than additive effects were obtained. Other experiments where varying concentrations of anti-β_2m and LPS were mixed in cultures of mouse spleen cells showed that suboptimal concentrations of both substances gave synergistic responses, whereas mixing of optimal concentrations or superoptimal concentrations of both substances resulted in inhibition. Therefore, we would conclude that those subsets of cells which are activated to antibody secretion by the different PBA's are partly overlapping. Definite proof that this is so will have to await direct studies using "priming" in vitro with one mitogen, followed by addition of hot tritiated thymidine to induce suicide of activated cells, and thereafter to study effects of a second B cell mitogen.

Synergy studies have also been performed with human cells. As has been stated above, both anti-β_2m and LPS give strong stimulation of human spleen cells, but only anti-β_2m stimulates blood lymphocytes (56). The variation in responsiveness to the two PBA's might reflect the fact that they activate partly distinct subsets of B cells, and that various lymphoid organs contain B cells belonging to different stages of maturation. Anti-β_2m induces increased DNA synthesis as well as antibody secretion in spleen cells, but, in our hands only DNA synthesis in blood lymphocytes. Therefore, one may speculate that antibody forming cell precursors are not recirculating cells. In the experiments with human spleen cells, we found that anti-β_2m and LPS mixed in optimal conditions gave almost additive responses, indicating that the two PBA s activate distinct B cell subsets. However, similar experiments with blood lymphocytes revealed that LPS, which in itself does not activate blood cells, clearly potentiated the effect of anti-β_2m. The reason for this finding is not clear. It could be caused by LPS reacting only at subthreshold stimulating concentrations with some blood lymphocytes, which could also react to anti-β_2m, or, alternatively, the activity of anti-β_2m is in some yet undefined way enhanced by LPS, perhaps by absorption on to LPS molecules.

Taken together, these findings show that the cells which respond with increased DNA synthesis to anti-β_2m are distinct from those that respond to DxS, LPS and to PPD, but partly overlapping with those that respond to LPS with antibody synthesis. The experiments reported above, which indicate that anti-β_2m but not LPS can activate human peripheral blood lymphocytes also argue in this direction.

Is cell collaboration necessary for the activation of B cells by antibodies against β_2 microglobulin?

In our earlier experiments with mouse spleen cells we reported that removal of adherent or phagocytic cells did not interfere with mitogenicity of anti-β_2m. This argues in favour of a direct effect of the antibody on B cells for triggerring to occur. Most other PBA s, with the possible exception of DxS (65) also directly activate B cells. A direct effect of anti-β_2m is also supported by findings that removal of Thy.1 positive T cells from spleen cells suspensions does not inhibit mitogenicity, nor does the addition of T cell mitogens to a mixture of spleen cells interfere positively or negatively with the mitogenicity caused by anti-β_2m on B cells. However, it was reported by Östberg et al. (61) that mitogenicity of anti-β_2m on human blood lymphocytes was optimal with certain mixtures of T and B cells, separated by centrifugation after formation of rosettes with SRBC and by passage through anti-Ig columns. Therefore, we have reinvestigated this problem and have again found that for activation of mouse spleen cells, anti-β_2m does not seem to need cell interaction (60). Anti-Thy.1 treated cells and anti-MBLA treated cells were mixed in various proportions. Mitogenicity was found to be directly related to the proportion of B cells present in the cell mixture. In experiments with human blood lymphocytes, however, mitogenicity can be induced by Fab fragments of anti-β_2m as well as of the IgG fraction of serum on "pure" SRBC-RFC separated cells containing around 90% surface Ig-positive cells, but we have noted on several occasions that even higher responses can be observed in mixtures of purified "T" and "B" cells. Further studies are needed to resolve the question whether cell collaboration might in fact augment the activation of human (or mouse) blood lymphocytes, but not of mouse (or human) spleen lymphocytes.

Which is the cell surface receptor which receives the activating signal by anti-β_2m?

It is unfortunate that there are many different cell surface structures, with presumed receptor activity whose function is unknown, and definite receptor sites for various functions, whose structure is unknown. As for polyclonal B cell activating substances, one of the best known systems is that of LPS activation. It is well documented that the receptor for activation of B cells by LPS is distinct from the specific cell surface immunoglobulin receptor (for ref. see 28). There is a mouse strain which has a genetic defect and lacks to receptor for LPS activation on B cells. Cells from this mouse strain (C3H/HeJ) can form normal amounts of antibodies against LPS under special circumstances, and therefore has a normal complement of LPS specific immunoglobulin receptors (66). Elegant experiments by Coutinho recently (67) demonstrated the existence of a specific receptor for LPS on the surface of a subset of B lymphocytes. Since different B cell subsets respond to different B cell mitogens, they most probably carry distinct cell surface receptors for triggerring to antibody synthesis. One possible way to define the structure of the cell surface receptor which is responsible for activation of a particular B cell subset is by the accidental definition of mice or men that lack the capacity to respond to that PBA, if it can be shown that this is due to a genetic defect. We have searched for high and low responders against anti-β_2m. But, there are some strains which regularly give a higher response than other strains, without showing only other signs of disturbed ratios or functions of T and B cells. We are currently trying to establish the genetic basis of this defect by the production of F_1, F_2 and backcross mice between high and low responder strains, but do not have any firm data yet.

Seemingly all cells, both T and B cells as well as other nucleated cells, carry $\beta_2$microglobulin. This is to be expected from the universal distribution of histocompatibility antigens, since β_2m is present in association with these structures on cells. It might be unexpected therefore that only a specific subset of B cells can be activated by antibodies against $\beta_2$microglobulin, should the activating signal to the cell be mediated through the reaction of the antibody with cell surface β_2m in association with HLA or H-2 structures. But, it is well known that other selective mitogens, such as Con A and LPS, which only activate T and B cells, respectively, can bind to cells of both subpopulations with equal strength. The next question therefore would be: is the presumed cell surface structure that reacts with Fab monomers of anti-β_2m for activation identical to $\beta_2$microglobulin, and if so, is that β_2m associated with H-2 and/or HLA molecules? We cannot answer any of these questions at present. We believe that since anti-β_2m activity was absorbed by soluble and by insoluble β_2m, and since anti-β_2m was the only antibody with mitogenic properties for B cells so far, and since Fab monomers of anti-β_2m are active, that the cell surface receptor which interacts with anti-β_2m is β_2m itself.

Calculations performed as to the total amount of β_2m on lymphocytes first indicated that there was an excess of β_2m on cells compared to the amount of H-2 or HLA antigenic molecules. Thus, it was believed that β_2m could be associated with other types of cell surface structures other than histocompatibility structures (for ref. see 50). However, recent evidence, which is primarily chemical, suggests that all cell β_2m is associated with MHS molecules, and therefore activation might be mediated through such a molecule.

We have studied the effect of antibodies directed against other cell surface components on mitogenicity by anti-β_2m. Antibodies against H-2 or HLA structures are not mitogenic. Neither have we found that antibodies against these antigens, in mice or man, can interfere with mitogenicity of anti-β_2m. Only excessive concentrations of combined multispecific anti-HLA sera interfere with mitogenicity of anti-β_2m and only partly, which could be due to steric effects on cell surfaces rather than specific blocking of activating receptor molecules. The fact that we have not yet been able to interfere with anti-β_2m mitogenicity with allogeneic sera does, however, not contradict the assumption that the activating receptor contains MHS antigenic structures. Since there is no allogeneic variation yet observed for β_2m itself, it might be necessary to repeat these experiments with heteroantibodies against MHS structures before definite conclusions can be drawn. Furthermore, we have performed several experiments to block activation by anti-β_2m by anti-Ig antibodies, anti-Thy.1, anti-MBLA, without any effects. Neither does the particular activation receptor seem to be identical to the Fc receptor or the C3 receptor, since anti-β_2m pretreatment of cells does not interfere with the formation of Fc- or C3-RFC. Blocking of Fc receptor function can be achieved by anti-Ia antibodies, but we have not yet observed any inhibition of anti-β_2m activation of mouse splenic lymphocytes by anti-Ia antibodies, neither in experiments where cells are first pretreated with anti-Ia in the absence of complement nor when it is mixed with the culture medium. Furthermore, antibodies against β_2m do not interfere with cytotoxicity against Ia specificities in the mouse or HLA-D products in the human (60). The purified Fc receptor seems to lack both Ia specificities and β_2m (68). Other indirect evidence also support the notion that the Fc receptor is not associated with β_2m.

In conclusion, the surface receptor with which anti-β_2m reacts to exert its mitogenic effect is not identical to any of the hitherto defined B cell receptors. Which it is remains to be elucidated. Let us only, in closing, state that we hoped that this antibody, which thus can activate resting lymphoid B cells to differentiate into mature antibody-secreting cells could be used as a handle to define the receptor molecule responsible for activation. But, as all cells carry β_2m and only few can be activated by this ligand, other approaches have to be searched for.

References

1. Möller, G. (ed.) Transplant. Rev. 16, 1973.
2. Coombs, R.R.A., Gurner, B.W., Wilson, A.B., Holm, G. and Lindgren, B. Int. Arch. Allergy, 39:658, 1970.
3. Frøland, S .S., Scand. J. Immunol. 1:269, 1972.
4. Jondal, M., Holm, G. and Wigzell, H., J. Exp. Med. 136:207, 1972.
5. Aiuti, F., Cerottini, J-C., Coombs, R.R.A., Cooper, M., Dickler, H.B., Frøland, S ., Fudenberg, H.H., Greaves, M.F., Grey, H.M., Kunkel, H.G., Natvig, J., Preud homme, J-L., Rabellino, E., Ritts, R.E., Rowe, D .S., Seligmann, M., Siegal, F .P., Stjernswärd, J., Terry, W.D. and Wybran, J. Scand. J. Immunol. 3:521, 1974.
6. Raff, M.C., Nature (Lond.), 224:378, 1969.
7. Aiuti, F. and Wigzell, H., Clin. Exp. Immunol., 13:171, 1973.
8. Toben, H.R. and Cooper, M.D., Clin. Res. (abstract), 20:797, 1972.
9. Möller, G., J. Exp. Med., 114:415, 1961.
10. Raff, M.C., Immunology, 19:637, 1970.
11. Unanue, E.R., Grey, H.M., Rabellino, E ., Campbell, P. and Schmidtke, J.R., J. Exp. Med. 133:1188, 1971.
12. Bianco, C ., Patrick, R. and Nussenzweig, V ., J. Exp. Med., 132:702, 1970.
13. Basten, A ., Miller, J.F.A.P., Sprent, J. and Pye, J., J. Exp. Med. 135:610, 1972.
14. Dickler, H.B. and Kunkel, H.G., J. Exp. Med., 136:191, 1972.
15. Haegert, D .G., Hallberg, T . and Coombs, R.R.A., Int. Arch. Allergy 46:525, 1974.
16. Möller, G., J. Exp. Med. 139:969, 1974.
17. Raff, M.C., Nase, S. and Mitchison, N.A., Nature (Lond.), 230:50, 1971.
18. Greaves, M.F. and Brown, G ., Nature New Biol., 246:116, 1973.
19. Möller, G. (ed.), Transplant. Rev., 11, 1972.
20. Coutinho, A. and Möller, G., Adv. Immunol., 20:113, 1975.
21. Nowell, P.C., Cancer Res. 20:462, 1960.
22. Janossy, G. and Greaves, M.F., Clin. Exp. Immunol., 8:483, 1971.
23. Stobo, J.D., Transplant. Rev., 11:60, 1972.

24. Andersson, J., Sjöberg, O. and Möller, G., Transplant. Rev., 11:131, 1972.

25. Greaves, M.F. and Janossy, G., Transplant. Rev. 11:87, 1972.

26. Waxdal, M.J. and Basham, T.Y., Nature (Lond.), 251:163, 1974.

27. Sultzer, B.M. and Nilsson, B.S., Nature New Biol., 240:198, 1972.

28. Coutinho, A. and Möller, G., Nature New Biol., 140:12, 1973.

29. Diamantstein. T., Blitstein-Willinger, E. and Schulz, G., Nature (Lond.), 250:596, 1974.

30. Gronowicz, E., Coutinho, A. and Möller, G., Scand. J. Immunol., 3:413, 1974.

31. Berggård, I. and Bearn, A.G., J. Biol. Chem., 243:4095, 1968.

32. Evrin, P-E., Peterson, P.A., Wide, L. and Berggård, I., Scand. J. Clin. Lab. Invest. 28:439, 1971.

33. Smithies, O. and Poulik, M.D., Science 175:187, 1972.

34. Peterson, P.A., Cunningham, B.A., Berggård, I. and Edelman, G.M., Proc. Nat. Acad. Sci. (Wash.), 69:1697, 1972.

35. Evrin, P-E. and Pertoft, M.J., J. Immunol., 111:1147, 1973.

36. Hütteroth, T.H., Cleve, H., Litwin, S.D. and Poulik, M.D., J, Exp, Med. 138:1608, 1973.

37. Nilsson, K., Evrin, P-E., Berggård, I. and Pontén, J., Nature (Lond.), 244:44, 1973.

38. Cresswell, P., Turner, M.J. and Strominger, J.L., Proc. Nat. Acad. Sci. (Wash.), 70:1603, 1973.

39. Tanigaki, N., Katagiri, M., Nakamuro, K., Kreiter, V.P. and Pressman, D., Fed. Proc., 32:1017, 1973.

40. Grey, H.M., Kubo, R.T., Colon, S.M. and Poulik, M.D., J. Exp. Med., 138:1608, 1973.

41. Nakamuro, K., Tanigaki, N. and Pressman, D., Proc. Nat. Acad. Sci. (Wash.), 70:2863, 1973.

42. Peterson, P.A., Rask, L. and Lindblom, B., Proc. Nat. Acad. Sci. (Wash.), 71:35, 1974.

43. Rask, L., Lindblom, B. and Peterson, P.A., Nature (Lond.), 249:833, 1974.

44. Silver, J. and Hood, L., Nature (Lond.), 249:764, 1974.

45. Lindblom, B.,Östberg, L. and Peterson, P.A., Tissue Antigen 4:186, 1974.

46. Solheim, B .G. and Thorsby, E., Tissue Antigen, 4:83, 1974.

47. Poulik, M.D., Bernoco, M., Bernoco, D. and Ceppellini, R., Science 182:1352, 1973.

48. Nilsson, K., Evrin, P-E. and Welsh, K.I., Transplant. Rev., 21:53, 1974.

49. Sanderson, A.R. and Welsh, K.I., Transplantation 17:281, 1974.

50. Möller, G. (ed.), Transplant. Rev., 21, 1974.

51. Uhr, J., personal communication.

52. Goodfellow, P.N., Jones, E.A., van Heymingen, V., Solomon, E., Bobrow, M., Miggiano, V. and Bodmer, W .F., Nature (Lond.), 254:267, 1975.

53. Jongsma, A ., Someren, H., van Westerfeldt, A., Hagemeier, A. and Pearson, P., Humangenetik 20:195, 1973.

54. Östberg, L., Rask, L., Nilsson, K. and Peterson, P.A., Eur. J. Immunol., 5:462, 1975.

55. Möller, E. and Persson, U., Scand. J. Immunol., 3:445, 1974.

56. Ringdén, O. and Möller, E., Scand. J. Immunol., 4:171, 1975.

57. Ringdén, O., Scand. J, Immunol., In press, 1976.

58. Forman, J. and Möller, G ., Transplant. Rev., 17:108, 1973.

59. Gronowicz, E, and Coutinho, A., Transplant. Rev., 24:3, 1975.

60. Möller, E., Persson, U., Rask, L. and Ringdén, O., in The role of products of the major histocompatibility gene complex in immune responses. edited by D. Katz and B. Benacerraf, Acad. Press, N.Y. 1976.

61. Östberg, L., Lindblom, B. and Peterson, P.A., Eur. J. Immunol., 6:108, 1976.

62. Ringdén, O. and Johansson, B.G., unpublished observations.

63. Sela, B-A., Wang, I.L. and Edelman, G.M., J. Exp. Med., 143:665, 1976.

64. Möller, G., Immunology, 19:583, 1970.

65. Persson, U., personal communication.

66. Coutinho, A. and Gronowicz, E., J. E xp. Med., 141:753, 1975.

67. Coutinho, A ., personal communication.

68. Rask, L., Klareskog, L., Östberg, L. and Peterson, P.A., Nature (Lond.), 257:231, 1975.

Chapter 23

ASSOCIATIVE CONTROL OF THE IMMUNE RESPONSE
TO CELL SURFACE ANTIGENS

P. Lake and N.A. Mitchison
ICRF Tumour Immunology Unit,
Department of Zoology, University College London,
Gower Street, London, WC1E 6BT

The immune response to cell surface antigens is regulated by the activity of helper and suppressor cells, probably of T_H and possibly of T_{CS} types. For this to operate, recognition must take place of two or more antigenic determinants, which may be carried on the same (intramolecular help) or different (intermolecular, intrastructural help) molecules. Regulation of this type has been shown to operate in the response to the murine allo-antigens H-2K, H-2D, Thy-1, and H-minors.

Regulation of the immune response to cell surface antigens matters from

a practical point of view in two ways. There is a need to increase the acti-

vity of weak antigens in cancer, and to decrease the activity of self antigens

in auto-immune disease. In order to achieve these aims it is logical to try

to use helper and suppressor T cells (T_H and T_{CS} cells) (1). The functions

of these cells is best understood with hapten-protein conjugated antigens;

the analogy between this type of response and possible control of the response

to cell surface antigens is illustrated in Fig. 1. In the response to conju-

gates, the size of the B cell response to the hapten is influenced by the

activity of T_H and T_{CS} cells reacting to carrier determinants: T_H cells in-

crease the response (2,3) and T_{CS} cells diminish it (4). It ought therefore

to be possible to increase or diminish the response to cell surface antigens

by inserting appropriate control determinants into the neighbouring surface.

407

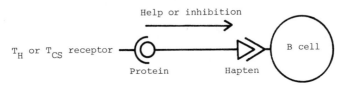

T_H or T_{CS} receptor

Protein   Hapten

B cell

Help or inhibition

Hapten-protein antigen

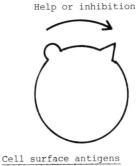

Help or inhibition

Cell surface antigens

FIGURE 1

The role of T cells in the response to haptens and cell surface antigens

Although much still remains unclear about the mode of action of these accessory T cells, one feature of the activity of T_H cells has been definitely established. In order for help to be delivered, the hapten and carrier determinants must be carried on the same physical structure, which is normally a single molecule (3). This implies that a bridge must be formed between T cell receptor and B cell, although it is still unclear whether in order for this to happen the T cell receptor must be shed as a 'factor' cytophilic for macrophages (5). If, therefore, in experiments with cell surface antigens it can be shown that an increase in the response depends on associating two determinants on the same surface, two conclusions can be drawn. One is that cell cooperation of the hapten-carrier type is probably operating. The other is that the two determinants are physically linked to an extent which enables

them to be co-processed in the immune system, even though they may be carried

on separate molecules. We term this type of help intermolecular, intra-

structural (6).

An important question which remains open is whether similar mechanisms

operate in the delivery of help, or suppression, to T cells themselves.

Although various T cell responses can certainly be helped by the activity of

T_H cells, it appears in most cases that a physical link between the helper

determinant and its target it not required (7). In a few instances, however,

the situation seems to resemble that of T-B help (8). The problem is that

the immunopathological effects of T cells are probably more important than

those of B cells, but help for B cells is easier to analyse. Perhaps we are

justified in concentrating for the time being on help for B cells: antibody

mediated effects may turn out to be quite important, and helper strategies

for B cells may apply also to T cells.

Various types of cell surface antigens could be employed for our pur-

pose. Previous studies have used viruses (9), haptens attached to the cell

surface (10), and antigens of different species combined in cell hybrids (11).

We concentrate on the use of allo-antigens, because, thanks to the develop-

ment of congenic (12) and inbred recombinant (13) mouse strains, it is now

possible to focus the immune response with exquisite specificity on a wide

range of well-defined antigens.

Experimental design and methods

We have used (i) primary responses, and (ii) secondary responses in

order to evaluate help and suppression. Primary responses are generated by

immunising mice with one or more injections of allogeneic cells, taken from

spleen or thymus. Secondary responses are generated either without transfer,

by priming with one type of cell and boosting with another, or by boosting

of cells adoptively transferred into irradiated recipients. Primary and
secondary responses can also be generated to Thy-1 in vitro (14). The responses are measured in terms either of antibody forming cells in the spleen
(14,15), or of cytolytic antibody by ^{51}Cr release.

Each design has its advantages and limitations. Primary immunisation
is more directly relevant to future practical applications. In order to
evaluate the contribution of help to a response, comparisons can be made between congenic strains or between congenic and normal inbred strains. Thus,
one can compare the response generated by Thy-1 in immunisation between A
and A.θAKR, two strains congenic for Thy-1, and between A and AKR which differ
for Thy-1, but are not congenic. This type of comparison yields information
about the net contribution to the response of helper activity directed at
all the antigens for which A and AKR differ. Clearly, experiments of this
type are limited in scope by the availability of suitable congenic strains.

Secondary responses permit more elaborate comparisons to be made. Thus,
for example, one can deprive spleen cells of T cells, and then reconstitute
with T cells primed against selected allo-antigens. In this way, it is possible to establish not only what types of help are necessary in order to
generate a response, but also what types are sufficient. In an example given
below in detail, a T_H response to H-minor antigens is sufficient but not
necessary in order to generate a secondary response to Thy-1. This greatly
broadens the scope of the analysis, and also permits analysis of the contribution of lymphocyte sub-populations.

Results

Figure 2 gives the result of a representative secondary response experiment. In this experiment AKR donors of B cells were immunised by eight
weekly injections of CBA thymocytes. Both strains are H-2^k, but AKR mice

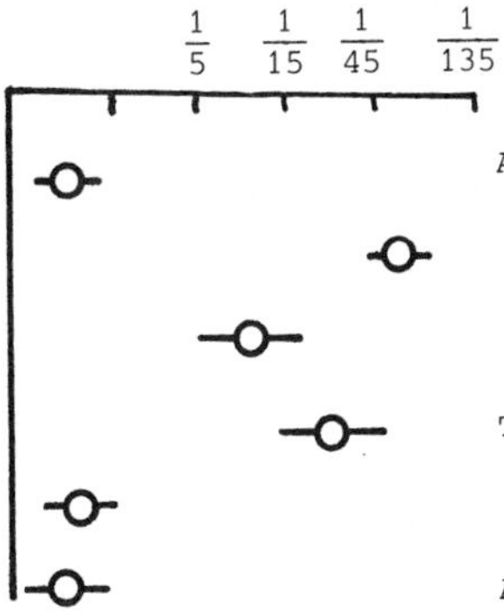

$\frac{1}{5}$ $\frac{1}{15}$ $\frac{1}{45}$ $\frac{1}{135}$

	B cells	T cells	Boosted
	Anti-Thy-1.2[1]	Unprimed	Thy-1.2 + H-minors[4]
	"	Anti-Thy-1.2[2]	"
	"	Anti-H-minors[3]	"
	T + B anti-Thy-1.2 unseparated[1]		"
		"	Nil
	Anti-Thy-1.2		Thy-1.2 + H-minors separately[5]

[1] AKR anti-CBA thymocytes x 8 (10^7 cells)
[2] AKR anti-CBA thymocytes x 2 (10^7 cells)
[3] AKR anti-AθAKR thymocytes x 2 (10^7 cells)
[4] CBAxAθAKR F_1 thymocytes (10^7 cells)
[5] CBA+AθAKR thymocytes (10^7 + 10^7 cells)

FIGURE 2

Intra- and intermolecular help for anti-Thy-1 response

are Thy-1.1 and CBA mice are Thy-1.2. The immunisation therefore generated
an anti-Thy-1.2 response. Spleen cell populations were trypsinised in order
to eliminate Thy-1.2 antigen retained on the cell surface, which would have
obscured the effects of the boost. B cells were isolated by treatment with
anti-Thy-1.1 plus complement. T cells were generated by similar but shorter
immunisations. They were not isolated, since controls not shown indicated
that these short-course immunisations generated negligible B cell activity.
The immunised cell populations were then mixed with the cells used as antigen
for boosting (which had been irradiated with 2000r in order to prevent any
response on their own part) and transferred intravenously into 600r irradiated AKR hosts. Sera were collected nine days post-transfer, and assayed by

[51]Cr release against CBA thymocyte targets. The results are expressed as
mean ± standard error titres, from groups of six hosts.

The data show that the response does not occur in the absence of the
boosting antigen. Isolated B cells do not respond. T cells primed against
either Thy-1.2 itself (i.e. providing intramolecular help) or H-minors (providing intermolecular help) reconstitute the response, but the intramolecular
help does so more effectively. Separation of the helper determinants from
the inducing determinants, by using a mixture of CBA plus A.θAKR cells instead
of F_1 cells, renders the antigen inactive. Other controls, not shown, indicate that unprimed T cells are much less effective in reconstituting the
response; they show also that the T cell activity is sensitive to treatment
with anti-Thy-1 plus complement.

Other results may be summarised as follows:

(1) In primary responses to H-2K and D, help from alloantigens outside the
major histocompatibility complex (MHC) make no detectable contribution. This
conclusion follows from the finding that responses in H-2 congenic strains are as
strong as those in unrelated inbred strains. The finding is not particularly
significant, because there is plenty of opportunity for intermolecular help
within the MHC, e.g. from I region antigens to K or D.

(2) One example is known of a requirement for intermolecular help in the
response to MHC antigens, namely in the IgG response of H-2[b] mice to D[d] (16).
This type of requirement is evidently limited, for the IgM response in the
same combination is independent of such help, and other immune response gene
phenotypes do not show the requirement (16).

(3) To judge from immunisations carried out with Thy-1 congenic A strain mice,
the primary response to Thy-1 can take place independently of intermolecular
help, but is increased by the provision of such help. Obviously, this conclusion needs qualification, to the extent that the production of the congenics

is complete. If, as is likely (17), the congenics still differ by additional H-minor loci linked to Thy-1, the requirement for intermolecular help may be absolute.

(4) In the secondary response no sign has yet been obtained of any benefit from intramolecular help in the response to K or D. This conclusion is qualified to the extent that these antigens are available only in a partially isolated form, through the use of the naturally occurring (A strain) or selected (B10.5R) recombinant strain (18). These recombinants are the best currently available for isolating the K or D end antigens, and no doubt may include other antigens; at any rate, the D antigen isolated in this way is every bit as effective in boosting as is the entire MHC.

A particularly interesting finding in this respect is that even in the $H-2^b$ phenotype, the anti-D^d secondary IgG response appears to proceed independently of help from outside the region isolated by the B10.5R recombination. This is contrary to the requirements evident in primary immunisation mentioned above. It prompts the speculation that responses which start dependent on outside help may be able to proceed independently: a matter of some significance in auto-immune disease.

(5) If, however, the secondary response to K and D is deprived of inter-MHC help, it can still be driven by help from H-minors outside. This is evident in experiments analogous to that illustrated in Fig. 2. Thus, for example, T cell help against minors generated either by CBA anti-B10.Br ($H-2^k$ congenics), or CBA anti-C3H, is able to drive the CBA anti-Balb (anti-$H-2^d$) B cell response. Here, then, we have an example of help which is sufficient but not necessary.

(6) The situation outlined above for K and D in the secondary response applies also to Thy-1.

(7) Thy-1 shows an interesting relationship to H-2 in the adoptive response: cells which carry Thy-1 plus an incompatible H-2 are less immunogenic than

cells which carry Thy-1 but are otherwise compatible. Thus, H-2 antigens
appear to suppress the response to Thy-1. They do so, apparently, through
a mechanism of intrastructural competition analogous to 'intramolecular
competition' as previously defined (19). A mechanism of this sort, in which
antigen is channelled off into a competing response is postulated for the
following reasons: (i) the inhibitory effect diminishes as the quantity of
antigen used for boosting is increased; (ii) the addition of an excess of cells
which carry the incompatible H-2 antigens but not Thy-1 diminishes the inhibition; and (iii) rendering of mice tolerant of H-2 antigens by standard
neonatal treatment interferes with the inhibition.

(8) Thy-1.1 is present not only on thymus cells of AKR mice, but also on those
of rats. Serologically the antigens from these two sources resemble one
another closely (20). Nevertheless, they do not cross-prime detectably. At
first sight, this finding contradicts the evidence, mentioned above, that the
secondary response to Thy-1 is driven under normal circumstances by intramolecular help. Once again, the explanation may be that these Thy-1 congenics retained differentiation antigens additional to Thy-1 itself, which
supplied intermolecular help and which rat cells lack. An alternative
explanation - in some ways a far fetched one - is that mouse and rat Thy-1
molecules differ by what is effectively a single determinant, which T_H
cells pre-empt and which therefore does not normally provoke an antibody response. The situation would then be analogous to that postulated for the mouse
polymorphic liver antigen F (21).

(9) Titration of helper activity has begun. Responses in the adoptive
secondary system are just sufficiently repeatable to make titration of T_H
cell populations worthwhile. This, in turn, makes it worthwhile to make
quantitative comparisons among different H-minor antigens, isolated either
in congenics or in back-crosses.

Discussion

The outcome of this reconnaissance into associative recognition of cell surface allo-antigens is in some ways disappointing. Examples of responses which depend on intermolecular help have been hard to find, and most responses thus appear to run optimally without outside aid. If this is always so, the chances of increasing a response by providing outside help are thin. However, the H-2 and Thy-1 systems, accessible as they are to experimental investigation on account of their strength, may for this very reason mislead us. Weaker antigens, which have so far attracted less attention, may well have greater need of outside help. It all depends on what makes weak antigens weak: to the extent that weakness results from lack of help, the approach remains valid.

On the other hand, this study has firmly established that distinct molecules on the cell surface are processed together in the immune system. This is not of course the first piece of information about the activity of T_H cells directed at cell surface antigens; their importance was first shown in the response to sheep erythrocytes (22). With an antigenic structure as complex as that of a foreign erythrocyte, however, and equally in the case of help directed at haptens located at random on the cell surface (23), the role of inter- versus intramolecular help would be hard to evaluate.

Because the molecules which are evidently co-processed in the immune system are also either known to presumed to possess independent lateral mobility in the plane of the cell surface, it is unlikely that co-processing is a sign of a physical link between molecules. More probably, co-processing involves larger units, up to the size of an intact cell. Clearly studies with cell fragments, and with gluteraldehyde-fixed cells, may provide further information on this point.

In the next stage of this approach to manipulation of the response, two
things will be needed. One is an explanation for the selective basis of the
working of control determinants. Why do some determinants stimulate T_H cells
while others stimulate T_{CS} cells? The other is a set of rules for constructing
effective determinants of either type. Both matters have been discussed
recently (24). The hypothesis was advanced that helper determinants associate physically with I molecules on the cell surface, while suppressor determinants associate with D or K, thus providing targets suitable for dual recognition respectively by T_H and T_{CS} cells. As yet, there is no direct evidence
for or against this possibility. As regards structural rules, the list of
key determinants which generate suppression is growing: it includes defined
amino acid sequences in lysozyme (25) , galactosidase (24), and myelin
basic protein (26) , as well as defined substitutes in arsanylated
human gammaglobulin (27), and tyrosylated glutamic acid-alanine copoly-
mer (2 8), As yet the common features of these structures have not
been identified, but one may hope that these will be revealed as the list grows

References

1. Medawar, P.B. and Simpson, E., Nature, 258:106, 1975.

2. Rajewsky, K., Schirrmacher, V., Nase, S. and Jerne, N.K., J. Exp. Med.,
    129:1131, 1969.

3. Mitchison, N.A., Eur. J. Immunol., 1:18, 1971.

4. Basten, A., Miller, J.F.A.P. and Johnson, P., Transplant. Rev., 26
    130, 1975.

5. Feldmann, M. and Nossal, G.J.V., Transplant. Rev., 13:3, 1972.

6. Lake, P. and Mitchison, N.A., Cold Spring Harbor Symp. Quant. Biol.,
    (in press), 1976.

7. Schendel, D.J., Alter, B.J. and Bach, F.H., Transpl. Proc., 5:1651, 1973.

8. Feldmann, M., Kilburn, D.G. and Levy, J., Nature, 256:741, 1975.

9. Lindenman, J., Biochim. Biophys. Acta, 355:49, 1974.

10. Bauminger, S. and Yachnin, S., Brit. J. Cancer, 26:77, 1972.

11. Watkins, J.F. and Chan, L., Nature, 223:1018, 1969.

12. Bailey, D.W., Immunogenetics, 2:249, 1975.

13. Bailey, D.W., Transplantation, 11:325, 1971.

14. Lake, P., Nature, 262:297, 1976.

15. Fuji, H., Zaleski, M. and Milgrom, F., Transplant. Proc., 3:852, 1971.

16. Wernet, D. and Lilly, F., J. Exp. Med., 141:573, 1975.

17. Flaherty, L.A. and Bennett, D., Transplantation, 16:505, 1973.

18. Sachs, D., David, D.C. Schreffler, D.C., Nathanson, S.G. and McDevitt, H.O., Immunogenetics, 2:301, 1975.

19. Schechter, I., J. Exp. Med., 127:237, 1968.

20. Letarte-Muirhead, M., Barclay, A.N. and Williams, A.F., Biochem.J. 151:685, 1975.

21. Lane, D. and Silver, D., Eur. J. Immunol., 6:480, 1976.

22. Miller, J.F.A.P. and Mitchell, G.F., J. Exp. Med., 128:801, 1968.

23. Janeway, C.A., Sharrow, S.O. and Simpson, E., Nature, 253:544, 1975.

24. Turkin, D. and Sercarz, E., Immunology (in press), 1976.

25. Hill, S.W. and Sercarz, E.E., Eur. J. Immunol., 5:317, 1975.

26. Swanborg, R.H., J. Immunol., 114:191, 1975.

27. Bullock, W.W., Katz, D.H. and Benacerraf, B. J. Exp. Med., 142:275,1975.

28. Kapp, J.A., Pierce, C.W., and Benacerraf, B. J. Exp. Med., 142:50,1975.

Chapter 24

EFFECT ON SYNTHETIC CAPACITY OF ANTIBODIES DIRECTED
AGAINST RECEPTORS OF ALLOTYPE-BEARING RABBIT CELLS

Louise T. Adler
Division of Immunology
St. Jude Children's Research Hospital
Memphis, Tennessee 38101

Abstract

Rabbit immunoglobulin molecules possess antigenically distinct
allotypic markers. Immunoglobulin molecules possessing such markers
which are attached to the surfaces of lymphocyte precursors of antibody-
forming cells are believed to serve as antigen receptors. Interactions
between allotypic determinants and antibodies directed against them can
provide useful models for the study of immunoregulatory phenomena. Exposure of a fetal or neonatal rabbit to antibodies against one of its
immunoglobulin allotypes results in a persistent defect in expression
of that type (allotype suppression). Some evidence indicates that this
phenomenon involves specific active suppression by lymphoid cells. By
examining the effects of anti-allotype antibodies on allotype-bearing
lymphocytes in vitro, certain specific events surrounding the processes
of lymphocyte ontogeny, and the differentiation and maturation of
antibody-forming cells are being analyzed.

The observation that there are genetically determined antigenic

differences among serum immunoglobulins of individual rabbits was first

made by Oudin in 1956 (1). Since that time the identity and in some cases

the location of a number of such antigenic determinants on the immunoglo-

bulin (Ig) molecule have been established (2). At the present time

rabbit allotypes have been identified on light chains of the kappa and

lambda classes, as well as on heavy chains of gamma, mu, and alpha

classes. Antibodies against a given allotype are generally obtained by

immunization of rabbits which do not synthesize Ig molecules with that

determinant. Allotypes encoded by the b locus (designated b4, b5, b6 and b9 and located on kappa chains) were among the earliest described and have probably been used more extensively than other allotypes as markers in immunological studies. Most of the work presented in this review deals with b locus allotypes.

Allotype Suppression in Vivo

A striking and far-reaching effect of antibodies against allotypic markers was first noted by Dray in 1962 (3). His study involved the immunization of female rabbits homozygous for the b4 determinant against another of the b locus allelic products, b5. These does were subsequently mated with homozygous b5 males. The ensuing offspring, genotypically b4b5, were, however, unable to make Ig of the paternal type, presumably as a result of their exposure to maternal antibodies against this gene product during intrauterine life. No adverse effects on the general well-being of the progeny were noted as a result of this onslaught by maternal antibodies. Total Ig levels were maintained at the normal level through a compensatory overproduction of the maternal type. One of the most fascinating aspects of Dray's observation was that this condition of specific "allotype suppression" persisted in some degree for at least a year, even though maternal antibody was not detectable in the serum of the young rabbits after the age of 8 weeks. Subsequently, it was shown that the same effect could be brought about by the injection of newborns with antibodies against one of their serum allotypes (4). Suppression involving other allotypic markers was soon demonstrated (5), and suppression of homozygotes also proved to be possible when experimental procedures were devised to circumvent the presence of passively derived maternal Ig in the circulation of homozygous neonates (6-8). The subject of allotype suppression in rabbits has recently been thoroughly reviewed by Dray (9) and Mage (10-12).

In Vitro Models for Allotype Suppression

Attempts to reveal the nature of the mechanism which maintains
suppression have occupied the center of interest in a number of laboratories
for many years. At the heart of the matter, of course, is the question of
the nature of the interaction between the anti-allotypic antibodies which
brought about suppression and their target in the fetal or neonatal rabbit.
One kind of approach to this problem has been to examine the effects of
such antibodies on rabbit lymphoid cells bearing allotypic markers in
various in vitro settings. These interactions have also been studied in
a more general context by those interested in the ontogeny of antibody-
forming cells or in fundamental principles of immune regulation rather
than in the phenomenon of allotype suppression per se. This section will
deal primarily with the results obtained by these groups of investigators.

Schuffler and Dray set out to develop an in vitro model in order to
analyze the mechanism of allotype suppression in the intact animal (13).
They exposed cultured spleen cells from normal b4b5 rabbits to anti-b4 or
anti-b5 antibody for the first 24 hr of culture, then washed the cells,
and determined the concentrations of b4 and b5 Ig formed by the cells after
a further 4 days of culture. Production of Ig bearing each marker was
determined by inhibition of precipitation in an indirect radioimmunoassay
system. Specific allotype suppression was observed, in that anti-b4 inhibited b4 production with no effect on b5 Ig levels, while the sole effect
of anti-b5 in these experiments was to suppress production of b5 Ig. In
analogous fashion antibodies against "a" locus determinants, located on
the Fd fragment of heavy chains were shown to inhibit specifically production of the corresponding allotypes only (14).

Inhibition was also observed when $F(ab')_2$ or Fab fragments of anti-b4
IgG were tested against b4b5 spleen cells (15). Attempts to induce allotype
suppression in vivo using $F(ab')_2$ fragments of anti-allotype antibody had

previously been unsuccessful (16), but this can be attributed to the
rapid elimination of such fragments in the intact animal (17). It has
been observed repeatedly that anti-allotype antibodies are not directly
cytotoxic for target cells marked with the corresponding allotypes (18-20).

We have devised an in vitro model for allotype suppression studies
which differs from that of Schuffler and Dray in that results are expressed
in terms of antibody responses by cultured spleen cells rather than total
Ig levels (21). Spleen cells of unprimed animals are stimulated with an
antigen obtained by solubilization of T2 bacteriophage, and the T2-neu-
tralizing antibodies formed can be both identified and titrated by means
of an amplification assay which uses specific anti-allotype sera (22, 23).
We feel justified in using the anti-T2 response as a valid indicator of
the total Ig production of a given allotype, since we have consistently
observed that the distribution of allotypes in antibodies formed by cells
of a heterozygous spleen donor roughly parallels the distribution of the
allotypes in the serum of that animal (23, 24). This finding reflects the
probable multi-determinant nature of our antigen, a subject discussed by Gell
(25), and contrasts strikingly with the pronounced tendency of certain
antigens to stimulate selectively cells of a given allotype (26, 27).
Using this system we have confirmed the basic finding of Schuffler and
Dray regarding the specificity of allotype suppression in vitro. That is,
anti-b5 serum prevented the formation of anti-T2 with the b5 marker by
spleen cells from b4b5 rabbits, but did not suppress production of b4
anti-T2. Anti-b4 exerted an entirely analogous specific effect on b4
anti-T2 formation. Additionally, an intriguing observation was made with
the cells of a few b4b5 rabbits which had unusually skewed ratios of b4 to
b5 in their serum. As noted earlier by Mage (28), Chou et al. (29-31), and
others, a natural "pecking order" exists in the relative serum concentrations of the two allelic types expressed by a heterozygous rabbit; for

example, b4 > b5 and b6 > b9. This suggests the occurrence of a naturally

occurring regulatory phenomenon, for which artificially induced allotype

suppression (in vivo) may serve as an exaggerated model. In our hands,

when spleen cells of rabbits with unusually high serum ratios of b4 to b5

were treated with anti-b4 serum, a highly significant elevation of the b5

anti-T2 response occurred (21). This seemed to be analogous to the compensa-

tory production of the alternate allotype seen in allotype suppression in

vivo, and contrasts with the absence of a compensatory effect seen in

Schuffler and Dray's experiments.

Probable Cellular Events During Allotype Suppression

Although it is clear from the above data that repression of allotype

synthesis can be readily induced in vitro, details of the inhibition mechanism

on a molecular and cellular level are not well understood. A number of

investigators have demonstrated the presence of allotype-bearing molecules

(presumably Ig or Ig-like receptors) on the surface of rabbit lymphoid cells

(32-34). The initiating event in allotype suppression could then be assumed

to be the attachment of anti-allotype antibody to the allotypic determinant

of a membrane-bound Ig receptor. The sequence of events following binding

of labeled anti-Ig to mouse spleen cells has been studied in some detail

(35-37). Uniform binding with some patchy areas of labeling was soon

followed by the aggregation of most or all of the label at one pole of the

cell ("capping"). Cap formation, which occurred at 37°C but not at 0°C,

was rapidly followed by pinocytosis of the antibody-Ig complex. This

reaction left the cell temporarily depleted of surface receptors. A

probably related phenomenon is that of "antigenic modulation." As observed

by Old and his colleagues (38, 39), this consists of the disappearance of

thymus-leukemia (TL) antigen from the surfaces of mouse lymphoid cells after

exposure to anti-TL antiserum.

Linthicum and Sell (40, 41) used an indirect immunoferritin labeling technique to study the interaction between lymphoid cells from homozygous b4 rabbits with anti-b4. Peripheral blood lymphocytes bound the antibody in sparse patches at $0°C$, and warming the cells to $37°C$ caused rapid endocytosis without cap formation. The authors suggested that the absence of cap formation might be attributable to the observed sparse distribution of allotype-bearing receptors on the cell membrane of these cells, a condition which would make cross-linkage into one large aggregate unlikely. A large percentage of splenic lymphocytes exhibited surface labeling properties similar to those of peripheral blood cells.

The disappearance of surface receptors after their union with anti-Ig reagents, followed by the generation of new receptors, accords well with the observed in vitro effects of anti-allotype serum treatment. It is clear from studying the effects of anti-allotype antibodies on cells that the single event of binding of antibodies to allotype-bearing Ig molecules (receptors) does not result in permanently blocking those activities of the cell in which receptors are involved. Prolonged exposure to anti-allotype antibodies seems to be required in order to cause complete shutdown of specific Ig or antibody synthesis (13, 21). It seems most likely from the microscopic observations cited that after binding of antibody and aggregation into patches or caps, followed by endocytosis or shedding of receptors, prompt regeneration of the receptors leads to a repetition of this entire process. In the continued presence of antibodies, an eventual exhaustion of the regenerative processes of the cell could very well occur, but such a breakdown may require a series of membrane regeneration cycles. Schuffler and Dray (13) found it necessary to culture b4b5 spleen cells with anti-b4 or anti-b5 antibody for 24 hours in order to achieve maximum inhibition of Ig synthesis. A 4 hour incubation period under their standard conditions for in vitro allotype suppression was only partially effective.

We have studied the kinetics of induction of allotype suppression in vitro
and have noted that exposure of homozygous b4 spleen cells to a potent
anti-b4 serum for as little as one hour may completely inhibit the capacity
of these cells to form neutralizing antibodies when they are subsequently
stimulated with solubilized T2 phage (21). With smaller doses or less
potent antisera, longer exposure times are required to effect complete
suppression.

The ultimate fate of cells subjected to an irreversible treatment
with specific anti-allotype serum is not known. As mentioned earlier in
this article, no direct evidence for a cytotoxic effect by anti-allotype
serum has been noted (18-20). Similar observations were made with regard
to antisera directed against mouse alloantigens by Pierce et al. (42), in
that prolonged exposure to such sera interfered with the responses of
mouse spleen cells to sheep erythrocytes in vitro, without any demonstrable cytotoxic action on the cells. However, it is possible that cells
which do not appear to be dead when the criterion of dye exclusion is used
could be metabolically inert due to the interference of anti-allotype antibodies with synthetic processes. A possible parallel may exist in the
induction of tolerance on a cellular level and the generation of the so-
called "tolerant cell." Considerable evidence indicates that exposure of
antigen-reactive cells to a high concentration of antigen leads to a
condition of immunological paralysis which is characterized by the inability
of these cells to regenerate receptors and the retention of antigen on the
membrane (43-46).

Entirely different criteria have been used by Sell and his colleagues
to assess the effects of anti-allotypic antibodies on allotype-bearing
rabbit lymphoid cells (47-52). They have repeatedly observed stimulation
of DNA synthesis and changes associated with typical "blast" cell morphology
as a result of such interaction. Anti-allotype stimulation of rabbit peri-

pheral blood lymphocytes was found to be completely reversible up to
36 or 42 hours of initiation by the addition of serum containing the
allotypic specificity toward which the antiserum was directed (53).
These observations may very well have bearing on the mechanism leading
to inhibition of Ig synthesis by anti-allotype serum. As postulated by
Gell (25) and others, stimulation of allotype-producing cells by anti-
allotype serum may cause inactivation of such cells through an irreversible and abortive blast transformation process.

Possible Differences Between Adult and Immature Lymphocytes

All of the in vitro studies described thus far have dealt with the
effects of anti-allotype serum on lymphoid cells of adult rabbits. However,
the in vivo induction of allotype suppression succeeds only during a critical
period of early life (54, 55). If one of the goals of the in vitro studies
under discussion is to shed some light on the mechanism of allotype
suppression in the intact animal, possible differences between adult and
fetal or neonatal cells need to be considered. It is well known that a
newborn rabbit possesses no detectable Ig of its own. Active Ig synthesis
begins on the first day of life (56), but for the first week or two this
is only a small fraction of the total concentration, most of which has been
obtained passively from the maternal circulation. Maternal Ig is rapidly
catabolized during the first month, and around that time allotypes encoded
by the young rabbit's genotype are easily detectable by immunological
methods (3, 57). Cells marked with these allotypes are present in low
numbers during the neonatal period and gradually reach adult levels during
the first 3 to 4 months of life (57, 58). The proliferative responses to
anti-allotype serum and to mitogens is present in lymphoid cells of newborns
but do not attain adult levels until about the age of one month (58). The
injection of anti-allotype serum into newborn heterozygotes causes a prompt

disappearance of cells marked with the corresponding allotype from the

lymphoid organs (57, 59). Furthermore, the lymphocytes from such allo-

typically suppressed rabbits exhibit a decreased ability to undergo

blast transformation when exposed to antiserum against the suppressed

type (50). It is tempting to speculate that allotype suppression may

be inducible in vivo only very early in life because of quantitative or

qualitative differences in the surface properties of adult as opposed to

immature cells. Recent evidence to suggest such differences comes from

an observation that spleen cells of infant mice regenerate surface Ig

poorly, if at all, after clearance by anti-Ig (60).

The Effects of Heterologous Anti-Allotypic Antibodies

Anti-allotype sera prepared in species other than the rabbit have

proved to be almost completely without effect in inducing allotype suppres-

sion in newborn rabbits. Lowe (18) reached this conclusion using a sheep

anti-b6 light chain serum, suitably absorbed to remove other anti-rabbit

specificities. This serum readily induced blast transformation in b6 lym-

phocytes in vitro, whereas rabbit antiserum specific for b6 light chains

induced both b6 suppression in vivo and blast transformation in vitro.

Goldman and Mage (54) achieved only marginal allotype suppression by

administering multiple injections of goat anti-b5 and anti-b4 sera to

newborn b4b5 rabbits. Such intensive treatment caused only slight tempor-

ary delays in attainment of the adult level of the affected allotype.

Purified goat anti-allotype antibodies were similarly ineffective.

Interestingly, when purified rabbit antibody fractions were tested for

their ability to induce allotype suppression in neonates, a total of 45 mg

of precipitable anti-b5 given over a period of several days was needed to

bring about complete b5 suppression. In contrast, the equivalent of less

than 5 mg of antibody achieved the same result when unfractionated serum

was used. The authors of this paper attributed their results to the low
stability of purified antibodies and to the more rapid degradation of
heterologous than homologous proteins in vivo. This conclusion is in
accord with Lowe's observation (18) that heterologous anti-allotype serum
does interact in a demonstrable way with allotype-bearing lymphocytes in
vitro, if not in vivo. Recently, Shek and Dubiski (61) have presented
evidence for a competitive binding effect on target lymphocytes between
heterologous antibodies or F(ab')$_2$ fragments of rabbit antibodies and
intact rabbit antibodies capable of inducing allotype suppression. This
strongly implies that the capacity to induce suppression in vivo does not
reside solely in ability to bind to target lymphocytes but that an
intact Fc portion of rabbit antibody is probably required.

Effects of Anti-Allotype Serum on Cells From Allotype-Suppressed Rabbits

A major part of our approach to the problem of allotype suppression
has centered on the reactivities of lymphoid cells from prenatally suppressed
rabbits in the culture system described in an earlier section. Ever since
allotype suppression was first described (3), workers in this area of
research have been struck by the long-lasting nature of the suppressive
effect in relation to the rapid elimination of the inducing anti-allotype
agent, an observation which made it difficult to accept continued target
cell elimination as the causative mechanism. The finding of other models
for allotype suppression in certain mouse strain combinations (62, 63), and
in particular the demonstration that specific suppression can be induced
in normal syngeneic recipients by the transfer of viable cells from
suppressed mice (64, 65) provided a firm base for the exciting concept of
an active suppression mechanism mediated by suppressor cells. The
scarcity of sufficient inbred rabbits has discouraged analogous direct
tests for suppressor cells. However, experimental observations made in

two laboratories independently provided justification for the idea of

an active suppression mechanism in rabbits as well. The first of these

was the finding by Harrison et al. (66), showing that during spontaneous

recovery from b5 suppression, lymphocytes marked with membrane b5 approach

normal numbers at the same time that serum Ig with allotype b5 is still at

a very low level. This observation led these investigators to postulate

the existence of a functional block in activation or maturation of b5-

bearing B lymphocytes in b5-suppressed rabbits which results in deficient

secretion of b5. A similar conclusion was reached by Adler and Fishman

(67), who noted that the pool of intracellular Ig in lymph node cells of

b4-suppressed b4b5 rabbits contained a much higher proportion of the

suppressed allotype than was indicated by the serum ratio of b4 to b5

concentrations. Further, their studies indicated that the block in

chronic suppression is located somewhere beyond synthesis of messenger

RNA relevant to the suppressed allotype. Other experiments done in our

laboratory, to be described below, have provided more direct evidence for

an active suppression mechanism in spleen cells of allotype-suppressed

rabbits.

We have routinely induced allotype suppression in rabbits which were

heterozygous at the b (kappa chain) locus by immunizing homozygous females

against the b locus allotype of their prospective homozygous mates. Most

of the progeny obtained by this method have been completely suppressed in

terms of the paternal type when first tested at weaning age (8 weeks),

using a test which detects as little as 1 µg of a given allotypic

specificity (23).

Table I lists the allotypic combinations we have utilized to breed

the rabbits for our studies. Data summarized in Table II show that, in

each instance, spleen cells taken from an allotype-suppressed rabbit

formed neutralizing antibodies which were marked with the maternal (non-

TABLE I

Breeding Schemes for Allotype Suppression

Dam	Sire	Progeny
b5, anti-b4	b4	(b4)b5
b6, anti-b4	b4	(b4)b6
b6, anti-b5	b5	(b5)b6
b5, anti-b6	b6	b5(b6)
b4, anti-b5	b5	b4(b5)
b4, anti-b6	b6	b4(b6)

The suppressed type of the offspring is shown in parentheses.

suppressed) allotype only, when they were stimulated with the soluble T2 antigen (S-T2) in culture. Cells of normal heterozygotes formed antibodies of both allotypic specificities encoded by their genotype.

In another series of experiments spleen cells of (b4)b5 (b4-suppressed b4b5) rabbits were exposed to anti-b5 serum (made in b4 rabbits) during the entire culture period of 4 to 5 days. As observed previously when normal b4b5 spleen were cultured in the presence of anti-b5, there was a marked inhibition of b5 anti-T2 production (21). Surprisingly, inhibition of the non-suppressed allotype (b5) was accompanied by the appearance of highly significant titers contributed by b4 anti-T2 (23). The distribution of allotypic specificities between b4 and b5 anti-T2 activities obtained under these conditions often resembled that formed by normal b4b5 spleen cells (Figure 1). These results provided direct evidence for the existence of an active suppression mechanism which continued to exert its effect in vitro on cells potentially capable of making the suppressed allotype, and which could seemingly be overcome by suppressing the activities of cells associated with production of the non-suppressed Ig type.

TABLE II

Allotype Distribution of T2 Neutralizing Antibodies Made in
Spleen Cell Cultures from Normal and Allotype-Suppressed Rabbits

Group	Allotype(s) of Spleen Donor	Reciprocal of 30% Neutralization Titer*		
		b4 anti-T2	b5 anti-T2	b6 anti-T2
Normal Homozygous	b4b4	64	<4	<4
	b5b5	<4	32	<4
	b6b6	<4	<4	32-64
Normal Heterozygous	b4b5	64	16-32	<4
	b4b6	16-32	<4	4-8
Allotype- Suppressed**	(b4)b5	<4	64	<4
	(b4)b6	<4	<4	128-256
	(b5)b6	<4	<4	16
	b5(b6)	<4	32-64	<4
	b4(b5)	64-128	<4	<4
	b4(b6)	32-64	<4	<4

*The highest dilution of culture fluid which neutralizes 30% of a test dose
of phage. The results shown are typical examples from a large number of
tissue culture experiments.

**The ratio of non-suppressed to suppressed Ig allotype in the serum of
these spleen donors at the time of sacrifice varied from 30:1 to >1000:1.

Closer examination of this phenomenon has revealed that two factors are
actually needed in order to bring about the reversal of suppression in
vitro, namely anti-allotype antibodies specific for the non-suppressed type
and also Ig marked with the suppressed type (68). The source of the second
factor could be either the serum containing antibodies against the non-
suppressed type (e.g., in the original experiments b4 suppression in

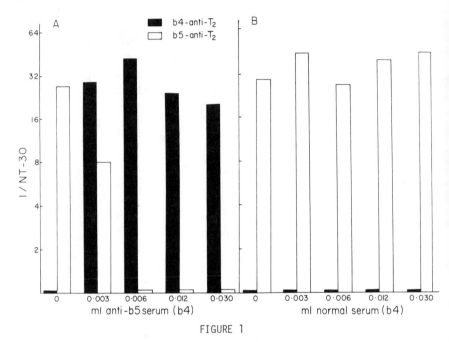

FIGURE 1

A. Effect of b4 anti-b5 serum on the response of (b4)b5 spleen cells to
S-T2. Antiserum remained in the cultures for the entire culture period
of 5 days.
B. Effect of normal b4 serum on the response of spleen cells from the
same donor.

Adler, L.T. (1974).

(b4)b5 cells had been overcome by b4 anti-b5 serum), or normal b4 serum could

be mixed with anti-b5 of another allotypic specificity (such as b6 anti-b5).

The principles learned in this first series of experiments were

applied to cells from rabbits of other allotypic combinations, suppressed

in their ability to synthesize b4, b5 or b6, and proved to operate in an

entirely analogous manner (24). Table III presents a list of the allo-

typic combinations of cells which have been tested to date and in which

suppression has been broken in vitro when antibodies against the non-

TABLE III

Release from Allotype Suppression In Vitro

Allotypic Specificity of Spleen Cells	Serum Components Needed to Release Suppression
(b4)b5	b4* and anti-b5
(b4)b6	b4* and anti-b6
(b5)b6	b5* and anti-b6
b5(b6)	b6* and anti-b5
b4(b5)	b5* and anti-b4

* This component can be present as a component of the anti-allotype serum or it may be added to antiserum of another allotypic specificity in the form of normal serum.

suppressed allotype and a source of Ig of the suppressed type (either the antiserum itself or normal serum) were applied to cultures.

Among alternate hypotheses which we have offered to account for these observations (24, 68), it is our impression that the following might most plausibly represent actual events occurring during the induction and maintenance of allotype suppression and its abrogation in vitro. It seems likely that the continued presence in the immature rabbit of antibodies against one of its allotypic determinants could completely inhibit expression of Ig marked with that determinant by cells potentially capable of synthesizing it. As the passively acquired antibody is catabolized, some cells should begin to express the suppressed type, but as they present themselves with membrane Ig of that type, they might be recognized as foreign cells by other lymphoid cells of the young animal, whose immunocompetence now would be steadily mounting. A specific cellular immune response might then be induced against cells marked with the suppressed type, presumably by cells of the non-suppressed type, and

these could be responsible, in some as yet unknown manner, for specifically
inhibiting synthesis or secretion of the suppressed Ig type. The recognition structure on the proposed "suppressor cell" might then be, in the case
of (b4)b5 cells as an example, a b5 anti-b4 entity. The demonstration
that precursors of b4-producing cells can be released from suppression
by inclusion of b4 anti-b5 serum in the culture medium could be explained
in terms of its supplying a twofold inhibition of anti-b4 suppression,
namely 1) the target cell product, b4, and 2) antibodies reactive with
the non-suppressed allotype, b5. This "double blow" might be needed if
the anti-b4 recognition structure were in fact a b5 anti-b4 entity on a
suppressor cell surface. A perhaps more trivial explanation for the anti-
b5 requirement would be that proliferation of b4-producing cells newly
released from suppression is favored by inhibiting b5 synthesis. Very
recently we have shown that normal Ig of the suppressed type alone is
sufficient to abrogate suppression in cultures of spleen cells from
rabbits in advanced states of recovery from chronic suppression (69).
This direct demonstration of the neutralizing effect of the target allotype upon the suppressive mechanism is consistent with our proposed scheme
to account for the perpetuation of allotype suppression.

Real support for the idea that auto-immunity against the suppressed
gene product is possible in allotype-suppressed rabbits has been provided
by the recent demonstration of Lowe et al. (70), showing that heterozygous
rabbits suppressed for b6 production could be made to raise a humoral
anti-b6 response when actively immunized against b6 Ig. This finding has
led Lowe and her colleagues (55) and also Sell (71) to postulate that
allotype suppression in vivo is dependent upon the secretion of small
amounts of antibody against the suppressed allotype by cells which express
the non-suppressed type. We have found no evidence for anti-b4 activity
in the globulin fraction of culture fluids made from cells of rabbits

completely suppressed for b4, even after 20-fold concentration. Similarly, highly concentrated extracts of disrupted spleen cells from such sources were totally devoid of anti-b4 activity (unpublished observations). There is little doubt of the part played by passively transferred serum antibodies in the induction of allotype suppression. However, the possible induction and participation of auto-immune humoral antibodies in the maintenance of chronic suppression is at the moment entirely speculative. A mechanism involving auto-induced cellular immunity deserves equal consideration.

Acknowledgements

The skilled secretarial assistance of Ms. Alva Moore is gratefully acknowledged. Support for those portions of the work cited which were conducted in our laboratory were supported by Grants No. AI 13159 and AI 13180 from the National Institutes of Health.

References

(1)   Oudin, J., C.R. Acad. Sci.,(Paris), 242: 2606, 1956.

(2)   Kindt, T.J., Adv. Immunol., 21: 35, 1975.

(3)   Dray, S., Nature, 195: 677, 1962.

(4)   Mage, R.G. and Dray, S., J. Immunol., 95: 525, 1965.

(5)   Mage, R.G., Young, G.O., and Dray, S., J. Immunol., 98: 502, 1967.

(6)   Dubiski, S., Nature, 214: 1365, 1967.

(7)   Vice, J.L., Hunt, W.L., and Dray, S., J. Immunol., 103:629, 1969.

(8)   Vice, J.L., Gilman-Sachs, A., Hunt, W.L., and Dray, S., J. Immunol., 104: 550, 1970.

(9)   Dray, S., in Ontogeny of Acquired Immunity, Ciba Foundation Symposium, p. 87, Associated Scientific Publishers, Amsterdam, 1972.

(10)   Mage, R.G., in Current Topics in Microbiology and Immunology, 63: 131, Springer-Verlag, Berlin, 1974.

(11) Mage, R.G., Fed. Proc., 34: 40, 1975.

(12) Mage, R.G., Transplant. Rev., 27: 84, 1975.

(13) Schuffler, C., and Dray, S., Cell. Immunol., 10: 267, 1974.

(14) Schuffler, C., and Dray, S., Cell. Immunol., 11: 377, 1974.

(15) Schuffler, C., and Dray, S., Cell. Immunol., 11: 367, 1974.

(16) Dubiski, S., and Swierczynska, Z., Int. Arch. Allergy, 40: 1, 1971.

(17) Spiegelberg, H., and Weigle, W.O., J. Exp. Med., 121: 323, 1965.

(18) Lowe, J.A., Immunology, 23: 591, 1972.

(19) Mond, J.J., Luzzati, A.L., and Thorbecke, G.J., J. Immunol., 108:
      566, 1972.

(20) Harrison, M.R., and Mage, R.G., J. Exp. Med., 138: 764, 1973.

(21) Adler, L.T., Fishman, M., and Adler, F.L., J. Immunol.,114: 1275,
      1975.

(22) Adler, F.L., Walker, W.S., and Fishman, M., Virology, 46: 797, 1971.

(23) Adler, L.T., J. Immunol., 113: 1107,1974.

(24) Adler, L.T., Transplant. Rev., 27: 3, 1975.

(25) Gell, P.G.H., in Regulation of the Antibody Response, edited by B.
      Cinader, p. 204, Charles C. Thomas, Springfield, Ill., 1968.

(26) Zimmerman, S., and Haurowitz, F., Immunochem., 11: 403, 1974.

(27) Kindt, T.J., J. Immunol., 112: 601, 1974.

(28) Mage, R.G., Cold Spring Harbor Symp. Quant. Biol., 32: 203, 1967.

(29) Chou, C.-T., Cinader, B., and Dubiski, S., Int. Arch. Allergy, 32:
      583, 1967.

(30) Chou., C.-T., Cinader, B., and Dubiski, S., Eur. J. Immunol., 2:
      391, 1972.

(31) Chou, C.-T., Cinader, B., and Dubiski, S., Cell. Immunol., 11: 304,
      1974.

(32) Coombs, R.R.A., Gurner, B.W., Janeway, C.A., Jr., Wilson, A.B.,
      Gell, P.G.H., and Kelus, A.S., Immunology, 18: 417, 1970.

(33) Pernis, B., Forni, L., and Amante, L., J. Exp. Med., 132: 1001, 1970.

(34) Davie, J.M., Paul, W.E., Mage, R.G., and Goldman, M.B., Proc. Nat. Acad. Sci. (Wash.), 68: 430, 1971.

(35) Taylor, R.B., Duffus, W.P.H., Raff, M.C., and de Petris, S., Nature, 233: 1971.

(36) Unanue, E.R., Perkins, W.D., and Karnovsky, M.J., J. Immunol., 108: 569, 1972.

(37) Loor, F., Forni, F., and Pernis, B., Eur. J. Immunol., 2: 203, 1972.

(38) Old, L.J., Stockert, E., Boyse, E.A., and Kim, J.H., J. Exp. Med., 127, 523, 1968.

(39) Stackpole, C.W., Jacobson, J.B., and Lordis, M.P., J. Exp. Med., 140: 939, 1974.

(40) Linthicum, D.S., and Sell, S., Cell. Immunol., 12: 443, 1974.

(41) Linthicum, D.S., and Sell, S., Cell. Immunol., 12: 459, 1974.

(42) Pierce, C.W., Kapp, J.A., Solliday, S.M., Dorf, M.E., and Benacerraf, B., J. Exp. Med., 140: 921, 1974.

(43) Aldo-Benson, M., and Borel, Y., J. Immunol., 112: 1793, 1974.

(44) Diener, E., and Paetkau, V.H., Proc. Nat. Acad. Sci. (Wash.), 69: 2364, 1972.

(45) Ault, K.A., Unanue, E.R., Katz, D.H., and Benacerraf, B., Proc. Nat. Acad. Sci. (Wash.), 71: 3111, 1974.

(46) Schaefer, A.E., Adler, L.T., and Fishman, M., J. Immunol., 114: 1281, 1975.

(47) Sell, S., and Gell, P.G.H., J. Exp. Med., 122: 423, 1965.

(48) Sell, S., and Gell, P.G.H., J. Exp. Med., 122: 923, 1965.

(49) Sell, S., J. Exp. Med., 127: 1139, 1968.

(50) Sell, S., J. Exp. Med., 128: 341, 1968.

(51) Sell, S., Lowe, J.A., and Gell, P.G.H., J. Immunol., 104: 103, 1970.

(52) Sell, S., Lowe, J.A., and Gell, P.G.H., J. Immunol., 104: 114, 1970.

(53) Sell, S., Lowe, J.A., and Gell, P.G.H., J. Immunol., 108: 674, 1972.

(54) Goldman, M.B., and Mage, R.G., Immunochemistry, 9:, 513, 1972.

(55) Catty, D., Lowe, J.A., and Gell, P.G.H., Transplant. Rev., 27: 157, 1975.

(56) Wainer, A., Robbins, J., Bellanti, J., Eitzman, D., and Smith, R.T., Nature, 198: 487, 1963.

(57) Harrison, M.R., and Mage, R.G., J. Exp. Med., 138: 764, 1973.

(58) Elfenbein, G.J., Harrison, M.R., and Mage, R.G., Cell. Immunol., 15: 303, 1975.

(59) Catty, D., Chambers, L., and Lowe, J.A., Immunology, 26: 331, 1974.

(60) Sidman, C.L., Fed. Proc., 34: 4604 (Abstr.), 1975.

(61) Shek, P.-N., and Dubiski, S., J. Immunol., 114: 621, 1975.

(62) Jacobson, E.B., and Herzenberg, L.A., J. Exp. Med., 135: 1151, 1972.

(63) Bosma, M.J., and Bosma, G.C., Nature, 259: 313, 1976.

(64) Jacobson, E.B., Herzenberg, L.A., Riblet, R., and Herzenberg, L.A., J. Exp. Med., 135: 1163, 1972.

(65) Herzenberg, L.A., Chan, E.L., Ravitch, M.M., Riblet, R.J., and Herzenberg, L.A., J. Exp. Med., 137: 1311, 1973.

(66) Harrison, M.R., Elfenbein, G.J., and Mage, R.G., Cell. Immunol., 11: 231, 1974.

(67) Adler, F.L., and Fishman, M., J. Immunol., 115: 129, 1975.

(68) Adler, L.T., and Adler, F.L., J. Exp. Med., 142: 332, 1975.

(69) Adler, L.T., and Adler, F.L., submitted for publication.

(70) Lowe, J.A., Cross, L.M., and Catty, D., Immunology, 28: 469, 1975.

(71) Sell, S., Transplant. Rev., 27: 135, 1975.

Chapter 25

IMMUNOLOGY OF THE ACETYLCHOLINE RECEPTOR

Vanda A. Lennon
Molecular Neuropathology Laboratory
The Salk Institute for Biological Studies
Post Office Box 1809
San Diego, California   92112

Abstract

Myasthenia gravis is a spontaneously occurring autoimmune disease in which antibodies and lymphocytes are specifically reactive with nicotinic ACh receptors of skeletal muscle. Antibodies reactive with junctional receptors of human muscle are found in 90% of patients with myasthenia gravis and not at all in other diseases. Their capacity to cross the placenta suggests their involvement in the pathogenesis of neonatal myasthenia. The role of the thymus in myasthenia gravis remains a mystery, but it has recently been established that the thymus contains nicotinic ACh receptors and that anti-receptor antibodies are present in myasthenic thymuses.

Antibodies of myasthenic patients detect only partial cross reactivity between ACh receptors of different species. However, greater antibody binding is observed with receptors isolated from denervated rat muscle than with receptors from normal rat muscle. This suggests that extrajunctional and junctional ACh receptors might express different antigenic determinants. Although human antibodies bind minimally to ACh receptors of the electric organs of eels and marine rays, lymphocyte reactivity to electric eel receptors is found in high incidence in myasthenic patients. This suggests that electric organ and mammalian muscle ACh receptors may share more lymphocyte-defined than serologically-defined antigenic determinants.

Both cellular and humoral immune responses to ACh receptors can be induced experimentally. Sufficient antigenic homology exists between receptors of different species that electric organ receptors are capable of inducing in mammals experimental autoimmune myasthenia gravis. Syngeneic muscle receptor also is immunogenic in rats. Induction of both myasthenia and antibodies to ACh receptor requires participation of thymus-derived lymphocytes. The majority of ACh receptors in myasthenic rat muscle exist complexed with antibody, but antibody is not bound directly to the receptor's ACh-binding site. Anti-receptor antibodies *in vitro* are capable of impairing the electrophysiological function of ACh receptors with minimal blocking of the ACh-binding site and in the absence of complement. Thus, myasthenia gravis and its experimental model provide unique biological tools for studying the structure, function and pathology of cell membrane receptors.

Introduction

Application of immunological techniques to analysis of the neuromuscular junction's structure and function is providing a new perspective
in the investigation of muscle physiology and pathology and cell-membrane
receptors in general. Identification, purification and analysis of the
nicotinic acetylcholine (ACh*) receptor molecule has been made possible
by the novel application of two exotic products of nature -- electric organs
of fishes, which provide a rich source of nicotinic ACh receptor, and α-
neurotoxins from the venom of elapid snakes, which bind tightly to the ACh-
binding site of the receptor molecule. Two practical dividends have arisen
from immunological studies of ACh receptors: the establishment of a valid
experimental animal model of the disease myasthenia gravis (1-5) and the
development of serological tests diagnostic for myasthenia gravis (6-8).

Experimentally induced myasthenia provides a unique model for studying
biological and therapeutic aspects of diseases mediated by autoimmunity to
cell-membrane receptors. Although myasthenia gravis is an uncommon disease,
it can now be regarded as a prototype immunopharmacological disease (9):
antibodies to ACh receptors of skeletal muscle are measurable in the serum
of 90% of patients with myasthenia gravis and are specific for that disease
(8), are capable of crossing the placenta to human offspring (8) and of
passively transferring disease across a species barrier when injected into
mice (10). This review will consider primarily immunological aspects of
nicotinic ACh receptors of skeletal muscle and electric organs of fishes.
Muscarinic ACh receptors and nicotinic ACh receptors of organs other than
muscle and thymus will not be discussed.

Chemistry, Structure and Pharmacology of Nicotinic ACh Receptors

Nicotinic ACh receptors are specialized glycoproteins which are integral
components of certain post-synaptic membranes. When the neurotransmitter
ACh binds with its receptor, the ionic conductance of the post-synaptic
membrane increases, resulting in depolarization of the adjacent membrane.
Different types of ACh receptors can be defined pharmacologically by their
specificities for agonists and antagonists and for permeating ions.

The small amounts of ACh receptors in mammalian tissues limit their
biochemical characterization using techniques available today. However,
the electric organs of the eel (*Electrophorus electricus*) and the marine
ray (*Torpedo*) contain large quantities of nicotinic ACh receptor in the

post-synaptic membranes of their electrogenic cells (electroplax) which
are homologous to neuromuscular junctions of skeletal muscle cells. By
mild detergent solubilization, ACh receptor is extractable from homogenized
electric organ in milligram quantities with preservation of structure, as
judged by its antigenicity and its capacity to bind ACh ligands. Isolated
receptor can be purified by affinity chromatography employing α-neurotoxins
coupled to agarose with elution by a competing ligand (11). Isolated
receptor protein is identified by its ability to specifically bind
cholinergic antagonists or agonists or affinity labeling reagents (12)
such as MBTA (4-[N-maleimido]benzyltrimethylammonium).

Purified ACh receptor is a hydrophobic complex of uniform size comprising several polypeptide subunits (12, and reviewed in 13). Its apparent
molecular weight (with associated detergent), as estimated by its sedimentation coefficient in a sucrose gradient (9.5S), is in the order of 250,000
daltons (13). Polyacrylamide gel electrophoresis of both native and
denatured eel receptor in non-dissociating conditions yields a single
band (14). *Torpedo* receptor appears to exist both as a monomer and a
dimer (15).

When reduced and saturated with SDS and heated 2 minutes at $100^{\circ}C$,
purified eel ACh receptor electrophoreses as two components of apparent
molecular weights 54,000 and 40,000 (11). The 40,000 dalton subunit bears
the ACh-binding site (12). Purified ACh receptor contains all the amino
acids of globular proteins and appears to bear a carbohydrate moiety
containing at least D-mannose and N-acetylgalactosamine (13). With negative
staining, purified eel receptor protein appears electron microscopically in
the form of rosettes of 8-9 nm diameter comprising 5-6 subunits (13). From
the appearance of freeze fracture images of *Torpedo* membranes (13) it has
been proposed that the ACh receptor protein is an "integral" molecule which
spans the lipid bilayer.

A water soluble form of ACh receptor has been prepared from eel electric
organ without use of detergents by dialysis of membrane fractions against a
low ionic strength buffer followed by controlled tryptic digestion (16).
Partial purification was obtained by affinity chromatography. The water
soluble receptor differed physicochemically from receptor solubilized by
Triton detergent, the toxin-binding component having a sedimentation coefficient of 32-35S with an estimated molecular weight $\approx$ 2-3 x 10^{6}. The
water soluble material retained the specificity of native receptor for ACh

ligands and, as will be discussed later, shared antigenicity with receptors prepared in detergent. Receptor solubilized without detergents may resemble more closely receptor in its native form. Moreover, avoidance of detergents is desirable in examining the chemical, immunochemical and biological properties of the receptor molecule.

Myoid Cells and Nicotinic ACh Receptors in the Thymus

Because of the frequency of pathological lesions (medullary lymphoid hyperplasia with germinal centers, and occasional tumors) in thymuses of patients with myasthenia gravis (17), it has long been suspected that involvement of that organ in myasthenia gravis might be due to its sharing an autoantigen(s) with skeletal muscle. This concept was supported by the finding of an antibody in the serum of 30% of myasthenic patients which reacts equally intensely, by indirect immunofluorescence, with striations of both mammalian skeletal muscle and thymic myoid cells (18). Because of its parasympathetic innervation (17), the thymus might be expected to have muscarinic ACh receptors, and indeed, muscarinic receptors have been demonstrated on thymus-derived lymphocytes (19). Recent evidence indicates that nicotinic ACh receptors are present in the thymus. Firstly, binding of ^{125}I-α-bungarotoxin to extracts of rat thymuses was specific as judged by its inhibition with benzoquinonium (20). Secondly, the thymus' toxin-binding material was precipitable by rat antibodies with specificity for rat muscle ACh receptor (20). Thirdly, sera and lymphocytes from rabbits immunized with eel receptor detected shared antigenic determinants in aqueous extracts of calf thymus (21). The location of nicotinic ACh receptors in the thymus has not yet been ascertained, but one would predict their association with myoid cells.

Junctional and Extrajunctional ACh Receptors of Muscle

In skeletal muscle of adult vertebrates, most ACh receptors are confined to the sub-synaptic region of the neuromuscular junction. They are concentrated at the crests of the junctional folds (22) where their density, which is remarkably similar among different species, appears to be a characteristic feature of vertebrate endplates (23,24). Extrajunctional ACh receptors are identifiable electrophysiologically by iontophoretic microapplication of ACh or by binding of labeled α-neurotoxins. They are found in muscle of fetuses and newborns (25) and in myotubes formed in tissue culture from developing muscle (26) or cloned muscle cell lines (27). In adult muscle, large numbers of extrajunctional receptors are found only after sectioning the innervating motor nerve (28) and in pathological states in which motor nerve degeneration occurs, e.g., amyotrophic lateral sclerosis (29). The appearance of extrajunctional receptors, which occurs in rats 2-3 days after denervation, is blocked by inhibitors of protein and RNA synthesis (30). That *de novo* synthesis of extrajunctional receptors occurs following denervation has been proven directly by their incorporation of a radioactive amino acid precursor (31). Reinnervation of denervated muscle results in disappearance of extrajunctional receptors and reestablishment of the normal adult distribution of receptors.

Extrajunctional and junctional ACh receptors exhibit similar general physiological properties, both mediating an increase in sodium and potassium conductance in response to ACh. When purified from normal and denervated rat diaphragms, toxin complexes of the two receptors were indistinguishable by gel filtration and by zone sedimentation (32). However, several quantitative differences between junctional and extrajunctional ACh receptors have been reported. These include differences in turnover rate (33), differences in sensitivity to ACh (34), differences in sensitivity to cholinergic antagonists (32,35-38) and differences in isoelectric point (32). Antigenic differences also have been described (39,40). The latter will be discussed in a later section of this review.

Experimental Autoimmune Myasthenia Gravis

An early observation in the course of raising antibodies for immunochemical characterization of purified ACh receptors was that rabbits immunized repeatedly with receptor in adjuvant developed profound muscular weakness (1,2,4). This observation led to development of an animal model of the disease myasthenia gravis (3). Experimental autoimmune myasthenia gravis (EAMG), an organ-specific autoimmune disease of skeletal muscle, is readily induced in mammals by a single intradermal inoculation of microgram quantities of purified ACh receptor with adjuvants. EAMG resembles spontaneously occurring human myasthenia gravis by every criterion examined so far. The symptoms and signs of both disorders are caused by a defect of neuromuscular transmission. Clinical observation reveals muscle weakness with easy fatigability and temporary improvement by rest and by anticholinesterase drugs. The weakness in EAMG (Figure 1), as in myasthenia gravis of man, predominantly affects muscles of the head and neck, upper

FIGURE 1

     Postures characteristic of EAMG in the Lewis rat (upper right and
lower). The upper left rat was immunized only with adjuvants. Severe
muscle weakness in EAMG predominantly affects the head and neck, fore-
limbs (note flexed digits) and respiratory muscles. The dark rings
around the eyes of the myasthenic rat are porphyrin stained tears
associated with endemic chronic respiratory disease which is exacer-
bated in EAMG.

limbs and respiration (3). Electrophysiologically, a decrementing response

of muscle occurs during repetitive motor nerve stimulation at low rates

(2-10/sec) and rapid repetitive stimulation induces post-activation facil-

itation followed by exhaustion (3,41) -- phenomena which occur in myasthenia

gravis. The electrical defect is repaired by anti-cholinesterases and sensitivity to curare is increased (41). The amplitude of miniature endplate potentials (mepp's) is reduced, but the number of transmitter quanta released by a nerve impulse are normal (42). Binding sites for ^{125}I-α-bungarotoxin are greatly reduced in numbers in neuromuscular junctions of myasthenic patients (43) and animals with EAMG (44,45). Ultrastructurally (Figure 2C), motor endplates of rats with chronic EAMG (46,47) are indistinguishable from those of patients with myasthenia gravis (48), the post-synaptic membrane being primarily involved. Immunologically, antibodies and cellular immunity to ACh receptors are demonstrable in both conditions. These data will be considered in detail in following sections of this review.

Rats develop two episodes of EAMG in response to a single inoculation with eel ACh receptor with complete Freund's adjuvant and *B. pertussis* vaccine (3). An acute phase occurs about eight days post-inoculation, is transient, and is characterized histologically (Figure 2B) by massive infiltration of the motor endplate region with mononuclear inflammatory cells which destroy the post-synaptic membrane (46,47). A human counterpart of the acute phase of EAMG has not been recognized as an entity, but the occurrence of "lymphorrhages" and the occasional association of polymyositis with myasthenia gravis (49) may be evidence of a similar cellular contribution to the pathogenesis of myasthenia gravis in man. After day 12, the rats recover and inflammatory cells disappear, but around day 30 a chronic phase of weakness ensues which usually progresses to death over several weeks. In the chronic phase (Figure 2C) the post-synaptic membrane is highly simplified with loss of its intricate folding and widening of the synaptic gap (46,47). This picture is identical with the appearance of human neuromuscular junctions in established myasthenia gravis.

Immunogenicity, Myasthenogenicity and Antigenic
Determinants of ACh Receptors

Isolated ACh receptors of eel and *Torpedo* are potently immunogenic in a variety of mammals when injected intradermally with adjuvants. Inoculation of rabbits (1,2,4,44), rats (3), guinea pigs (3,50), monkeys (5), and goats (51) leads to development of EAMG and antibodies to receptors. Cellular immune responses to receptor also have been demonstrated in rats (52), guinea pigs (50) and monkeys (5). In Lewis rats, isolated syngeneic muscle ACh receptors, with adjuvants, have proven as immunogenic as eel receptor for inducing EAMG and autoantibodies to receptor (53). Rats (20) and rabbits (15)

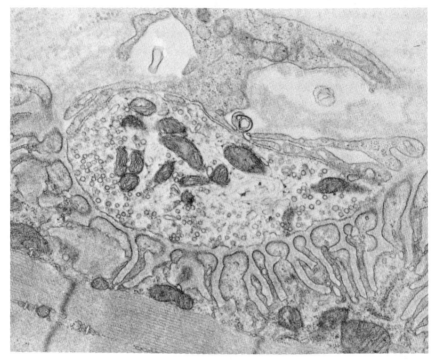

FIGURE 2

    A. Ultrastructure of a normal rat's neuromuscular junction (x 21,800).
Note the vessicles in the nerve terminal (presumed to contain ACh), the
narrow synaptic space and intricate folding of the post-synaptic membrane.
              (Courtesy of Dr. A.G. Engel, Ann. N.Y. Acad. Sci., in press.)

immunized, respectively, with eel and *Torpedo* receptors denatured in SDS
produced antibodies reactive with native ACh receptors, but no signs of
disease ensued. Rabbits (20) and rats (54) immunized repeatedly with
purified polypeptide chain subunits of eel ACh receptor likewise showed no
signs of EAMG. These observations suggest that the antigen(s) required for
EAMG induction has a complex structure. This contrasts with the immunogenic
requirements for induction of experimental autoimmune encephalomyelitis
which, for the guinea pig (55), rabbit (56) and rat (57), consist of a short
linear sequence of amino acids.

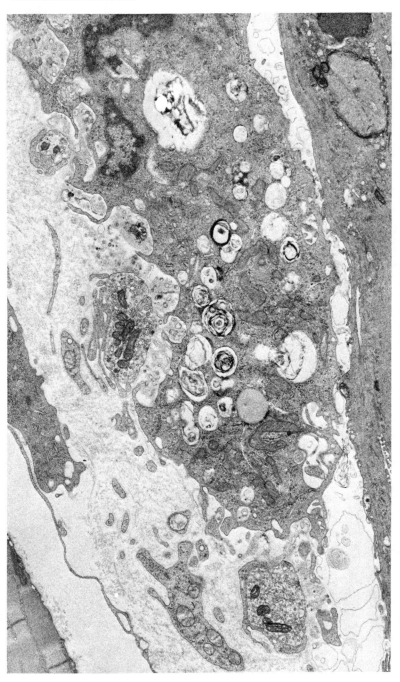

FIGURE 2B. Neuromuscular junction of a rat with acute EAMG (x 8,400). A macrophage containing phagocytotic vacuoles, separates the overlying intact nerve terminals from the underlying muscle fiber which has been denuded of post-synaptic membrane. (Courtesy of Dr. A. G. Engel, J. Neuropath. Exp. Neurol., in press.)

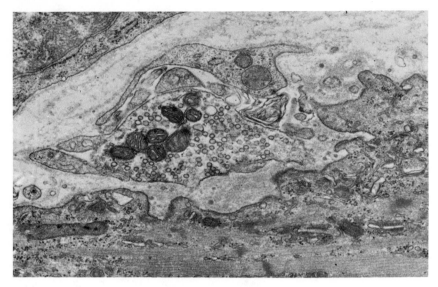

FIG. 2C. Neuromuscular junction of a rat with chronic EAMG (x 21,400).
Note widened synaptic space and paucity of post-synaptic folds.
            (Courtesy of Dr. A.G. Engel, Ann. N.Y. Acad. Sci., in press.)

Eel ACh receptor prepared without detergent induced EAMG in rabbits
(16).   Antibodies to both water-soluble and detergent-soluble receptors
bound aqueous receptor.   These data suggest that aqueous extracted receptor
exists in a conformation similar to that in the intact membrane.   When
tested by immunodiffusion with antiserum to crude aqueous receptor, affinity
purified aqueous receptor gave a single precipitin line with a reaction of
identity to that formed with a crude detergent-solubilized extract, indicating the sharing of a major antigenic determinant.

Experimentally induced antibodies to receptor form a single precipitin
line on double diffusion in agar with purified receptor (2,51,58,59).   Other
standard immunological methods have been employed to demonstrate antibodies

to ACh receptor: (a) precipitation of radiolabeled receptor, either directly
with immune serum (44) or indirectly after addition of an anti-antibody (60);
(b) complement fixation (7); (c) indirect hemagglutination (61); and (d)
indirect immunofluorescence (59). Anti-receptor antibodies have been demonstrated also by their ability to inhibit (a) binding of radiolabeled α-
neurotoxins to receptors in electroplax (51), and (b) *in vitro* electrophysiological responses of electroplax (2,20,60) and skeletal muscle (44,62)
to ACh and its analogues.

Antibodies to ACh receptors show partial species cross-reactivity.
For example, rabbit antiserum to eel receptor precipitated the toxin
binding components of eel membranes > *Torpedo* > chick embryo muscle (2);
rabbit antibodies against *Torpedo* receptor cross-reacted with ACh receptor
in frog neuromuscular junctions (44). The extent of serologically detectable cross-reactivity between receptors of different species appears to be
small. For example, the titer for *Torpedo* receptor of goat anti-eel antibodies was only 3% of the titer for eel receptor (51). Absorption of that
antiserum with purified *Torpedo* receptor removed most *Torpedo* reactivity
but not the reactivity with eel receptor (51). Rat antibodies to eel
receptor cross react with rat muscle ACh receptor (3,20,52), the titers
for isolated junctional receptor of rat muscle being about one one-hun-
dredth of the titer for eel receptor (20). In this instance absorption
with eel membranes efficiently removed the autoreactive antibody (20).
Titers of antibody to eel receptor in rat antisera raised against syngeneic
muscle ACh receptor were about one-fourth the titers against rat muscle
receptor (53).

Antibody binding sites on the ACh receptor molecule have not yet
been defined, but a foreign molecule the size of the eel receptor would be
expected to contain numerous antigenic determinants, many of which would
be species specific. Brockes and Hall (32) reported that rabbit antiserum
to eel receptor bound equally well to junctional and extrajunctional rat
ACh receptors complexed with ^{125}I-toxin. This implies that junctional and
extrajunctional receptors are very similar antigenically. However, as will
be discussed later, human myasthenic antibodies appear able to distinguish
antigenic differences between junctional and extrajunctional receptors (39,40).

Published reports concerning the existence of antibodies directed to the ACh-binding site of receptor molecules are confusing. This confusion has arisen in part because of employment by different investigators of different preparations and biological sources of ACh receptors and of antibodies from different species, including man. Substrates used to test the capacity of anti-receptor antibodies to inhibit binding of ^{125}I-α-neurotoxins include eel electroplax, purified electric organ receptors, crude membrane fractions of muscle and putative junctional and extra-junctional receptors prepared respectively from normal and denervated muscle. Studies with experimentally induced anti-receptor antibodies suggest that a serologically-defined antigenic determinant(s) of electric organ receptors is situated close to, but probably does not include, the ACh-binding site. Patrick *et al.* (60) reported that binding of rabbit antibodies to eel receptor which was ^{3}H-acetylated, leaving the toxin-binding site accessible, was inhibited to a small extent in the presence of unlabeled cholinergic ligands. Valderrama *et al.* (15) showed further that rabbit anti-*Torpedo* antisera bound more receptor which was alkylated with [^{3}H]MBTA (a small molecule [372 daltons] with affinity for the ACh-binding region) than receptor labeled with ^{3}H-toxin (7,800 daltons). Those authors reported also that rabbit anti-eel antibodies cross-reacting with *Torpedo* receptor were directed predominantly to the 40,000 dalton subunit which contains the ACh-binding site. Anti-eel antibodies cross-reacting with *Torpedo* receptor can be absorbed by *Torpedo* membranes and eluted with 3M KSCN (51). Lindstrom reported that cross-reacting goat anti-eel antibodies purified in this way were able to bind to ^{3}H-acetylated eel receptor in the presence of unlabeled toxin (51). This indicates that antibody cross-reacting with eel and *Torpedo* receptors was not directed to the toxin-binding site. Evidence from studies with eel electroplax combining electrophysiology and labeled-toxin-binding, suggests that antibody can impair the function of the ACh receptor molecule without blocking its ACh-binding site. Thus, Lindstrom *et al.* (63) reported that binding of ^{125}I-toxin to isolated eel electroplax was inhibited by only 10-15% in the presence of goat antibody concentrations which decreased by 70% the electrophysiological responses of the electrogenic cell (see below). Studies of the capacity of anti-receptor antibodies to block toxin-binding sites of muscle have for the most part employed human myasthenic sera and will be discussed in a later section.

Bioelectrical Effects of Experimentally Induced
Antibodies to ACh Receptors

Sugiyama *et al.* (2) reported that application of rabbit anti-receptor serum to isolated eel electroplax reduced by 70% the electrophysiological response to bath-applied carbamylcholine. This effect was not observed with normal rabbit serum. Moreover, less than 10% of the initial response was recovered one hour after prolonged rinsing with physiological solutions. Increasing antiserum concentration and exposure for longer periods did not significantly increase the degree of electrophysiological block. Patrick *et al.* (60) demonstrated that gamma globulin of rabbit anti-eel serum was responsible for inhibiting carbamylcholine induced depolarization of the eel electroplax. A similar effect of immune goat gamma globulin was not abolished by heating at $56^{o}C$ for 30 min (20), implying that complement was not necessary for impairment of bioelectrical function by antibody. Green and coworkers (44) reported that application to isolated frog muscle of rabbit antiserum to eel receptor reduced the amplitude of mepp's. Within 30-60 minutes the mean mepp amplitude fell to 15% of the original and after 3 hours to 7% or less. Deep fibers were affected later, presumably due to diffusion delays, but eventually all endplates ceased to transmit impulses and reversal of the antibody-mediated block was very slow. Bevan *et al.* (62) found that sera of rats with EAMG, but not sera from rats immunized only with adjuvants, reduced the extrajunctional ACh sensitivity of denervated rat diaphragm to iontophoretically applied ACh. This provides further evidence that experimentally induced antibodies to ACh receptors, like human myasthenic antibodies, are directed to extrajunctional as well as junctional receptors. Impairment of mammalian ACh receptor function by antibody was not affected by heating the antisera at $56^{o}C$ for 30 minutes.

Immunoglobulin Classes of Experimentally Induced
Antibodies to ACh Receptors

Antibodies to both eel and muscle receptors were detected in serum of rats as early as three days following a single inoculation of eel receptor (53). By the tenth day (52) more than 50% of the antibody to eel receptor sedimented as 7S (presumably IgG), the rest as 19S (presumably IgM). After day 20 the titers of antibodies reacting with muscle receptors rose abruptly, and by day 25 antibodies to both eel and muscle receptors sedimented exclusively as 7S molecules. The IgG class of this antibody was substantiated by its chromatographic properties on DEAE cellulose (54) and by its precipitation with rabbit antibody to rat IgG (3,52,53). Rabbit antibodies to eel receptor also were reported to be IgG (59).

Interaction of Antibodies with ACh Receptors *In Situ*

Zurn and Fulpius (64) showed in mice that the post-synaptic membranes of intact neuromuscular junctions were readily accessible *in vivo* to circulating molecules the size of IgG and IgM. Lindstrom *et al.* found that the bulk of ACh receptor extracted quantitatively (by Triton solubilization) from rats with EAMG did not sediment on sucrose gradients at 9.5S like muscle receptor from normal rats (20), but sedimented at 18-20S, corresponding to a molecular weight of 600-800,000 daltons. Furthermore, receptor isolated from rats with EAMG was precipitable directly by rabbit anti-rat immunoglobulin. Addition of exogenous rat anti-receptor antibody during extraction of receptor from normal rat muscle did not yield antibody-receptor complexes. These data indicate that ACh receptors (250,000 daltons) in muscles of rats with EAMG exist complexed with a few molecules of antibody (150,000 daltons). The fact that extracted antibody-receptor complexes were still capable of binding ^{125}I-α-bungarotoxin, is further evidence that anti-receptor antibody does not bind directly to the ACh-binding site. Receptors extracted from muscle of rats immunized with eel receptor were complexed with antibody as early as six days post-inoculation (53). Thus, binding of antibody to receptor *in situ* is detectable before the appearance of inflammatory cells in the motor endplate region in the acute phase of EAMG (46,47).

Experimentally Induced Cellular Immunity to ACh Receptors

Rats (52), guinea pigs (50) and monkeys (5) develop cellular immunity to receptor after immunization with eel receptor. Lennon *et al.* (52) reported that in rats inoculated once with eel ACh receptor, cell-mediated immunity to ACh receptor appeared before the clinical, electrophysiological and histological events of acute EAMG. Delayed cutaneous reactivity (a classic property of thymus-derived lymphocytes [T-cells] 65), was demonstrated with eel receptor 4 days after immunization. As EAMG progressed into its chronic phase, rats lost their specific delayed skin reactivity to receptor and became anergic also to purified protein derivative of tuberculin (a mycobacterial component of the adjuvant used in the induction of EAMG). Because of the complexity of factors which can influence cellular

immune responses *in vivo* in the later stage of EAMG, detailed analysis of
cell-mediated immunity to receptor will require investigation *in vitro*.
Abramsky *et al.* (5) reported the successful application of lymph node
lymphocyte transformation for *in vitro* demonstration of the cellular
immunity of immunized monkeys to eel receptor.

Clinical and electrophysiological signs of EAMG were transferred to
normal rats (52) and guinea pigs (50) using viable cells prepared from
lymph nodes of syngeneic animals immunized with receptor. In neither
instance was antibody to receptor demonstrated in sera of recipients,
but delayed skin reactivity to receptor was demonstrated in recipient
guinea pigs (50). It is nevertheless possible that the bone marrow-
derived (B-) lymphocytes present in the transferred population of lymph
node cells (approximately 25%) produced sufficient autoantibody in
recipient animals to impair neuromuscular transmission.

Rats depleted of T-cells (by a combination of thymectomy, X-irradiation
and reconstitution with spleen cells treated with anti-thymocyte serum and
complement) did not develop EAMG after appropriate challenge with eel ACh
receptor (52). Moreover, in the absence of T-cells, increased curare sensitivity did not ensue, and no detectable antibody to ACh receptor was produced. The presence of an intact thymus or transfusion of thymocytes or
splenic T-cells was necessary for both production of antibody to receptor
and for immunopharmacological impairment of motor endplate function (52).
The acute phase of EAMG was inhibited when normal rats were treated, early
after immunization with eel receptor, with rabbit antiserum to rat thymocytes
(52). However, the chronic phase occurred at the expected time, and anti-
receptor antibody was demonstrable in the serum in titers comparable to
those of rats treated with normal rabbit serum. These data suggest that
T-cells mediating a cellular immune response to ACh receptor might be
involved in the pathogenesis of the acute phase of EAMG. Thymectomy after
the onset of chronic EAMG did not alter the course of the disease (52).
Thus it appears that T-cells with specificity for ACh receptor provide help
for B-cells in their production of antibody to ACh receptor and also give
rise to delayed-type hypersensitivity reactions. Cellular immune responses
to receptor may be involved in part in amplifying destruction of the post-
synaptic membrane by macrophages in the acute phase of EAMG (46,47).

Antibodies to ACh Receptors in Patients
with Myasthenia Gravis

Several investigators have reported detection of anti-ACh receptor antibodies in sera of patients with myasthenia gravis. Antibody cross reacting with ACh receptor of *Torpedo* electric organ was demonstrated in a complement fixation assay in 9 of 15 myasthenic sera (7). However, with immunoprecipitation (20) and indirect hemagglutination (61) assays, negligible amounts of antibody to eel (20) and *Torpedo* (61) receptors were detected. These different results might reflect methodologic differences in preparing receptors, or performance or sensitivity of the assays. However, if this observation has an immunologic basis, it suggests that the antigenic determinant(s) of electric organ receptor which cross-reacts serologically with mammalian muscle ACh receptor might be near the ACh-binding site (which is occupied by the ^{125}I-labeled toxin molecule in the radioimmunoassay).

Four distinct reactivities with ACh receptor have been ascribed to myasthenic antibodies -- one blocking the α-neurotoxin binding site of receptors (6,29); another reacting with antigenic determinant(s), outside the toxin-binding site (8); a third reacting selectively with extrajunctional receptors (39); and a fourth reacting with a carbohydrate moiety of the receptor (66). However, these distinct reactivities may not reflect antibodies with specificities for four separate antigenic determinants.

Almon *et al.* (6) reported that 5 of 15 sera from patients with myasthenia gravis caused up to 48% inhibition of binding of ^{125}I-labeled α-bungarotoxin to ACh receptors prepared from rat muscle denervated 10 days earlier. The serum inhibitory factor was shown to be IgG and, unlike cholinergic ligands, antibody did not compete with toxin for cholinergic binding sites, but it reduced the number of sites available for binding without affecting binding affinity (39). Myasthenic sera did not significantly block binding of ^{125}I-toxin to ACh receptors prepared from normal human (20) or rat muscles (39), or *Torpedo* electric organ (61). Mittag *et al.* (66), employing assays with denervated rat muscle receptor as described by Almon and his colleagues (6,39), found that only 2 of 28 myasthenic sera inhibited binding of ^{125}I-toxin. However, an immuno-electron microscopic technique has demonstrated antibodies capable of blocking the toxin-binding sites of both human and rat neuromuscular

junctions (29), with a higher incidence of positivity in myasthenic sera
(75%) using denervated human muscle as substrate, than with neuromuscular
junction (44%). Sera from patients with other neurologic diseases did not
block toxin binding. Thus, extrajunctional receptors, preferably in the
membrane environment of denervated muscle, appear to be the required substrate for demonstrating the capacity of myasthenic antibody to block the
binding of toxin to ACh receptor. It is probable that blocking of toxin
binding is an allosteric effect of anti-receptor antibodies. Reproducible
demonstration of this antibody effect using isolated extrajunctional receptors
could depend on extraneous factors which might affect the antigenicity or
conformation of the receptor preparation such as the detergent and the conditions of storing isolated receptor.

Immunoprecipitation assays employing solubilized receptors labeled with
^{125}I-toxins have proven very sensitive for detecting antibodies directed to
determinants outside the ACh-binding site (8,39,66). Using human junctional
ACh receptor complexed with ^{125}I-α-bungarotoxin, Lindstrom *et al.* (8)
demonstrated antibodies in 87% of sera from 71 myasthenic patients.
Antibody was detected also in sera from two babies of a myasthenic
mother but was not detected in any of 175 sera from individuals
without myasthenia, including patients with other neurologic or autoimmune
diseases.

Human myasthenic sera cross react only slightly with junctional muscle
receptors of rat (20,39). However, antibodies reacting with extrajunctional
rat receptors have been detected in up to 85% of myasthenic sera (66,67).
Almon and Appel (40) reported that a change occurred in the antigenicity
of rat muscle receptors 3 days after denervation which was reflected as a
difference in their reaction with myasthenic IgG. Cross absorption studies
should establish whether a subset of myasthenic antibodies is directed to
antigenic determinant(s) expressed only on extrajunctional ACh receptors of
mammalian skeletal muscle.

Concanavalin A (Con A) binds to specific carbohydrate residues of
both junctional and extrajunctional ACh receptors of rat muscle, without
affecting their capacity for binding α-bungarotoxin (32). Mittag *et al.*
(66) reported that 10 of 15 myasthenic sera blocked Con A binding to
solubilized rat extrajunctional receptors. Some sera with high titers of
anti-receptor antibody (as assessed by immunoprecipitation of ^{125}I-toxin-

receptor complexes) did not appreciably block Con A binding, while others, with low anti-receptor titers, did block Con A binding. These data suggest that an antigenic determinant at or near a carbohydrate moiety of extrajunctional receptors might be recognized by a subpopulation of myasthenic antibodies. Alternatively, apparent blocking by antibody of "specific" sites on the receptor molecule could be determined largely by the conformation of the antigen, which itself might be influenced by bound antibody. That is, binding activities may not equate with specific antigenic determinants.

It is not yet known whether other anti-muscle antibodies defined serologically in myasthenic sera are directed to determinants of the ACh receptor. Anti-myoid antibody, which is found in approximately 30% of myasthenic sera, is higher in incidence in patients with detectable anti-ACh receptor antibodies (68,69). However, its presence in myasthenic sera did not correlate with anti-receptor titers (69) except that none was found in sera without demonstrable anti-receptor antibodies (68,69). Unlike myoid antibody, anti-receptor antibody was not found in patients with thymoma without myasthenia gravis (8,68). Absorption studies with human ACh receptor are needed to establish whether myoid antibody has specificity for ACh receptor.

Little is yet known of the biological properties of human antibodies to ACh receptor. Some are capable of crossing the human placenta (8) and are presumably responsible for the transient myasthenia which occasionally occurs in newborn infants of myasthenic mothers (17). Some are capable of activating the complement system on interacting with antigenic determinants of ACh receptor (17); serum complement levels have been reported to fluctuate with disease activity in patients with myasthenia gravis (70). Evidence for the capacity of human myasthenic antibody to impair neuromuscular transmission was reported by Toyka and coworkers (10). After repeated intraperitoneal injections of pooled myasthenic gamma globulin, 4 of 14 mice exhibited decrementing electromyograms on repetitive nerve stimulation. Improvement occurred in only one following administration of neostigmine, but all recipient mice displayed diminution of mepp amplitudes and reduction of toxin-binding sites at their neuromuscular junctions.

The stimulus and site of autoantibody production in patients with myasthenia gravis is a mystery. The presence in myasthenic thymuses of

increased numbers of B-lymphocytes (71) and germinal centers, the hallmark
of antibody production, suggests that the thymus might be a source of antibody. Myasthenic thymuses in culture synthesize more antibody than normal
thymuses (72), and antibody to ACh receptor has been extracted from
myasthenic thymuses (66). Drainage of the thoracic duct of patients with
myasthenia gravis was reported to alleviate signs of myasthenia before
blood lymphocyte numbers fell (73). Improvement was pronounced, but
temporary, and rapid return of the myasthenic state followed reinfusion
of cell-free lymph. These observations suggest that the lymph was rich
in antibodies to ACh receptor. If this were so, peripheral lymph nodes
also might be synthesizing antibodies to ACh receptor.

In summary, the occurrence of antibodies to ACh receptor in sera of
patients with myasthenia gravis provides sensitive and specific tests for
diagnosing that disease. Serial evaluation of a patient's titer of anti-
receptor antibody might prove valuable in monitoring the response to
immunotherapy. Mammalian muscle ACh receptor molecules appear to have
several antigenic determinants. Subtle differences in the specificities
of anti-receptor antibodies, e.g., for junctional and extrajunctional
receptors or for carbohydrate residues, may provide valuable molecular
probes for analyzing the antigenicity, structure and function of the ACh
receptor molecule.

Cellular Immunity to ACh Receptors in Patients
With Myasthenia Gravis

Peripheral blood lymphocytes from patients with myasthenia gravis have
been reported to exhibit *in vitro* sensitivity to both muscle antigens (74)
and autologous thymus cells (71,75). Abramsky *et al.* (76,77) demonstrated
that blood lymphocytes from 100% of non-immunosuppressed myasthenic patients
were specifically stimulated when cultured with eel ACh receptor. Lymphocytes from patients with other neurologic diseases did not react. Moreover,
when myasthenic patients were treated with corticosteroids, clinical improvement was accompanied by decreased *in vitro* responsiveness of blood lymphocytes to ACh receptor (77). Richman *et al.* (75) reported in myasthenic
patients a significant correlation between the severity of clinical disease
and the degree of *in vitro* responsiveness of blood lymphocytes to eel
receptor. The high incidence of myasthenic patients with lymphocyte

reactivity to eel receptors suggests that perhaps more lymphocyte-defined than serologically-defined antigenic determinants are shared by ACh receptors of electric organs and human muscle, since little cross-reactivity has been detected with myasthenic sera (20,61). It might be noted that in experimental autoimmune encephalomyelitis, separate antigenic determinants are responsible for induction of cellular immunity (and disease) and antibodies to myelin basic protein (78). The apparent association of lymphocyte responsiveness and disease activity suggests that cellular immunity to ACh receptor may be important in mediating or perpetuating impairment of neuromuscular transmission in myasthenia gravis.

Acknowledgements

I thank Drs. R. Almon, E. Lambert and J. Lesley for their helpful comments in the preparation of this manuscript. The author's laboratory is supported by grants from the National Institutes of Health (NS 11719-01) and the Los Angeles Chapter of the Myasthenia Gravis Foundation.

References

*Abbreviations used in this review: ACh, acetylcholine; B-cells, bone marrow-derived lymphocytes; Con A, concanavalin A; EAMG, experimental autoimmune myasthenia gravis; MBTA, 4-(N-maleimido)benzyltrimethylammonium; mepp's, miniature endplate potentials (small transient depolarizations of the post-synaptic membrane resulting from random spontaneous release of single quanta of ACh from a nerve ending); T-cell, thymus-derived lymphocyte.

1.   Patrick, J. and Lindstrom, J., Science, 180:871, 1973.
2.   Sugiyama, H., Benda, P., Meunier, J.-C. and Changeux, J.-P., FEBS
      Letters, 35:124, 1973.
3.   Lennon, V.A., Lindstrom, J.M. and Seybold, M.E., J. Exp. Med., 141:1365,
      1975.
4.   Heilbronn, E., Mattson, C.H., Stalberg, E. and Hilton-Brown, P.,
      J. Neurol. Sci., 24:59, 1975.
5.   Tarrab-Hazdai, R.A., Aharonov, I., Silman, S., Fuchs, S. and Abramsky,O.
      Nature (Lond.), 256:128, 1975.
6.   Almon, R.R., Andrew, C.G. and Appel, S.H., Science, 186:55, 1974.
7.   Aharonov, A., Abramsky, O., Tarrab-Hazdai, R. and Fuchs, S., Lancet,
      2:340, 1975.
8.   Lindstrom, J.M., Seybold, M.E., Lennon, V.A., Whittingham, S. and
      Duane, D.D., Neurology (Minneap.), in press.
9.   Lennon, V.A. and Carnegie, P.R., Lancet, 1:630, 1971.
10.   Toyka, K.V., Drachman, D.B., Pestronk, A. and Kao, I., Science, 190:397,
      1975.

11. Lindstrom, J. and Patrick, J., in Synaptic Transmission and Neuronal Interaction, M.V.L. Bennett, editor, p. 191, Raven Press, New York, 1974.

12. Karlin, A. and Cowburn, D.A., Proc. Nat. Acad. Sci. USA, 70:3636, 1973.

13. Changeux, J.-P., Benedetti, L., Bourgeois, J.-P., Brisson, A., Cartaud, J., Devaux, P., Grunhagen, H., Moreau, M., Popot, J.-L., Sobel, A. and Weber, M., in The Synapse, Cold Spring Harbor Symposia on Quantitative Biology, in press, 1976.

14. Klett, R.P., Fulpius, B.W., Cooper, D., Smith, M., Reich, E. and Possani, L.D., J. Biol. Chem., 248:6841, 1973.

15. Valderrama, R., Weill, C.L., McNamee, M.G. and Karlin, A., Ann. N.Y. Acad. Sci., in press, 1976.

16. Aharonov, A., Kalderon, N., Silman, I. and Fuchs, S., Immunochem., 12:765, 1975.

17. Goldstein, G. and Mackay, I.R., in The Human Thymus, W. Heinemann Ltd., London, 1969.

18. Strauss, A.J.L., van der Geld, H.W.R., Kemp, P.G., Exum, E.D. and Goodman, H.C., Ann. N.Y. Acad. Sci., 124:744, 1965.

19. Strom, T.B., Stytkowski, A.J., Carpenter, C.B. and Merrill, J.P., Proc. Nat. Acad. Sci. USA, 71:1330, 1974.

20. Lindstrom, J.M., Lennon, V.A., Seybold, M.E. and Whittingham, S., Ann. N.Y. Acad. Sci. USA, in press, 1976.

21. Aharonov, A., Tarrab-Hazdai, R., Abramsky, O. and Fuchs, S., Proc. Nat. Acad. Sci. USA, 72:1456, 1975.

22. Fertuck, H.C. and Salpeter, M.M., Proc. Nat. Acad. Sci. USA, 71:1376, 1974.

23. Albuquerque, E.X., Barnard, E.A., Porter, C.W. and Warnick, J.E., Proc. Nat. Acad. Sci. USA, 71:2818, 1974.

24. Porter, C.W. and Barnard, E.A., J. Memb. Biol., 20:31, 1975.

25. Diamond, J. and Miledi, R., J. Physiol., 162:393, 1962.

26. Fischbach, G.D. and Cohen, S., Dev. Biol., 31:147, 1973.

27. Steinbach, J.H., Harris, A.J., Patrick, J., Schubert, D. and Heinemann, S., J. Gen. Physiol., 62:255, 1973.

28. Axelsson, J. and Thesleff, S., J. Physiol. (Lond.), 149:178, 1959.

29. Bender, A.N., Ringel, S.P., Engel, W.K., Daniels, M.P. and Vogel, Z., Lancet, 1:607, 1975.

30. Fambrough, D.M., Science, 168:372, 1970.

31. Brockes, J.P. and Hall, Z.W., Proc. Nat. Acad. Sci. USA, 72:1368, 1975.

32. Brockes, J.P. and Hall, Z.W., Biochem., 14:2100, 1975.

33. Berg, D.K. and Hall, Z.W., J. Physiol. (Lond.), 252:771, 1975.

34. Feltz, A. and Mallart, A., J. Physiol. (Lond.), 218:101, 1971.

35. Beranek, R. and Vyskocil, F., J. Physiol. (Lond.), 188:53, 1967.

36. Lapa, A.J., Albuquerque, E.X. and Daly, J., Exp. Neurol., 43:375, 1974.

37. Chiu, T.H., Lapa, A.J., Barnard, E.A. and Albuquerque, E.X., Exp. Neurol., 43:399, 1974.

38. Almon, R.R., Andrew, C.G. and Appel. S.H., Biochem., 13:5522, 1974.

39. Almon, R.R. and Appel, S.H., Biochim. Biophys. Acta, 393:66, 1975.

40. Almon, R.R. and Appel, S.H., Biochem., in press, 1976.

41. Seybold, M.E., Lambert, E.H., Lennon, V.A. and Lindstrom, J., Ann. N.Y. Acad. Sci., in press, 1976.

42. Lambert, E.H., Lindstrom, J.M. and Lennon, V.A., Ann. N.Y. Acad. Sci., in press, 1976.

43. Fambrough, D.M., Drachman, D.B. and Satyamurti, S., Science, 182:293, 1973.

44. Green, D.P.L., Miledi, R. and Vincent, A., Proc. R. Soc. Lond., 189:57, 1975.

45. Engel, A.G., Lindstrom, J.M., Lambert, E.H. and Lennon, V.A.,
    Amer. Acad. Neurol. Abstracts, Neurology (Minneap.), 26:371, 1976.
46. Engel, A.G., Tsujihata, M., Lindstrom, J.M. and Lennon, V.A.,
    Ann. N.Y. Acad. Sci., in press, 1976.
47. Engel, A.G., Tsujihata, M., Lambert, E.H., Lindstrom, J.M. and
    Lennon, V.A., J. Neuropath. Exp. Neurol., in press, 1976.
48. Engel, A.G. and Santa, T., Ann. N.Y. Acad. Sci., 183:46, 1971.
49. Johns, T.R., Crowley, W.J., Miller, J.Q. and Campa, J.F.,
    Ann. N.Y. Acad. Sci., 183:64, 1971.
50. Tarrab-Hazdai, R., Aharonov, A., Abramsky, O., Yaar, I. and Fuchs, S.,
    J. Exp. Med., 142:785, 1975.
51. Lindstrom, J., J. Supramol. Struc., 4:389 (349), 1976.
52. Lennon, V.A., Lindstrom, J.M. and Seybold, M.E., Ann. N.Y. Acad. Sci.,
    in press, 1976.
53. Lindstrom, J., Einarson, B., Lennon, V.A. and Seybold, M.E.,
    submitted for publication.
54. Lindstrom, J., Lennon, V.A. and Seybold, M., unpublished data.
55. Westall, F.C., Robinson, A.B., Caccam, J., Jackson, J. and
    Eylar, E.H., Nature, 229:22, 1971.
56. Shapira, R., Chou, F.C.-H., McKneally, S., Urban, E. and Kibler, R.F.,
    Science, 173:736, 1971.
57. Westall, F.C., Thompson, M. and Lennon, V.A., Fed. Proc., 35:436, 1976.
58. Heilbronn, E. and Mattson, C., J. Neurochem., 22:315, 1974.
59. Penn, A.S., Chang, H.W., Lovelace, R.E., Niemi, W. and Miranda, A.,
    Ann. N.Y. Acad. Sci., in press, 1976.
60. Patrick, J., Lindstrom, J., Culp, B. and McMillan, J., Proc. Nat. Acad.
    Sci. USA, 70:3334, 1973.
61. Aarli, J.A., Mattson, C. and Heilbronn, E., Scand. J. Immunol. 5:849,
    1976.
62. Bevan, S., Heinemann, S., Lennon, V.A., and Lindstrom, J.,
    Nature (Lond.), 260:438, 1976.
63. Lindstrom, J., Francy, M. and Einarson, B., J. Supramol. Struc.,
    in press.
64. Zurn, A.D. and Fulpius, B.W., Clin. Exp. Immunol., in press, 1976.
65. Miller, J.F.A.P. and Osoba, D., Physiol. Rev., 47:437, 1967.
66. Mittag, T., Kornfeld, P., Tormay, A. and Woo, C., New Engl. J. Med.,
    294:691, 1976.
67. Appel, S.H., Almon, R.R. and Levy, N., New Engl. J. Med., 293:760, 1975
68. Ringel, S.P., Bender, A.N., Engel, W.K. and Smith, H.J., Lancet,
    1:1388, 1975.
69. Lennon, V.A., Whittingham, S. and Lindstrom, J., unpublished data.
70. Nastuk, W.L., Plescia, O.J. and Osserman, K.E., Proc. Soc. Exp. Biol.,
    105:177, 1960.
71. Abdou, N., Lisak, R.P., Zweiman, B., Abrahamson, I. and Penn, A.S.,
    New Engl. J. Med., 291:1271, 1974.
72. Smiley, J.D., Bradley, J., Daly, D. and Ziff, M., Clin. Exp. Immunol.,
    4:387, 1969.
73. Bergstrom, K., Franksson, C., Matell, G. and von Reis, G.,
    Europ. Neurol., 9:157, 1973.
74. Knott, E., Genkins, G. and Rule, A.H., Neurol. (Minneap.), 23:374, 1973
75. Richman, D.P., Patrick, J. and Arnason, B.G.W., New Engl. J. Med.,
    294:694, 1976.
76. Abramsky, O., Aharonov, A., Webb, C. and Fuchs, S., Clin. Exp. Immunol.
    19:11, 1975.
77. Abramsky, O., Aharonov, A., Teitelbaum, D. and Fuchs, S., Arch. Neurol.
    32:684, 1975.
78. Lennon, V.A., Wilks, A.V. and Carnegie, P.R., J. Immunol., 105:1223, 19

Chapter 26

IMMUNOLOGY OF THE THYROTROPIN RECEPTOR

B. Rees Smith
Departments of Medicine and Clinical Biochemistry,
Wellcome Research Laboratories, University of
Newcastle upon Tyne, Newcastle upon Tyne, NE1 4LP
England

Abstract

Antibodies to the thyrotropin receptor appear to be
responsible for hyperthyroidism in Graves' disease. The
antibodies, described as thyroid-stimulating antibodies (TSAb)
mimic the effects of thyrotropin (TSH) by binding to the TSH
receptor and activating adenylate cyclase. TSAb consist of an
electrophoretically heterogeneous population of IgG and the
thyroid-stimulating site is formed by combination of heavy and
light chains in the Fab part of the molecule. Binding studies
indicate that the TSAb molecule interacts monovalently with
membrane bound TSH receptors and that TSAb consists of an
antibody population which shows a restricted heterogeneity
with regard to TSH receptor affinity. Studies in patients
with Graves' disease and hyperthyroidism indicate that the
levels of TSAb correlate well with thyroidal iodine uptake and
the absence of pituitary control of thyroid function. However
in some patients with ophthalmic Graves' disease or autoimmune
thyroiditis there is evidence of serum antibodies which interact
with the TSH receptor but are unable to stimulate thyroid
function.

Introduction

There is now considerable evidence that the hyperthyroid-

ism associated with Graves' disease is due to antibodies to

the thyrotropin receptor. The antibodies, sometimes described

as thyroid-stimulating antibodies (TSAb) appear to mimic the

effects of thyrotropin (TSH) by binding to the TSH receptor

and activating adenylate cyclase. This paper reviews concis-

ely current knowledge concerning the role of TSAb and the TSH

receptor in Graves' disease.

461

Properties of TSAb

The presence of thyroid-stimulating antibodies in the serum of patients with Graves' disease was first described by Adams and Purves in 1956 (1). The antibody activity is associated with a heterogeneous population of IgG (2,3) and the thyroid-stimulating site is formed by combination of heavy and light chains in the Fab part of the molecule (Table 1 and Ref. 4).

The overall effect of TSAb on the thyroid is identical to TSH (6) but the time course of action of the antibody appears to be prolonged relative to TSH in vivo and in some in vitro systems (2,5,7,8).

TABLE 1

Thyroid-stimulating activity of TSAb subunits

Sample	Thyroid-stimulating activity measured by an in vivo bioassay (5) (% increase in blood ^{131}I + S.E.)
intact TSAb 1.0 mg/ml	1983 ± 478
reduced and alkylated TSAb 1.0 mg/ml	2041 ± 227
Heavy chain 6.0 mg/ml	477 ± 130
Light chain 2.0 mg/ml	146 ± 19
Recombined heavy and light chains 6.4 mg/ml	1707 ± 184
Fab fragment 1.5 mg/ml	1156 ± 171
Buffer only	124 ± 20

The effects of both TSH and TSAb are mimicked by adenosine 3':5' cyclic monophosphate (cyclic AMP) (9) and both TSAb and TSH stimulate cyclic AMP production in thyroid tissue slices (7) and thyroid membranes (10,11,12,13).

The first step in thyroid cell stimulation by TSAb appears therefore to be contact with an adenyl cyclase linked site on the cell surface. Leading from this we can postulate that interaction of this site with the immune system is responsible for initiation of TSAb synthesis.

The interaction of TSAb with thyroid tissue has been studied in considerable detail. The TSAb receptor sites appear to be localized in the thyroid cell membrane (6) and solubilization of the receptor activity has been effected by freezing and thawing (14,15). The characteristics of the TSAb receptor and its interaction with TSAb are summarized in Table II. The receptor appears to have a molecular weight of about 30,000 and isoelectric point of 4.5 (3). The complex formed between the solubilized receptor and TSAb has a molecular weight of about 200,000 and, in the presence of excess receptor, probably contains two molecules of receptor and one molecule of antibody (3).

TABLE II

The TSAb-soluble TSAb receptor interaction

	TSAb + receptor $\rightleftharpoons$		TSAb-receptor
isoelectric point	7-10	4.5	4-7
molecular weight	150,000	30,000	200,000

Subsequent to characterization of the TSAb receptor techniques have been developed to study the binding of ^{125}I-labelled TSH to TSH receptors in thyroid membranes (11,13,16, 17). With the TSH-TSH receptor system it was possible to demonstrate that TSAb inhibited the binding of labelled TSH to the TSH receptor (12,16,17,18) and this indicated that the two receptors were closely related. Furthermore, studies with soluble receptors indicate that TSAb and TSH bind to the same receptor molecule (Table III and Ref. 12) suggesting that TSAb is an antibody to the TSH receptor. The effects of TSAb on the binding of labelled TSH to thyroid membranes are localized in the Fab part of the TSAb molecule (Fig. 1), in good agreement with the localization of the thyroid stimulating activity (Table I).

TABLE III

The effect of soluble TSAb receptor on the interaction
between ^{125}I-labelled TSH and thyroid membranes[†]

Test material	Membrane bound ^{125}I-labelled TSH (%)	
	experiment 1	experiment 2
Bovine serum albumin	32.2	30.0
Purified soluble TSAb receptor	14.2	11.1
0.1 units of TSH	6.7	5.0

[†] ^{125}I-labelled TSH was incubated with Sephadex G-200 purified soluble TSAb receptor (3), bovine serum albumin solution or TSH. Thyroid membranes were than added and the amount of labelled TSH bound determined after centrifugation.

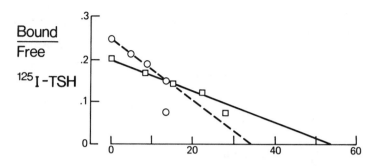

FIGURE 1

Plots of $\frac{bound}{free}$ labelled TSH vs amount of TSAb (□) and
TSAb-Fab (O) bound for the interaction between TSAb or
TSAb-Fab, ^{125}I-labelled TSH and human thyroid membranes. The
amount of TSAb bound was calculated by use of equation 4. The
proportion by weight of TSAb and TSAb-Fab in the total serum
was expressed as a constant (m).

The effects of TSAb on the binding of labelled TSH to

thyroid membranes provide a convenient way of studying the

TSAb-membrane interaction. If we consider the interaction of

TSAb (Ab) and TSH (H) with the same membrane receptor (R) to

be represented by the simple reactions:-

$Ab + R \rightleftharpoons AbR$ and $H + R \rightleftharpoons HR$ (equation 1)

with association constants K_{Ab} and K_H respectively then it is

possible to make a quantitative estimate of the amount of TSAb

bound to the membranes in terms of K_{Ab}, K_H, the amount of

labelled TSH bound and the amount of TSAb added (Ab_i). This

expression can be developed as follows:-

Application of the law of mass action to equation 1

gives:-

$$\frac{[AbR]}{[Ab]K_{Ab}} = \frac{[HR]}{[H]K_H} = [R] \qquad \text{(equation 2)}$$

and by rearrangement,

$$\frac{[AbR]}{[Ab]} = \frac{K_{Ab}}{K_H} \times \frac{[HR]}{[H]} \qquad\qquad \text{(equation 3)}$$

Substitution of $\quad [Ab] = [Ab_i] - [AbR] \quad$ and further rearrangement

gives:-

$$[AbR] = \frac{K_{Ab}}{K_H} \times \frac{[HR]}{[H]} \times \Big([Ab_i] - [AbR]\Big)$$

When small amounts of membranes are used it can be shown

experimentally that $[Ab_i]$ is much greater than $[AbR]$ and

consequently the equation reduces to:-

$$[AbR] = \frac{K_{Ab}}{K_H} \times \frac{[HR]}{[H]} \times [Ab_i] \qquad\qquad \text{(equation 4)}$$

Scatchard (19) rearrangement of equation 2 gives:-

$$\frac{[AbR]}{[Ab]} = K_{Ab} \Big([Ro] - [AbR]\Big)$$

where $[Ro] = $ total receptor concentration

Substitution for $\dfrac{[AbR]}{[Ab]}$ from equation 3 and $[AbR]$ from

equation 4 gives:-

$$\frac{K_{Ab}}{K_H} \times \frac{[HR]}{[H]} = K_{Ab} \left([Ro] - \frac{K_{Ab}}{K_H} \times \frac{[HR]}{[H]} \times [Ab_i] \right)$$

and by rearrangement:-

$$\frac{[HR]}{[H]} = K_{Ab} \left([Ro] \times \frac{K_H}{K_{Ab}} - \frac{[HR]}{[H]} \times [Ab_i] \right) \qquad \text{(equation 5)}$$

Consequently, a plot of $\dfrac{[HR]}{[H]}$ vs $\dfrac{[HR]}{[H]} \times [Ab_i]$

should be linear with slope $- K_{Ab}$ and intercept (p) on the

$\dfrac{[HR]}{[H]} \times [Ab_i]$ axis of $[Ro] \times \dfrac{K_H}{K_{Ab}}$

i.e. $\quad p = [Ro] \times \dfrac{K_H}{K_{Ab}} \qquad\qquad\qquad \text{(equation 6)}$

Plots of $\dfrac{[HR]}{[H]}$ vs $\dfrac{[HR]}{[H]} \times [Ab_i]$ for the effects of TSAb

and TSAb-Fab on the binding of labelled TSH to human thyroid

membranes are shown in Figure 1.   Both plots are approximately
linear as predicted in equation 5.

The linearity of the plots indicates that K_{Ab} (given by
the negative slope of the curve) is independant of [AbR].   This
suggests that TSAb contain a population of antibody molecules
which show a considerably restricted heterogeneity with regard
to receptor binding affinity.

Comparison of the slopes of the TSAb and TSAb-Fab plots
after adjustment for the difference in molecular weight
between IgG and Fab suggests that the association constant of
intact TSAb is 1.8 times that of the Fab fragment.   Similarly
comparison of the capacities of the membranes for TSAb by use
of equation 6 indicates that they have a similar capacity for
Fab as intact TSAb.   This suggests that TSAb interacts monovalently with the membranes.

Estimation of the absolute values of K_{Ab}, and Ro for
TSAb requires a knowledge of the proportion of TSAb-IgG in the
total serum IgG.   Methods for the accurate determination of
this value have not yet been described.

The binding of TSAb and TSAb-Fab to thyroid membranes
activates the membrane adenyl cyclase.   Comparison of the
amount of TSAb or TSAb-Fab bound (calculated from equation 4)
with the amount of cyclic AMP produced is shown in Figure 2.
The slope of the Fab and intact IgG plots are very similar
which suggests that the binding of one molecule of IgG to the
TSH receptor has the same effect as the binding of one molecule
of Fab.

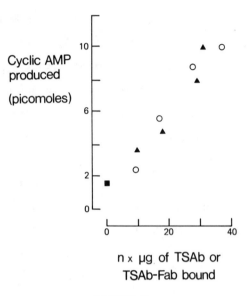

FIGURE 2

Comparison of receptor binding and cyclic AMP production
in human thyroid membranes. The amounts of TSAb (O) and
TSAb-Fab (▲) bound were calculated using equation 4. Conjoint
points are shown (■). The proportion of TSAb molecules in
the total serum IgG was expressed as a constant (n).

TSAb and thyroid function

The effect of TSAb on the binding of ^{125}I-labelled TSH

to human thyroid membranes has been used to develop a radio-

receptor assay for TSAb (16). With this assay Mukhtar et al

(10) were able to detect TSAb in 57 out of 57 patients with

untreated Graves' disease and hyperthyroidism. They were also

able to demonstrate a highly significant correlation between

the level of TSAb measured in the receptor assay and early

^{131}I uptakes. This data provided good evidence for a causative

role of TSAb in the hyperthyroidism of Graves' disease.

Thyroid function in health is controlled by TSH. The

serum level of TSH is principally regulated at the pituitary

level by the actions of triiodothyronine (T_3), possibly
thyroxine (T_4) and thyrotropin releasing hormone (TRH) (20).
Thyroid hormones reduce serum TSH levels and TRH increases
serum TSH levels. The effect of exogenous TRH on serum TSH
provides the basis of the TRH test of thyroid function.
Administration of T_3 results in a fall in serum TSH with subsequent reduction in thyroidal radioiodine uptake and this
effect provides the basis of the T_3 suppression test.

Thyroid-stimulators of non-pituitary origin such as TSAb
clearly cannot be controlled by the TRH-thyroid hormone-TSH
system. Recently, we have compared the levels of TSAb with
the results of T_3 suppression and TRH tests in a group of
patients who had been treated for Graves' disease and hyperthyroidism (21). This study found an excellent agreement
between the presence of TSAb and absence of TSH control of
thyroid function and provided further evidence of a causative
role for TSAb in the hyperthyroidism of Graves' disease.

The effect of treatment for Graves' disease on

serum TSAb

Virtually all patients with untreated Graves' disease
and hyperthyroidism have serum TSAb levels detectable by
receptor assay (10). Patients, treated by partial thyroidectomy, antithyroid drugs or radioiodine showed 17%, 53% and
50% detectable TSAb respectively. The values for the frequency
of detectable TSAb correspond with the frequency of remission
in similar groups of patients.

The reduction in detectable serum TSAb following treatment
by antithyroid drugs or radioiodine probably represents spontaneous remission but in the case of partial thyroidectomy the

operation itself has a dramatic effect on serum TSAb levels.
The mechanism of the effect of surgery is not understood at
present but clearly this observation implicates a major role
for the thyroid in the control of TSAb synthesis.

Thyroid-stimulating antibodies in other thyroid diseases

Ophthalmic Graves' disease is a condition characterized
by the eye signs of Graves' disease in the absence of hyperthyroidism. Recent studies in our laboratory (22) showed
that the serum of 23 out of 57 patients with ophthalmic Graves'
disease contained TSAb levels detectable in the receptor assay.
The existence of these thyroid-stimulating antibodies in
patients who show no symptoms of hyperthyroidism is difficult
to explain. Further, in this group of patients the presence
or absence of TSAb did not correlate well with the results of
T_3 suppression and TRH tests. This was in marked contrast to
our observations in a group of patients who had been treated
for hyperthyroid Graves' disease (21).

At least two factors can be introduced to rationalize the
apparently anomalous behaviour of TSAb in patients with
ophthalmic Graves' disease. Firstly, the results of the T_3
suppression and TRH tests indicated that in some patients with
high levels of TSAb thyroid function could be stimulated by
TSH but not by the patient's circulating TSAb (Table IV).

TABLE IV

TSAb and thyroid function in a patient with ophthalmic

Graves' disease

serum TSAb index (10)	6 hr ^{131}I uptake (%)		serum TSH (uU/ml)	
	before T_3	after T_3	before TRH	20 min after TRH
48	30	27	1	1

This suggested that the patient's serum contained receptor binding antibodies which were unable to stimulate the thyroid. Presumably the receptor-antibody interaction, because of inappropriate antibody affinity or specificity, was unable to induce the changes in membrane structure necessary for cell stimulation. Secondly, many patients with ophthalmic Graves' disease suffer from some degree of autoimmune destruction of the thyroid. Consequently even if TSAb was capable of stimulating the thyroid, destruction could have been so extensive that the thyroid was unable to produce elevated levels of thyroid hormone.

About 14% of patients with autoimmune thyroiditis have levels of serum TSAb detectable by receptor assay (10). The TSAb activity is quite distinct from the antithyroglobulin and antimicrosomal antibodies characteristic of these patients (16) but the presence of TSAb in patients with autoimmune thyroiditis emphasizes its close relationship to Graves' disease.

The reasons for the inability of TSAb to induce hyperthyroidism in patients with autoimmune thyroiditis are probably similar to those implicated in ophthalmic Graves' disease.

Preliminary experiments have indicated that the serum of 1 out of 6 patients with carcinoma of the thyroid contained TSAb activity detectable in the receptor assay (16). More recently, an extensive study has shown that 24 out of 84 patients with thyroid carcinoma contained detectable serum TSAb activity (23). There was no clear association between tumor type and the presence of TSAb. It is possible that the production of TSAb in these patients was the result of tumor induced changes in the thyroid or immune system. Alternatively

these patients may have Graves' disease quite independantly of
the carcinoma.

Phylogenetic specificity of TSAb

The partial phylogenetic specificity of TSAb was first
demonstrated by Adams and Kennedy (24), and subsequently confirmed in several laboratories (7,25,26). Present evidence
suggests that some thyroid-stimulating antibodies are capable
of stimulating the human thyroid but not the thyroids of
experimental animals. This is due, at least in part, to the
inability of human specific antibodies to recognize the animal
TSH receptor (16). However, the recent observations in
patients with ophthalmic Graves' disease and autoimmune
thyroiditis discussed in an earlier section of this review
suggest that variations in antibody affinity and more subtle
changes in specificity could also be involved.

Direct stimulation of the thyroid by lymphocytes

There has been a report of direct stimulation of isolated
bovine thyroid cells by peripheral blood lymphocytes from
patients with Graves' disease (27). This could possibly occur
by cell-cell contact involving an interaction between lymphocyte surface bound TSAb and the TSH receptor. However, we
have been unable to demonstrate a direct interaction between
Graves' lymphocytes and the TSH receptor in thyroid membranes
(20). Furthermore, detailed studies with a new cytochemical
bioassay (8,28), which has a sensitivity for TSH and TSAb in
the region of 10^4 times that of any other biological or
immunological system, have been unable to demonstrate a direct
effect of Graves' lymphocytes on the thyroid cell.

Synthesis of thyroid-stimulating antibodies

Cultures of peripheral blood lymphocytes from patients with Graves' disease have been reported to synthesize TSAb (29-31). In these experiments an in vivo mouse bioassay (5) was used to measure TSAb production and the levels of TSAb reported in the lymphocyte cultures were close to the limit of sensitivity of the assay. However, use of the highly sensitive and precise cytochemical bioassay (8) should permit a detailed investigation of TSAb synthesis in lymphocyte cultures.

Attempts have been made to prepare antibodies to the thyrotropin receptor by immunizing experimental animals with thyroid membranes (32). It is possible that anti-receptor antibodies prepared in this way could stimulate the thyroid but the presence of TSAb activity in the antisera has not been definitely established.

One of the principal problems involved in raising antisera to the TSH receptor is the minute amount of receptor present in the thyroid. For example, crude guinea pig thyroid membranes prepared from 1 gram of tissue have a TSH binding capacity of less than 50 ng of hormone (17).

TSAb and the complement system

The binding of TSAb to the TSH receptor does not appear to result in any cytotoxic effect (6). This suggests that the complement system is not activated by the antibody-receptor interaction.

Thyroid-stimulating antibody activity is always associated with IgG (2) and Fc bridging by the first component of complement would not be expected to occur unless TSAb molecules were

bound to closely adjacent TSH receptor sites (33). The serum
concentration of TSAb is extremely low (3) and the density of
TSH receptors on the thyroid cell surface also appears to be
low (3,11,13,16,17). Consequently binding of TSAb to closely
adjacent TSH receptor sites would appear to be unlikely and
this provides a possible explanation for the failure of the
TSAb-TSH receptor interaction to activate the complement
system.

Conclusions

Graves' disease is characterized by an interaction
between the endocrine and immune systems which results in the
formation of antibodies to the thyrotropin receptor. The
anti-receptor antibodies cause the hyperthyroidism of Graves'
disease by binding to the receptor and mimicking the
stimulatory effects of the hormone.

References

(1)   Adams, D. D. and Purves, H. D., Proc. Univ. Otago Med.
      Sch., 34:11, 1956.

(2)   Smith, B. R., Munro, D. S. and Dorrington, K. J., Biochim.
      Biophys. Acta, 188:89, 1969.

(3)   Smith, B. R., Biochim. Biophys. Acta, 229:659, 1971.

(4)   Smith, B. R., Dorrington, K. J. and Munro, D. S., Biochim.
      Biophys. Acta, 192:277, 1969.

(5)   McKenzie, J. M., Endocrinology, 63:372, 1958.

(6)   Dorrington, K. J. and Munro, D. S., Clin. Pharm. Therap.
      7:788, 1966.

(7)   Kendall-Taylor, P., Brit. Med. J., iii:72, 1973.

(8)   Petersen, V. B., Rees Smith, B. and Hall, R., J. Clin.
      Endocr. Metab., 41:199, 1975.

(9)   Kendall-Taylor, P., J. Endocr., 52:533, 1972.

(10) Mukhtar, E. D., Rees Smith, B., Pyle, G. A., Hall, R. and Vice, P., Lancet, i:713, 1975.

(11) Manley, S. W., Bourke, J. R. and Hawker, R. W., J. Endocr., 61:419, 1974.

(12) Manley, S. W., Bourke, J. R. and Hawker, R. W., J. Endocr., 61:437, 1974.

(13) Verrier, B., Fayet, G. and Lissitzky, S., Eur. J. Biochem. 42:355, 1974.

(14) Smith, B. R., J. Endocr., 46:45, 1970.

(15) Dirmikis, S. and Munro, D. S., J. Endocr., 58:577, 1973.

(16) Smith, B. R. and Hall, R., Lancet, ii:427, 1974.

(17) Smith, B. R. and Hall, R., FEBS Letts., 42:301, 1974.

(18) Mehdi, S. Q. and Nussey, S. S., Biochem. J., 145:105, 1975.

(19) Scatchard, G., Ann. N. Y. Acad. Sci., 51:660, 1948.

(20) Hall, R., Smith, B. R. and Mukhtar, E. D., Clin. Endocr., 4:213, 1975.

(21) Clague, R., Mukhtar, E. D., Nutt, J., Clark, F., Scott, M., Evered, D., Pyle, G. A., Rees Smith, B. and Hall, R., J. Clin. Endocr. Metab., In press.

(22) Teng, C. S., Rees Smith, B., Clark, F., Evered, D. and Hall, R., In preparation.

(23) Teng, C. S., Ross, W., Rees Smith, B. and Hall, R., In preparation.

(24) Adams, D. D. and Kennedy, T. H., J. Clin. Endocr. Metab., 33:47, 1971.

(25) Shishiba, Y., Shinizu, T., Yoshimura, S. and Shizume, K., J. Clin. Endocr. Metab., 36:517, 1973.

(26) Onaya, T., Kotani, M., Yamada, T. and Ochi, Y., J. Clin. Endocr. Metab., 36:859, 1973.

(27) Edmonds, M. W., Row, V. V. and Volpe, R., J. Clin. Endocr. Metab., 31:480, 1970.

(28) Petersen, V. B., McLachlan, S., Davies, T. F., Rees Smith, B. and Hall, R., In preparation.

(29) McKenzie, J. M. and Gordon, J., in Current Topics in Thyroid Research, edited by C. Cassano and M. Andreoli, p. 445, Academic Press, New York, 1965.

(30) Miyai, K., Fukuchi, M., Kumahara, U. and Abe, H., J. Clin.
     Endocr. Metab., 27:855, 1967.

(31) Wall, J. R., Good, B. F., Forbes, I. J. and Hetzel, B. S.,
     Clin. Exp. Immunol., 4:14, 555.

(32) Beall, G. W., Daniel, P. M., Pratt, O. E. and Solomon,
     D. H., J. Clin. Endocr. Metab., 29:1460, 1969.

(33) Müller-Eberhard, H. J., Adv. Immunol., 8:1, 1968.

Chapter 27

THE IMMUNOLOGY OF THE INSULIN RECEPTOR

J.S. Flier, C.R. Kahn, D.B. Jarrett, and J. Roth
Diabetes Branch, NIAMDD
National Institutes of Health
Bethesda, Maryland 20014

Abstract

We have detected and characterized anti-insulin-receptor autoantibodies which circulate in several patients with insulin resistant diabetes.
These antibodies are predominantly IgG and are polyclonal. They inhibit
insulin binding to its receptor on a variety of tissues from widely
separated species. Antibodies obtained from different patients appear to
bind to different determinants on the receptor and alter receptor function
in several ways. Some anti-receptor antibodies are capable of stimulating
insulin-like effects on target tissues, while others block insulin-
stimulated effects. Direct labeling of anti-receptor antibody with ^{125}I
permits use of these antibodies as an assay and probe of insulin receptors.

The first step in insulin action is the binding to a specific plasma

membrane receptor, the molecular structure which serves to recognize

biologically active insulin and insulin analogues (1,2). The insulin

receptor interaction has been intensively studied in recent years and

is, in all cases, defined by several functional criteria. Thus, binding

occurs to a finite number of sites (i.e., is saturable), is rapid and

reversible, and most importantly, is specific for insulin and other

peptides in proportion to their insulin-like biological activity.

Using direct measurement of ^{125}I-insulin bound to cells and membranes,

the properties of insulin receptors from a wide variety of species and

tissues have been investigated. These include species as divergent as man,

rodent, bird, and fish and such tissues as adipocyte, hepatocyte, fibroblast, and erythrocyte (3-7). When insulin receptors from such varied sources are carefully studied for such parameters as binding specificity, kinetics, pH optimum and site-site interactions, it is clear that few changes in binding properties have occurred with evolution. This remarkable evolutionary stability indicates the importance of the functions subserved by this receptor.

Most of the information characterizing the insulin receptor involves analysis of the functional properties of insulin binding. On another level, efforts to study the physical biochemistry of the receptor molecule have advanced, employing a variety of techniques of protein isolation and purification. These studies demonstrate that the receptor is an asymmetric protein with molecular weight about 300,000 possibly composed of several subunits (8).

Recent work in our laboratory has employed a new tool for the study of the insulin receptor. We have used a group of naturally occurring autoantibodies directed at this membrane component as a unique probe of receptor structure and function. In this review, we will discuss the setting in which these antibodies develop, their preliminary immunologic characterization, the nature of their interaction with the insulin receptor, and their utility in further analysis of the insulin receptor.

Syndrome of Insulin Resistance With Anti-Insulin Receptor Antibodies

Alterations of insulin receptor concentration or affinity play an important role in some states of altered insulin sensitivity (9). This fact is well illustrated by a recently described clinical syndrome of insulin resistance and acanthosis nigricans (10). In this syndrome there is extreme resistance to both endogenous and exogenous insulin. Thus, these patients have insulin concentrations that are 10 to 100-fold increased in both the basal and stimulated states and are resistant to

the effects of up to 1000 times the usual dose of exogenous insulin.
Known causes of insulin resistance, such as obesity, lipoatrophy, anti-
insulin antibodies, acromegaly or Cushing's disease are not present.
However, each of these patients has a defect at the level of the insulin
receptor, with markedly reduced insulin binding to their own cells
studied in vitro (Figure 1, top). One group of these patients has a
variety of "autoimmune" features suggestive of a generalized immune
disease, including increased gamma globulin and sedimentation rate,
anti-DNA and anti-nuclear antibodies, leukopenia, alopecia and arthralgias.
In these latter patients the insulin resistance is associated with
circulating inhibitors which specifically bind to the cell membrane and
interfere with insulin receptor function (11).

The Antibody Nature of The Insulin Binding Inhibitor

The assay for this inhibitor is simple (11). Cells or membranes
with insulin receptors are exposed to serum or serum fractions, usually
for 1 hour at 22°C., washed extensively to remove unbound antibody or
hormone, and then exposed to ^{125}I-insulin for binding assay. Serum from
the most insulin resistant patients inhibits up to 95% of specific insulin
binding in this assay. The titer of anti-receptor activity varies over
a wide range in our patients, from as high as 1/4000 to as low as 1/4,
and correlates well with the observed clinical insulin resistance.

We have used a variety of standard techniques to prove that the
circulating inhibitor of insulin receptor binding is an immunoglobulin
(12). Firstly, the inhibitory activity is fully preserved in a 33%
ammonium sulfate precipitate of these sera. In addition, the inhibitor
migrates with the immunoglobulins on G-200 Sephadex gel filtration and
DEAE cellulose chromatography. Immunoprecipitation experiments further
support the immunoglobulin nature of the binding inhibitor. The inhibitory

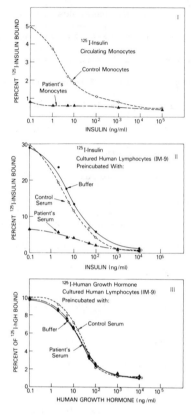

FIGURE 1

TOP: Binding of ^{125}I-insulin to circulating monocytes from a normal
control and a patient with anti-receptor antibodies and insulin resistance.
The monocyte concentration was about 12 X 10^6/ml, the ^{125}I-insulin
concentration was 200 pg/ml, and the unlabeled insulin concentrations were
as indicated. Incubation was conducted at 22° for 3 hours and the cell
bound ^{125}I-insulin separated from the free insulin by centrifugation.

CENTER: Effect of serum preincubation on insulin binding to cultured
lymphocytes. Human lymphocytes in culture (IM-9 line) were exposed to a
1:2 dilution of serum from the above patient from 60 min. at 4°C., and
washed X 3. ^{125}I-insulin (100 pg/ml) and unlabeled insulin at the
indicated at 15° for 90 min. The percent ^{125}I-insulin bound was determined
above.

BOTTOM: Effect of serum preincubation on growth hormone binding to
cultured lymphocytes. The cells were exposed to serum as above then
incubated with 100 pg/ml of ^{125}I-hGH in the presence of increasing
amounts of unlabeled hGH for 90 min. at 30°. The percent ^{125}I-hGH
was determined is above (Adapted from Reference 11).

activity of serum is completely removed by appropriate immunoprecipitation

with anti-human immunoglobulin antisera. Further studies reveal that the

antibody populations in these sera are polyclonal, since activity is

associated with both kappa and lambda light chains. In addition most

of these antibodies are of the IgG class, although one patient clearly

has activity in the IgM globulins as well. Finally, by preparing an $F(ab)_2$

fraction of IgG by peptic digestion, we have been able to demonstrate

that these blocking antibodies act through F_{ab} binding determinants

rather than through the F_c portion of the molecule. Thus, the inhibitor

is, by multiple criteria, an antibody.

Specificity of The Anti-Receptor Antibody

Having demonstrated that the inhibitor of insulin receptor binding

is an immunoglobulin, we may address the question of the specificity of

its effect. We have developed several lines of evidence that the binding

inhibition is not the result of a non-specific membrane toxicity, but

rather, is the consequence of binding to determinants on, or closely

linked to the insulin receptor. First, exposure to these immunoglobulins

does not adversely affect cell viability as measured by trypan blue dye

exclusion or by Cr^{57} labeling techniques (13). More importantly, the

inhibition is entirely specific for the insulin receptor.

Cells of the IM-9 lymphoblastoid cell line have specific membrane

receptors for both insulin and human growth hormone. When these cells

are exposed to immunoglobulins from our patients insulin binding is

markedly reduced, while growth hormone binding remains intact (Figure 1).

Similarly, these same immunoglobulins impair insulin binding to its

receptor on liver cell membranes, while causing no reduction in the

binding of NSILA-s, a closely related insulin-like peptide, to its

specific receptor on the same membranes (14). This data provides
information on the specificity of the antibody, as well as indicating
the unique ability of such antisera to distinguish closely related but
distinct receptors.

Much can be learned from the species and tissue specificity of anti-
receptor antibodies obtained from different patients. These sera
variably inhibit insulin binding to insulin receptors on all species
tested, including man, rodent, bird, and fish - species separated by
hundreds of millions of years of evolution. In addition, this effect is
seen with receptors from every tissue tested, including adipocytes,
hepatocytes, monocytes, fibroblasts, and nucleated red cells. Specificity
for the insulin receptor, together with effectiveness on cells from widely
varying species and tissues, strongly suggest that the antibodies are
binding to determinants on or very closely linked to the insulin receptor.

Although these sera may inhibit insulin binding to cells from a
variety of species, a given serum may not be equipotent against all
species tested. With one serum, inhibitory titers were very similar
using receptors from widely varying species, including mammals and fish
(15). It is likely that the antibodies in this serum see a portion of
the receptor that has been tightly preserved through evolution suggesting
that these antibodies bind to a region of the receptor with an important
functional role. With another anti-receptor serum, the inhibitory titer
against human and rodent tissues was 2 orders of magnitude higher than
that against receptors on avian cells and fish cells. This suggests
that the binding determinants for these antibodies (or a significant
fraction of them) may be somewhat removed from those parts of the receptor
which are critical for insulin binding specificity. Thus, modification
of receptor structure in regions unimportant for insulin binding
might be detected through the use of anti-receptor antibodies.

Anti-receptor antibody binding is rapid and temperature dependent, with a K_d of 10^9 M^{-1} estimated by indirect studies. When cultured human lymphocytes are pretreated with these sera, and then ^{125}I-insulin is allowed to bind to a steady state at 15°C., Scatchard analysis of the binding data suggests several mechanisms by which insulin binding may be reduced. One anti-receptor serum markedly reduces binding affinity, without a significant reduction in total receptor number. Kinetic data demonstrate that insulin association rate is reduced, and dissociation rate may be accelerated, thereby explaining the reduced affinity. With another serum, both steady state and kinetic data suggest that much or all of the binding defect is due to a reduced receptor number with normal affinity of remaining receptors. A third serum shows a mixed effect. This fits with previous data that indicate heterogeneity of these anti-receptor antisera.

Effects of Anti-Insulin-Receptor Antibodies On Isolated Adipocytes

Isolated rat adipocytes provide a system in which the effect of anti-insulin-receptor antibodies on both insulin binding and biological activity can be studied (16). When isolated adipocytes are exposed to normal serum and then washed, neither ^{125}I-insulin binding nor subsequent basal or stimulated glucose oxidation is altered. In contrast, exposure of these cells to sera containing anti-receptor antibodies alters all three of these parameters.

Anti-receptor sera which inhibit insulin binding to human lymphocytes and monocytes, inhibit insulin binding to rat adipocytes with almost identical potency. With these cells at 37° C., the effect of all sera appears to be a decrease in receptor affinity rather than a decrease in receptor number, consistent with the notion that the anti-receptor antibodies act as competitive inhibitors of insulin binding. This inhibitory activity is retained in purified IgG fractions and in the

isolated F(ab)$_2$ fragment of the antibody.

Fat cells preincubated with two of the anti-receptor antisera also
show a marked stimulation of basal glucose oxidation, the most potent
serum producing maximal stimulation even at dilutions of 1 to 1000
(Figure 2). At submaximally effective concentrations, this effect of the
serum on glucose oxidation is further augmented by insulin, while at
maximally effective concentrations the serum and insulin effect were

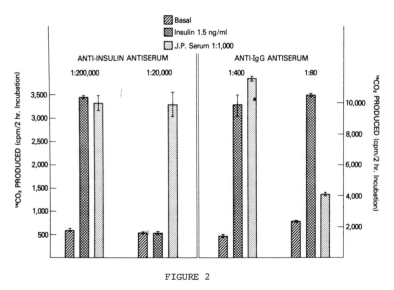

FIGURE 2

Effect of anti-IgG and anti-insulin serum on the insulin-like
activity of anti-receptor antibodies. Rat adipocytes were isolated by
collagenase digestion and incubated for 2 hours at 37° in a Krebs-Ringer
bicarbonate buffer containing 0.3 mM glucose and [U-^{14}C] glucose. At
the end of the incubation the samples were acidified and the evolved
^{14}CO$_2$ collected in hyamine hydroxide in hanging wells. In some flasks
(indicated by cross-hatching) 1.5 ng/ml of insulin was added; this concentration produces maximal glucose oxidation. Serum J.P. which contains
anti-receptor antibodies produced similar stimulation of glucose oxidation
at 1:1000 dilution (stippled bar at left). When the incubation was
conducted in the presence of a high concentration (1:20,000) of anti-
insulin antiserum, the effect of insulin was abolished, while the effect
of serum J.P. was unaffected. In contrast, anti-Ig antiserum at high
concentration (1:80) abolished the stimulation produced by serum J.P.,
but had no effect on insulin-stimulated glucose oxidation.

similar in magnitude and not additive, suggesting that insulin and anti-
receptor antibodies stimulate glucose oxidation via a similar pathway.
Like the effect on insulin binding, the effect of anti-receptor anti-sera
on glucose oxidation is blocked by incubation of serum with anti-human
IgG, but not anti-insulin antibodies (Figure 2). Furthermore, the
stimulatory activity is retained in partially purified immunoglobulin
fractions and in the $F(ab)_2$ fragment of the IgG.

One of the three sera thus far studied showed only a small effect
on basal glucose oxidation. This serum was also unique in that cells
preincubated with a concentration which inhibited insulin by 80%
showed a 5-fold shift to the right in the curve for insulin stimulated
glucose oxidation. Taken together with the data presented above, these
studies further suggest that the anti-receptor antibodies act at or very
near the site on the insulin receptor important for both insulin binding
and generation of biological effect.

Direct Studies of Antibody Binding

Using these indirect approaches, we have shown that the anti-receptor
autoantibodies have a unique specificity for the insulin receptor. This
specificity has been further accentuated in direct studies of the
interaction between the antibodies and the insulin receptor (17). Standard
techniques were used to prepare IgG fractions from the sera of patients
with anti-receptor antibodies. This fraction was labeled with [125]Iodine
and by selective cytoadsorption and elution from cells rich in insulin
receptors, a selectively enriched ^{125}I-anti-receptor antibody preparation
was obtained. The ^{125}I-antibody bound to a wide variety of cells in direct
proportion to the insulin receptor concentration present (Figure 3).
Modulation of the insulin receptor concentration by either tryptic digestion
of the cell surface, or specific insulin-mediated-loss of insulin receptors,

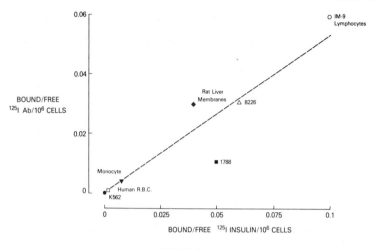

FIGURE 3

The binding of ^{125}I-anti-receptor antibody and ^{125}I-insulin. Seven cell types were each incubated with the ^{125}I-labeled ligand under conditions of relative receptor excess. The Bound/Free ratio for both ^{125}I-anti-receptor antibody and ^{125}I-insulin was calculated from the difference in binding measured in the presence and absence of 100 μg/ml unlabeled insulin.

was associated with a reduction in both labeled insulin and antibody binding. In contrast, insulin and antibody binding were unaffected when human growth hormone was used to specifically induce a loss of its receptors on these cells. Anti-receptor antibody binding was also specifically competed for by insulin and insulin analogues. Thus insulins which ranged over 300-fold in biological potency inhibited the binding of both labeled insulin and anti-receptor antibody in proportion to their ability to bind to the insulin receptor. The only substances, other than insulin which inhibited the binding of the anti-receptor antibody were whole sera, IgG and F(ab)$_2$ fractions from other patients with antibodies to the insulin receptor. Human growth hormone, which has specific receptors on these cells, as well as unrelated peptide hormones like ACTH, were without effect upon the binding of either insulin or the anti-receptor autoantibodies. Thus by both indirect and direct studies these anti-receptor autoantibodies appear

to interact specifically with cell surface determinants which are intimately

associated with the insulin receptor.

Summary

We have detected and characterized anti-insulin receptor auto-

antibodies in a group of patients with extreme insulin resistance. These

antibodies are predominantly IgG and are polyclonal in nature. Using

these unusual antibodies as probes, we have been able to approach the

insulin receptor for the first time from an immunological prespective.

The reason for the development of these anti-receptor antibodies is

unknown, but there are precedents in myasthenia gravis and Graves' disease,

in which antibodies to the acetylcholine receptor and thyroid stimulating

hormone receptor, respectively, have been found. In myasthenia gravis,

antibodies bind to the acetylcholine receptor and result in resistance

to the effect of this neurotransmitter, but this effect seems not to be

dependent upon inhibition of ligand binding (18). In Graves' disease,

antibodies inhibit TSH binding and also produce a TSH-like effect on

target cells, resulting in a state of thyroid over-activity and autonomy

from the usual trophic stimulus, TSH (19,20). The antibodies to the

insulin receptor possess some properties of each of these systems. Thus,

these antibodies in vivo are associated with a state of insulin

resistance, and they appear to produce the effect by interfering with

insulin binding to its membrane receptor. However, upon acute exposure

to at least one target tissue in vitro, binding is associated with

production of an insulin-like effect, analogous to the effect of the

stimulatory immunoglobulin in Graves' disease.

It is unknown whether the observed autoantibodies develop in

response to a primary or acquired receptor abnormality, or whether normal

receptors become the targets of an independent immune dysfunction.

Perhaps, as suggested by Lennon and Carnegie (21), the distinct structural and functional requirements of membrane receptor renders them likely to become the targets of an immune response.

Apart from their role in the insulin resistant syndrome which led to their discovery, these antibodies, as a group, provide a new tool for investigation of the insulin receptor. Receptors from different species and tissues that have indistinguishable insulin binding properties may now be separated on the basis of affinity for a heterogeneous population of antibodies directed at different regions of the receptor molecule. In this manner, the phylogenetic history of the receptor may be more clearly defined. In addition, use of the antibodies to perturb, and in some cases activate the receptor, may be useful in studies of the mechanism of action of insulin. Finally, the antibodies may be used as a unique reagent for the detection and quantitation of insulin receptors, without the need to bind hormone, labeled or unlabeled, as part of the assay procedure. Using this technique, structural variants of the receptor which have altered hormone binding properties might be found. Thus, these antibodies to the insulin receptor may be used to define new disorders of this important membrane compartment.

References

1. Kahn, C.R., in Methods in Membrane Biology, edited by E.D. Korn, p. 81, Plenum Press, New York, 1975.

2. Roth, J., Metabolism, 22:1059, 1973.

3. Gliemann, J., and Gammeltoft, S., Diabetologia, 10:105, 1974.

4. Kahn, C. R., Freychet, P., Roth, J., and Neville, D.M., Jr., J. Biol. Chem., 249:2249, 1974.

5. Freychet, P., Rosselin, G., Rancon, F., Fouchereau, M., and Broer, Y., Horm. Metab. Res., Supplement 5, 72, 1974.

6. Rechler, M.M., and Podskalny, J.M., Diabetes, 25:250, 1976.

7. Ginsberg, B.H., Kahn, C.R., and Roth, J., Endocrinology, in press.

8. Ginsberg, B.H., Cohen, R.M., and Kahn, C.R., Diabetes, 25:322, 1976.

9. Roth, J., Kahn, C.R., Lesniak, M.A., Gorden, P., De Meyts, P., Megyesi,
    K., Neville, D.M., Jr., Gavin, J.R., Soll, A.H., Freychet, P.,
    Goldfine, I.D., Bar, R.S., and Archer, J.A., Rec. Prog. Horm., Res.,
    31:95, 1975.

10. Kahn, C.R., Flier, J.S., Bar, R.S., Archer, J.A., Gorden, P., Martin,
    M.M., and Roth, J., New Eng. J. Med., 294:739, 1976.

11. Flier, J.S., Kahn, C.R., Roth, J., and Bar, R.S., Science, 190:63,
    1975.

12. Flier, J.S., Kahn, C.R., Jarrett, D.B., and Roth, J., submitted for
    publication.

13. Maratos-Flier, E., unpublished results.

14. Megyesi, K., Kahn, C.R., Roth, J., Neville, D.M., Jr., Nissley, P.S.,
    Humbel, R.E., and Froesch, E.R., J. Biol. Chem., 250:8990, 1975.

15. Muggeo, M., Ginsberg, B.H., Kahn, C.R., De Meyts, P., and Roth, J.,
    manuscript in preparation.

16. Kahn, C.R., Baird, K., Flier, J.S., and Jarrett, D.B., Diabetes, 25:
    322, 1976.

17. Jarrett, D.B., and Roth, J., Clin. Res., 24:427, 1976.

18. Lindstrom, J.A., Seybold, M.E., Lennon, V.A., Whittingham, S., and
    Duane, D.O., Neurology, in press.

19. Smith, B.R., and Hall, R., Lancet ii:427, 1974.

20. Mukhtar, E.D., Smith, B.R., Pyle, G.A., and Hall, R., Lancet i:713,
    1975.

21. Lennon, V.A., and Carnegie, P.R., Lancet i:630, 1971.

    D.B.J. is the recipient of a Clinical Sciences Fellowship from
    the National Health and Medical Research Council of Australia.

Chapter 28

IDENTIFICATION OF RECEPTORS FOR ANIMAL VIRUSES

Wm. R. Gallaher and C. Howe
Department of Microbiology
Louisiana State University Medical Center
New Orleans, Louisiana 70112

Abstract

Present knowledge of cell surface receptors for animal viruses is reviewed. The methods used for enumeration and identification of receptors are critically examined with respect to particular advantages and disadvantages. Specific controls and alternative interpretations are suggested in connection with the reactions of lectins which block viral attachment to cells. Currently available information for each group of animal viruses is summarized in order to define the extent to which the corresponding receptors have been identified. It is concluded that the full range of virus-receptor interactions has not yet been explored even for those viruses of which there is the most detailed knowledge. For some groups, moreover, the receptor is totally uncharacterized. Six areas in which future investigative effort might be productive are identified, including the isolation of membrane components and the immunochemical definition of viral receptors.

In the mid-1960's two divergent roads developed in the field of virology. One has led to our present detailed knowledge of molecular genetics and viral replication. The other has concerned what might be called the cellular biology of animal viruses, encompassing such topics as the mechanisms by which viruses kill cells, and the interaction of viruses with specific receptors on or in the plasma membrane. Most virologists have taken the former road, with the result that significant disproportion now exists between our knowledge of the molecular biology of viruses and our still rudimentary understanding of the nature of those interactions essential for initiation of viral infection.

491

To this day the detailed chemical structure of only one receptor substance for animal viruses is known, and that due largely to the fact that
myxoviruses were found to contain an enzyme, neuraminidase, which cleaves
the cell surface receptor in a precise and measurable manner (1).

The purpose of this review is to indicate that ample opportunity
exists to rapidly gain a more substantial understanding of viral receptors.
It will also become clear that immunologists, with their extensive experience in the molecular aspects of ligand research, are in good position to
contribute to this area.

The scope of this review will be restricted to three specific topics:
an overview of the methods available for studies of receptors, a brief
summary of the current status of information about receptors for each group
of animal viruses, and a final section discussing certain approaches or
methods which could prove to be productive in the future. For specific
information on the composition of cell membranes or on early virus-cell
interactions one may rely on several recent reviews (2,3,4,5,6). Likewise,
reasonably current information regarding viral components and products can
be obtained elsewhere (7).

Methodology of Receptor Research

Viruses attach to cells by simple adsorption, envisaged as noncovalent
bonding to the three-dimensional structure of a receptor site. "Receptor",
here is defined as a functional group on the surface of cells which shows
specific attachment to a complete infectious virion. The avidity and conditions for maximal adsorption vary widely among the major groups of
viruses. It is therefore important first to determine the optimal virus-
host system, and then to establish what conditions govern viral binding (2).
Attachment of virus is usually quantified by one or more of four general
methods: hemagglutination, plaque formation, uptake of radioactively

labeled virus, and electron microscopic visualization of virions attached
to receptor complexes. Identification of receptor activity is usually
accomplished through the demonstration of interference with attachment. A
prime example is the capacity of soluble sialopeptide to bind preferentially
to influenza A virions (1), inhibiting attachment of virus to cell surfaces.

The preferred system, wherever applicable, is the use of competent
erythrocytes in a viral hemagglutination reaction, in which the specificity
of receptor analogs may be evaluated by hemagglutination inhibition. An
advantage of this system lies in the fact that viruses do not replicate in
erythrocytes, so that interference is only at the level of attachment. Also,
the assay system is relatively simple and sensitive, and erythrocytes are
both available in large quantity and relatively constant in composition.
However, hemagglutination may be limited by more specialized conditions
than is infection of cultured cells. Such divergence may reflect the
existence of two different sets of receptor substances, one for hemagglutination, the other for initiation of infection. Moreover, hemagglutination
measures virions as particles, only a small proportion of which may also
be infectious units.

Determination of viral attachment by plaque formation is directly
dependent on infectious units, and can therefore be applied to analysis of
receptor systems which influence viral tropisms in vivo, e.g. binding
studies with explants or cell lines from target tissues (8). Quantitation
of virus–cell interactions over a wider range of multiplicities of infection is likewise possible. A qualitative estimate of interference can
be obtained in measuring plaque reduction by pretreatment of a cell mono-
layer with an inhibitor, prior to superinfection by a constant inoculum of
an appropriate challenge virus. More accurate quantitative interference
data can be obtained by transferring infected cells to fresh monolayers
in an infective center assay (9), or by measuring removal of virus from

suspension (10). The disadvantage of such a system is that it must be

checked for interference with viral penetration or replication, possibil-

ities not always unambiguously ruled out.

Measuring uptake of radioactively labeled virus is most useful (11),

particularly where neither a hemagglutination or plaque assay is available.

It combines technical ease and quantitative accuracy. However, no distinc-

tion can be made between non-functional viral particles and infectious

units.

Electron microscopic techniques, particularly those entailing the for-

mation of surface replicas of erythrocytes or cultured cells (12,13), are

useful in determining both distribution density and topology of receptors

for a variety of ligands, as these relate to morphological components of

membranes, e.g. intramembranal particles. Electron microscopy demands,

however, the use of very high miltiplicities of infection, not always

attainable, which may introduce artifacts difficult to interpret.

In all such systems, impurities or particle heterogeneity in the viral

preparation can seriously affect results. Therefore, such studies are best

done with infectious virus purified by at least two cycles of rate zonal

or isopycnic ultracentrifugation, and shown by isotope dilution or

immunological methods to be substantially free of contaminating cellular

material.

Evaluating Inhibition of Viral Attachment

There are three general classes of inhibitors of viral attachment:

soluble or particulate receptor analogs which compete with receptors for

viral attachment, lectins which compete with virions for attachment to

receptors, and enzymes which alter or destroy receptors. Receptor analogs

occur as natural components of many animal sera (see 2). Lectins with a

variety of substrate specificities have been isolated from both plant   and invertebrate sources (14). In certain systems, viral particles, either

defective virions or of a different type or strain, may interfere with

viral attachment (10,15,16,17). In some systems the interference may not

be reciprocal. For example, influenza A virus can attach to Sendai virus-

treated cells, but the reverse does not hold (17). Also, one must be sure

interference does not involve interferon or other intracellular interfering

mechanisms (9). While this approach does not identify viral receptors, it

does define groups of viruses sharing a receptor, and their relative

avidity for that receptor, and permits the choice of a virus for more

detailed study against a broader biological background.

The only completely unambiguous result is obtained when a substance

is found not to inhibit attachment of one type of virus, while in concurrent

assays it is found capable of inhibiting an unrelated virus. Any inhibition

of viral attachment should be subject to additional studies to determine

whether the inhibition is meaningful. In general, data showing a dose re-

sponse tend to substantiate the interpretation that specific competition

between virus and inhibitor over a broad range of inhibitor concentration

has occurred. Even when the effect of an inhibitor appears to be specific,

other factors must be considered in the interpretation of results. For

instance, neuraminidase, in denuding erythrocytes of N-acetyl neuraminic

acid (NANA), drastically reduces their negative surface charge (18). This

in turn, by changing the structure of the remaining glycoprotein, exposes

new reactive sites, e.g. for concanavalin A (Con A) (19), and may obscure

or abolish others. In the case of influenza A, enzyme data were confirmed

by the demonstration that soluble substances containing NANA inhibit viral

attachment (1). Enzyme data alone constitute suggestive rather than con-

clusive evidence of the structure of a receptor.

While lectins can be powerful probes of specific functional groups on

cell surfaces, their application to studies of receptors requires careful

control. Any inhibition of attachment may be due to steric hindrance, from
a site vicinal to the viral receptor, by multimeric forms of the lectin,
which may or may not be overcome by use of high multiplicities of virus (20).
Many lectins also bind to viral glycoproteins, thus neutralizing viral
infectivity (21). Were one to coat cells with antibody reactive with both
cell and viral surfaces, one might expect "inhibition" of subsequent viral
attachment even if the antibody was not specific for the receptor, since the
bound antibody would be available for direct viral neutralization. The same
possibly may pertain to inhibition by lectins. As noted below, the attachment of several viruses is inhibited by Con A, but Con A also agglutinates
and neutralizes many of those same viruses. Until the alternative explanations we suggest are ruled out, it is best to reserve judgment on such results. One way to eliminate the ambiguity would be to demonstrate that an
oligosaccharide ligand of the lectin, e.g. α-methyl-mannoside for Con A,
would serve as a competitive inhibitor or receptor analog, preventing viral
attachment to indicator cells, similar to the use of sialopeptides against
attachment of influenza A virus. This would confirm the specificity of the
lectin and unambiguously identify the receptor substance. However, to our
knowledge such studies have yet not been done in any of the lectin experiments reported to date.

Analysis of Host Systems for Viral Attachment

Attachment of bacteriophage is beyond the scope of this review. It is
useful to note, however, that a large body of precise data has been derived
from the analysis of bacterial mutants in which surface structures could be
examined biochemically and immunochemically (22). While the same degree of
genetic manipulation is not yet possible in animal cells, exact identification of receptors will almost certainly require dissection of the surface
membrane with similar precision. A major effort in this direction has for

some time been made with the human erythrocyte membrane, the principal

features of which are reviewed elsewhere (5). Most studies of receptors

for viruses have been done with human erythrocytes of group O, although

only the reoviruses seem to differentiate among the ABO blood groups (23).

The effect of other allogeneic variations on viral attachment is, unfortu-

nately, either nonexistent or unknown. However, allotypes serve as useful

markers for isolation of receptor substances. The lectin affinties of the

major erythrocyte glycoproteins are also known (5) and are therefore avail-

able for comparison with lectin inhibition of viral attachment.

Genetic manipulation of animal cells in order to investigate surface

receptors is still under development. Some attempt has been made to use

somatic cell hybrids to investigate viral host range (24), but permissiveness

for a given virus may be dependent on several steps in replication. In some

cases, partial isolation of receptor substances has been achieved, as will

be noted below. The major cell surface glycoprotein of chick embryo cells

has been isolated, but has not yet been characterized with respect to viral

receptor activity (25).

Current Status of Receptor Identification

We briefly summarize below the information on receptors for each major

group of animal viruses. Insofar as such studies involve virus-erythrocyte

interactions, more detailed descriptions may be found elsewhere (2).

Papovaviruses. The principal receptor for polyoma virus seems to con-

tain NANA, since receptor analogs of NANA prevent its attachment (26,27).

To our knowledge, similar information is not available for either SV40 or

papilloma viruses. Interference among papovaviruses has not been demon-

strated.

Adenoviruses. Three groups of adenovirus strains can be distinguished

on the basis of hemagglutination (28). Each group, and possibly some types within each group, has a different receptor specificity. The receptor for

adenovirus type 9 of group 2 seems to be neuraminidase-sensitive (29),

while for serotypes of group 3 attachment is enhanced by the same enzyme

(30). In the case of group 3 strains, a protein-protein interaction with

carboxylic residues on erythrocytes is the most probable mode of viral

attachment. This interpretation is based on the finding that the receptor

is sensitive to subtilisin and carbodiimide (11,31). Attachment inter-

ference between group 3 strains can be demonstrated by cross-saturation

experiments with viral penton fibers. It has been similarly shown that

group 3 adenoviruses and enteroviruses do not share a common receptor (11).

Herpesviruses. Little is known about receptors for this group of

viruses, in part due to the lack of any hemagglutinating factor in the

virion. However, a broad spectrum of cell lines from both human and animal

sources are susceptible to herpes simplex virus (HSV). Attachment of EB

virus is restricted to lymphoid cells, or hybrids of lymphoid and

epithelial cells (32), and the receptor may involve sialopeptide (33).

In case of HSV, attachment is promoted by thyroid hormone and inhibited by

parathyroid hormone (34,35). Interference between strains of HSV has not

been demonstrated. In mixed infections one phenotype may predominate, but

the progeny virions are genotypically mixed (36,37). Heterophile agglu-

tination has been shown on the surface of HSV-infected cells, and may

provide a probe of receptor activity (38). An oligosaccharide fraction of

BHK-21 cells has been shown to inhibit attachment of HSV (39). Con A pre-

vents attachment of HSV, but this lectin can bind to the virion as well

(21). The use of receptor analogs to confirm the lectin results has not

been reported. Cell fusion induced by semipurified extracts of HSV-infected

cells cannot be suppressed by a variety of lectins, but studies with HSV

specific antisera suggest that the fusion factor may not be a structural

component of HSV virions (40).

Poxviruses. Except for the fact that enteroviruses do not interfere with attachment of vaccinia virus (41), the character of the receptor for vaccinia virus is totally unknown. Surface interference among poxviruses has not been studied in detail. In mixed infections the progeny are genotypically mixed even when one phenotype predominates (42).

Picornaviruses. As might be expected for a group of viruses containing both enteroviruses and rhinoviruses, with well over a hundred serotypes, the evidence suggests that there are several types of receptors for these viruses. This diversity is reflected by differential sensitivity to enzymes and other reagents, as well as a variety of interference patterns. With rare exception (43), receptors for picornaviruses are insensitive to neuraminidase (44,45,46). Several echoviruses and Coxsackie B viruses share the same chymotrypsin and papain sensitive receptor, distinct from that for poliovirus (47). Prior infection with these viruses does not prevent attachment of Coxsackie A viruses (15). Con A inhibits attachment of poliovirus to cells (21), but the inhibition is reversed by high multiplicities of infection (20). There is more than one receptor for rhinoviruses, and the one for some types is blocked by Con A (46). Lectins do not bind to picornaviruses. It is therefore possible that certain lectins will prove useful in identifying receptors for this heterogeneous group of viruses.

Reoviruses. Each of the three reovirus serotypes agglutinates human erythrocytes, cells of blood group A having the highest reactivity, followed in order by allotypes AB, O and B (23). However, the implication from these findings that A allotypic determinant saccharides may have some specific relation to reovirus receptor activity has not been confirmed with receptor analogs. Experiments to explore this possibility are clearly feasible with highly purified blood group substances available from several human and animal tissues and secretions. Hemagglutination of human erythrocytes is not affected by sialopeptide receptor analogs. How-

ever, reovirus type 3 also reacts with bovine erythrocytes, an interaction
which is sensitive to inhibition by sialopeptides (48). Rather unexpectedly,
hemagglutination is affected by treatment of virus, but not cells, with
agents affecting carbohydrates (see 49), even though no carbohydrate is
present in virions. These rather mixed results indicate much future work
is required to define reovirus receptors.

Togaviruses. Hemagglutination of goose erythrocytes by togaviruses
has been studied with respect to inhibitory material from various sources,
which have in common a high lipid content (50). Sindbis virus will react
with lipid membranes in the absence of protein or carbohydrate (51). However, a distinction must be made between the rather specialized conditions
required for togavirus hemagglutination or interaction with model membranes,
and the less stringent conditions under which these viruses avidly infect
cultured cells in the presence of serum. Also, the number of receptors on
cultured cells has been found to be no more than 10^5 per cell (13), a
figure more consonant with glycoprotein than with lipid receptors. Studies
with lectins have been restricted to agglutination of viruses and virus-
infected cells, with no attention to their potential as receptor probes
(52). Apart from a limited study showing that neuraminidase has no interfering action (53),there is very little firm information regarding
receptors either for togaviruses, or for the unrelated rubella virus
which shares similar requirements for hemagglutination.

Myxoviruses. It has been firmly established that NANA is part of the
receptor site on erythrocytes with the highest and, in some cases, the only
reactivity for myxoviral hemagglutinin (1,2,5,54). The receptor on human
erythrocytes forms a part of a major glycoprotein, glycophorin, which also
bears MN allotypic specificity (55), and which requires terminal NANA
residues for viral receptor activity (56). Artificial membranes can also
have receptor activity if gangliosides with terminal NANA are present (57).

All of these findings make the role of NANA nearly incontrovertible in this

receptor specificity, which has been used to define this taxonomic group

of viruses. However, far less information is available on the interaction

of myxoviruses with cultured cells, the receptors on which have only been

assumed to be similar to that on glycophorin (see 58). Also, definite

differences exist among myxoviruses with respect to receptor specificity.

Some strains of influenza A retain avidity for erythrocytes extensively

treated with neuraminidase (54) or after total removal of cell surface

sialopeptide by trypsin (56). In contrast to influenza viruses of groups

A and B, group C viruses lack neuraminidase, and their attachment to

erythrocytes is unaffected by the enzyme (59,60). Some other oligosaccharide

may be involved in this hemagglutinin receptor, and enzyme(s) other than

neuraminidase in elution of virus from erythrocytes. Even allowing for the

possibility that NANA bound to gangliosides may not be sensitive to

neuraminidase in situ (61), it is still perhaps best not to elevate the

role of NANA in myxovirus receptors to the status of dogma. Other as yet

unidentified macromolecules in mammalian cell membranes may also posses

significant receptor activity.

   Paramyxoviruses. As with myxoviruses, the role of NANA in receptor

activity for paramyxoviruses borders on dogma. Indeed, Sendai virus and

some strains of Newcastle disease virus (NDV) have a more stringent

requirement for this sugar than myxoviruses (17). Neuraminidase prevents

exogenous cell fusion induced by NDV (62). It has also been shown that

attachment of Sendai virus to liposomal membranes requires ganglioside

receptors (63). However, neuraminidase effects only a five-fold maximal

reduction in the capacity of chick embryo cells to adsorb NDV (10).

Neuraminidase has no effect on attachment at low multiplicities of infection

with some strains of NDV (64,65, Duffy and Bratt, in preparation). It

would therefore appear that there is a distinct heterogeneity among paramyxoviruses with respect to their interaction with receptors, especially insofar as it involves NANA.

Measles and similar viruses have been termed "pseudomyxoviruses" solely because they lack neuraminidase and resist any effect of the enzyme on their attachment. However, as the molecular biology of these viruses becomes increasingly clear, they will be identified with paramyxoviruses. Measles virus broadens the receptor specificity of the group to include interactions which require NANA not on the cell surface but on the surface of the virion for maximum infectivity (Duffy and Howe, in preparation).

The attachment of Sendai virus and NDV has been shown to be reduced by pretreating cells with a variety of lectins (21,66). However, with NDV cell fusion rather than attachment was assayed. Yet it might be predicted that lectins would inhibit massive alteration of the cell through steric effects. The aggregate specificity of the lectins is so broad as to confirm that they function from neighboring sites or by binding to the viral envelope itself. Therefore, the spectrum of receptor specificities for paramyxoviruses has not been fully defined.

Oncornaviruses. The type C tumor viruses of mice and fowl are classified on the basis of host range restrictions in inbred lines of each species (67,68). In the avian system the genes for resistance to viruses are recessive (67), while in the mouse system somatic cell hybrids have been used to show that resistance is dominant (24). Both sensitive and resistant cell types have been shown to adsorb virus, but it has not been shown that the receptor for excluded strains is the same as that which permits infection of the other group (67). Within each group, interference between closely related viruses can be demonstrated, which is exerted at least partly at the attachment step (16,69). Avian tumor viruses react with lectins (70), but the reaction of lectins with viral receptors has not been reported. There is therefore virtually no information concerning receptors for oncornaviruses.

Conclusions and Future Prospects

As is plain from the foregoing, a review of viral receptors is not a
recitation of well-established facts, but rather an enumeration of salient
possibilities. Even for the best studied viral groups, the full range of
receptor interactions has not been fully explored. For other groups the
potential for expansion is limitless. One is left with the impression that
individual viruses may react to a range of receptors broader than that
initially supposed in conformity to the "one-virus-one receptor" concept
set by group A influenza viruses and bacteriophage. Animal cells present
to viruses a more irregular and heterogeneous surface than either the cell
coats of bacteria, with their regular polymers of lipopolysaccharide etc.,
or the surface membrane of erythrocytes. Perhaps the virus which can attach
to any of several specific biochemical configurations in a variable complex
of glycoproteins has a commensurate survival advantage. We also conclude
that there are too few reagents now in use as receptor probes, and those
in use are used too uncritically, to dissect the biochemistry of virus-
receptor interactions with precision.

Nevertheless, the prospects are good for a definitive explanation of
the specific nature of receptors for animal viruses. For each group of
viruses, there are at least the following six approaches toward this goal.

1.) Isolation of membrane components sufficiently pure to allow
precise chemical characterization, while retaining the capacity to interfere
with receptors bound to cell surfaces, should be attempted. This approach
should be extended to include cultured cells with a broad range of susceptibility to viruses, such as BHK-21 cells (39), and limited to those erythrocyte systems for which the cell species and hemagglutination conditions
are comparable to the host range of active infection.

2.) The role of allotypic variation in governing susceptibility to
viruses should be explored. Accordingly a large variety of reagent

antibodies to cell surface components can now be prepared and used to determine which surface allotypic markers participate in, or are vicinal to, viral receptors. It is perhaps in this area that immunologists can have maximum impact with moderate effort, since the only cell-specific sera tested for their effect on viral infection have been made against whole cells (71,72,73), obviously with mixed results and interpretations limited in their validity.

3.) The use of lectins should be expanded but the observed reactions carefully controlled for artifacts. Lectin inhibition of viral attachment is best used to select oligosaccharides which might serve as receptor analogs, and to define the specific lectins useful in affinity chromatography for isolation of receptors from solubilized cell membranes.

4.) The pattern of competition for attachment among animal viruses, both within and across taxonomic boundaries, should be expanded. The number of receptor types is almost certainly finite. Accordingly, some determination of the degree to which different viral groups share receptors would greatly facilitate the choice of virus-cell systems for detailed study.

5.) Artificial membrane systems should be developed further, either liposomal systems or lipid monolayers containing specific ligands (57,63), or erythrocytes specifically sensitized to viral attachment by interaction with lectins or other ligands (74).

6.) Our capacity for genetic manipulation of the host should be increased, both by screening existing variants for differences in ability to attach virus, and also by selecting mutant lines for resistance to viral infection or by allotypic cell surface changes. This approach might yield cell strains with mutations in receptor genes. Somatic cell hybrids should be constructed for the purpose of determining whether a receptor phenotype is dominant, as for example, A and B blood group determinants. Perhaps even mapping of human receptor genes, e.g. by following chromosomal deletions from mouse-human hybrids, could be attempted, similar to the mapping of

human genes for enzymes, HL-A and the immune response.

The ability to understand, and perhaps then to prevent, virus-receptor

interactions is of obvious significance in our overall knowledge of viral

infection. The opportunity to rapidly expand this understanding should not

be ignored.

This study was supported by NIH grant AI 10945, and by grants from the Edward G. Schlieder Educational Foundation and the Cancer Association of Greater New Orleans. We are grateful to Paula Duffy for permission to refer to her unpublished findings concerning paramyxovirus receptors.

References

1. Gottschalk, A., in The Viruses, vol. 3, edited by F.M. Burnet and W.M. Stanley, p. 51, Academic Press, New York, 1959.

2. Howe, C. and Lee, L.T., Adv. Virus Res., 17:1, 1972.

3. Dales, S., Bacteriol. Rev., 37:103, 1973.

4. Lonberg-Holm, K. and Philipson, L., Monographs in Virology, vol. 9, edited by J.L. Melnick, S. Karger, New York, 1974.

5. Bächi, T., Deas, J. and Howe, C., in Cell Surface Reviews, edited by G. Poste and G.L. Nicolson, No. Holland Div., ASP Biological and Medical Press, Amsterdam, in press.

6. Beers, R.F., Jr., and Bassett, E.G., editors, Cell Membrane Receptors for Viruses, Antigens and Antibodies, Polypeptide Hormones, and Small Molecules, Raven Press, New York, 1976.

7. Fraenkel-Conrat, H. and Wagner, R.R., editors, Comprehensive Virology, vols. 2-4, Plenum Press, New York, 1974.

8. Holland, J.J., Virology, 15:312, 1961.

9. Bratt, M.A. and Rubin, H., Virology, 33:598, 1967.

10. Bratt, M.A. and Rubin, H., Virology, 35:395, 1968.

11. Philipson, L., Lonberg-Holm, K. and Petterson, U., J. Virol., 2:1064, 1968.

12. Bächi, T., Aguet, M. and Howe, C., J. Virol., 11:1004, 1973.

13. Birdwell, C.R. and Strauss, J.H., J. Virol. 14:672, 1975.

14. Lis, H. and Sharon, N., Annu. Rev. Biochem., 42:541, 1973.

15. Crowell, R. L., J. Bacteriol., 91:198, 1966.

16. Steck, F.T. and Rubin, H., Virology, 29:642, 1966.

17. Woodruff, J.F. and Woodruff, J.J., J. Immunol., 112:2176, 1974.

18. Eylar, E.H., Madoff, M.A., Brody, O.V. and Oncley, J.L., J. Biol. Chem.,
     237:1992, 1962.

19. Bächi, T. and Schnebli, H.P., Exp. Cell Res., 91:285, 1975.

20. Lonberg-Holm, K., J. Gen. Virol., 28:313, 1975.

21. Okada, Y. and Kim, J., Virology, 50:507, 1972.

22. Lindberg, A.A., Annu. Rev. Microbiol., 27:205, 1973.

23. Brubaker, M.M., West, B. and Ellis, R.J., Proc. Soc. Exp. Biol. Med.,
     115:1118, 1964.

24. Scolnick, E.M. and Parks, W.P., Virology, 59:168, 1974.

25. Yamada, K.M., Yamada, S.S. and Pastan, I., Proc. Nat. Acad. Sci. USA,
     72:3158, 1975.

26. Hartley, J.W., Rowe, W.P., Chanock, R.M. and Andrews, B.E., J. Exp. Med.,
     110:81, 1959.

27. Crawford, L.V., Virology, 18:177, 1962.

28. Rosen, L., Amer. J. Hyg., 71:242, 1960.

29. Wadell, G., Proc. Soc. Exp. Biol. Med., 132:413, 1969.

30. Boulanger, P.A., Houdret, N., Scharfman, A. and Lemay, P., J. Gen.
     Virol., 16:429, 1972.

31. Neurath, A.R., Hartzell, R.W. and Rubin, B.A., Virology, 42:789, 1970.

32. Glaser, R., De Thé, G., Lenoir, G., Desgranges, C., and Ho, J.H.C.,
     Abstr. Annu. Meet. Am. Soc. Microbiol., p. 248, Washington, D.C., 1976.

33. Patel, P., Menezes, J. and Joncas, J., Abstr. Annu. Meet. Am. Soc.
     Microbiol., p. 247, Washington, D. C., 1976.

34. Roizman, B., Proc. Nat. Acad. Sci., USA, 48:795, 1962.

35. Roizman, B., Proc. Nat. Acad. Sci., USA, 48:973, 1962.

36. Roizman, B., Proc. Nat. Acad. Sci., USA, 49:165, 1963.

37. Roizman, B. and Aurelian, L., J. Mol. Biol., 11:528, 1965.

38. Watkins, J.F., Nature (London), 202:1364, 1964.

39. Lemaster, S.L. and Blough, H.A., Abstr. Annu. Meet. Am. Soc. Microbiol., p. 249, Washington, D.C., 1976.

40. Levitan, D.B. and Blough, H.A., J. Virol., 18:1081, 1976.

41. Crowell, R. L., J. Bacteriol., 86:517, 1963.

42. Ichihashi, Y. and Dales, S., Virology, 46:533, 1971.

43. Enegren, B.J., Duffy, E.M., Angel, M.A. and Burness, A.T.H., Abstr. Annu. Meet. Am. Soc. Microbiol., p. 211, Washington, D.C., 1976.

44. Holland, J.J., and McLaren, C.C., J. Exp. Med., 109:487, 1959.

45. Zajac, F. and Crowell, R.L., J. Bacteriol., 89:574, 1965.

46. Lonberg-Holm, K., and Korant, B.D., J. Virol., 9:29, 1972.

47. Philipson, L. and Bengtsson, S., Virology, 18:457, 1962.

48. Gomatos, P.J. and Tamm, I., Virology, 17:455, 1962.

49. Lerner, A.M. and Miranda, Q.R., Virology, 36:277, 1968.

50. Nicoli, J., Ann. Inst. Pasteur, Paris, 109:472, 1965.

51. Mooney, J.J. Dalrymple, J.M., Alving, C.R., and Russell, P.K., J. Virol., 15:225, 1975.

52. Birdwell, C.R. and Strauss, J.H., Virol., 11:502, 1973.

53. Karabatsos, N., J. Immunol., 91:76, 1963.

54. Choppin, P.W. and Tamm, I., in CIBA Foundation Symposium, Cellular Biology of Myxovirus Infection, edited by G.E.W. Wolstenholme, p. 218, Little Brown, Boston, 1964.

55. Marchesi, V.T., Tillack, T.W., and Scott, R.E., J. Exp. Med. 135:1209, 1972.

56. Winzler, R.J., in Cellular Receptors, edited by R.T. Smith, and R.A. Good, p. 11, Appleton, New York, 1969.

57. Tiffany, J.M. and Blough, H.A., Virology, 44:18, 1971.

58. Schultze, I., in The Influenza Viruses and Influenza, edited by E.D. Kilbourne, p. 53, Academic Press, New York, 1975.

59. Kendal, A.P., Virology, 65:87, 1975.

60. Loughlin, M., Labat, D., O'Callaghan, R.J. and Howe, C., Abstr. Annu. Meet. Am. Soc. Microbiol., p. 264, Washington, D.C., 1975.

61. Weinstein, D.B., Marsh, J.B., Glick, M.C. and Warren, L., J. Biol. Chem., 245:3923, 1970.

62. Kohn, A. Virology, 26:228, 1965.

63. Haywood, A.M., J. Mol. Biol., 83:427, 1974.

64. Bratt, M.A. and Gallaher, W.R., in Membrane Research, edited by
     C.F. Fox, p. 383, Academic Press, New York, 1972.

65. Poste, G. and Waterson, A.P., in Negative Strand Viruses, vol. 2,
     edited by B.W.J. Mahy, and R.D. Barry, p. 905, Academic Press, New
     York, 1973.

66. Poste, G., Alexander, D. J., Reeve, P. and Hewlett, G., J. Gen. Virol.
     23:255, 1974.

67. Crittenden, L.B., J. Nat. Cancer Inst., 41:145, 1968.

68. Hartley, J.W., Rowe, W.P., and Heubner, R.J., J. Virol., 5:221, 1970.

69. Vogt, P.K. and Ishizaki, R., Virology, 30:368, 1966.

70. Ishizaki, R. and Bolognesi, D.P., J. Virol., 17:132, 1976.

71. Quersin-Thiry, L., J. Immunol., 81:253, 1958.

72. Habel, K., Hornibrook, J.W., Gregg, N.C., Silverberg, R.J. and
     Takemoto, K.K., Virology, 5:7, 1958.

73. Apostolov, K. and Waterson, A.P., Microbios., 11A:85, 1974.

74. Yamamoto, K., Inoue, K., and Susuki, K., Nature (London), 250:511,
     1974.

Chapter 29

IMMUNOLOGICAL ASPECTS OF SPERM RECEPTORS ON THE
ZONA PELLUCIDA OF MAMMALIAN EGGS

Bonnie S. Dunbar and C. Alex Shivers
Department of Zoology
University of Tennessee
Knoxville, Tennessee 37916

Abstract

Sperm receptor sites on the surface of the zona pellucida are
believed to be necessary for fertilization on the mammalian egg. Ovary
specific antibodies are known to precipitate the zona pellucida and are
also responsible for inhibition of sperm binding and subsequent
fertilization. The nature of the proposed receptor sites which are
blocked by antibodies are discussed relative to comparable studies in
which phytoagglutinins and other agents are used to determine the
chemical nature of sperm receptor sites.

Introduction

The zona pellucida of the mammalian egg is a thick transparent,

envelope of noncellular structure surrounding the egg, is formed in

the ovary during development of the oocyte and is present until the time

of implantation of the fertilized egg (1, 2, 3). Because it is the

external investment of the egg, many investigators have speculated the

zona pellucida to be important in initial stages of fertilization as the

sperm must attach and pass through it before making contact with the

plasma membrane of the egg. This association of the sperm with the zona

has been demonstrated to be a species-specific event and has consequently

stimulated the search for specific sperm receptor sites as well as

possible antigenic sites on the surface of the zona (4, 5, 6, 7, 8).

In addition to its role in sperm recognition, the zona is believed to be involved in the block to polyspermy, (4, 9) in maintaining the normal cleavage pattern and prevention of egg fusion (10) and allowing movement of cleaving eggs in the oviduct (11, 12).

Both physical and biochemical properties of the zona pellucida have been studied in order to obtain an insight into the interaction of the egg and sperm as well as to define the relationship of the egg and embryo with the uterine environment prior to implantation. The permeability properties of the zona have been of interest particularly with respect to sperm penetration and to the zona's possible role in immunological protection of the egg from its environment. The zona pellucida has been demonstrated to be permeable to such large molecules as immunoglobulins (mo. wt. 150,000 - 900,000) and ferritin (mo. wt. 480,000) but is impermeable to some smaller molecules such as heparin (mo. wt. 16,000). The ability of molecules to pass through the zona apparently does not necessarily depend on molecular weight but also on molecular configuration, degree of aggregation or charge characteristics of the molecule in relation to the zona (13, 14, 15, 16, 17). Although the zona surface has been observed with the light and scanning electron microscope to be an irregular, fenestrated (netlike) structure, it is generally considered that the pores in the net are consistently smaller than the sperm head and it is therefore not possible for the sperm to penetrate the zona without somehow altering the latter's structure (18).

Although the zona pellucida varies biochemically among species, it has generally been characterized to be a complex of sulfated, acid and neutral mucopolysaccharide and protein possibly in the form of glyco-peptide units, which are stabilized by disulfide bonds, hydrophobic interactions, or salt bridges (3, 19, 20, 21, 22, 23). Since both of

these components are known to be antigenic, the zona pellucida has been
an attractive target for immunological studies particularly with respect
to sperm binding and fertilization (24, 25).

The biochemical properties of the zona have been studied to determine
the nature of surface receptors which might determine species specificity
of sperm recognition in fertilization. Hamster surface receptors can be
destroyed or masked by trypsin or by a trypsin-like enzyme from the egg
cortical granules (26, 27). Sperm attachment and penetration through the
zona pellucida of the rabbit, however, was not altered by trypsin or
neuraminidase (28). These studies indicate that in some species, the
zona surface receptors that are involved in sperm attachment are protein
components although it has been suggested that trypsin treatment results
in zona modifications which are not found after fertilization in vivo (29).

The recent popularity for use of plant lectins in cell recognition
systems has given some insights into the process of sperm attachment and
binding to the zona pellucida. Studies with phytoagglutinins aid in the
analysis of biochemical properties responsible for specificity in the
process of sperm attachment which might be responsible for blocking hybrid
fertilization. By treating hamster eggs with wheat germ agglutinin
(WGA), which is known to bind N-acetyl-D-glucosamine, sperm attachment
to the zona pellucida could be reduced as well as prevention of
sperm penetration through this structure (30). More recent studies have
shown that the agglutinins Ricinis communis (RCA), Dolichos biflorus (DB)
and concanavalin A (Con A) (specific for terminal residues similar to
D-galactose, N-acetyl-D-glucosamine and a-D-mannose or a-D-glucose
respectively) were effective in preventing fertilization although sperm
binding was not prevented. The effective concentrations required for
different lectins were quite different (7). Ultrastructural localization

of these lectin binding sites, using ferritin-lectin conjugates, demonstrated
that the binding sites for RCA and WGA were localized in the outermost
regions of the zona pellucida while the Con A receptors appeared to be
distributed sparsely throughout the zona (22). The conclusion drawn from
these studies with phytoagglutinins, in conjunction with the studies
involving the trypsinization experiments described above was that the
attachment or binding sites on the zona pellucida involve the polypeptide-
saccharide complexes, but that the saccharide moieties may be the actual
binding sites. Such alterations of the zona which are probably due to
extensive cross-linking of saccharide chains may then be responsible for
the failure of the sperm to recognize the surface binding sites. Another
possibility is that the sperm-borne lysine, an enzyme which is thought to
facilitate penetration of the sperm head through the zona (28), can no
longer depolymerize the zona material (7).

Immunological Studies of the Zona Pellucida

Immunological techniques have been used with increasing popularity
to look at the process of fertilization, particularly with studies which
pertain to diagnostic tests relative to fertility and sterility (8, 18, 31).
The ovary has been studied as a possible source of egg antigens and
several workers have been using heterologous antisera produced against
saline extracts of whole ovary to study ovary specific antigens in the
hamster, mouse and rabbit. The rabbit was used to produce antibodies
against hamster (25) and mouse (8) and pig   (unpublished) ovaries and the
sheep was used to produce antibodies against rabbit ovary (32). In
addition to some antigenic components that are also present in blood serum
and somatic tissues, the ovary contains organ specific antigens. Antibodies
which react with the specific ovarian antigens were isolated by absorbing
the antiovary serum with somatic tissues. Within the sensitivity of agar
gel diffusion and immunoelectrophoretic tests, it was concluded that the

hamster, mouse and rabbit ovaries contained 3, 2, and 2 ovary specific

antigens respectively. These numbers probably indicate the minimum number

of specific antigens in ovary (8).

Because of the location of the zona pellucida, its composition, and

the large number of zona coated eggs found in the ovary it was of interest

to find if the zona was responsible for any of the ovarian specific

antigens. With the use of immunofluorescent techniques it was possible

to show that the ovary antibodies were restricted to certain areas of the

ovary with evidence that some antigens are associated with the zona

pellucida. Ovary specific antigens of the hamster and rabbit (8, 32)

were found in the zona pellucida, theca interna and atretic follicles.

These results are similar to results reported for gonad specific antigens

in the ovary of the guinea pig (33). The binding of antiovary antibody to

zona of a pig egg is demonstrated in Figure 1.

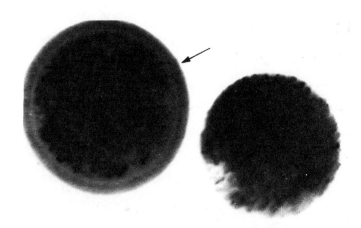

FIGURE 1

Zona coated pig eggs treated with immunoglobulin fraction of rabbit anti-
pig ovary (left) and rabbit preimmune serum (right) followed by treatment
with fluorecein conjugated sheep anti-rabbit immunoglobulin. Note the
fluorescent zona (arrow) in the anti-ovary serum treated egg. The egg
cytoplasm autofluoresces.

Mouse zona was demonstrated to be antigenic when antiserum was produced in rabbits against freshly ovulated eggs with their cumulus cells and against embryos at two preimplantation stages (morula and blastocyst). Immunofluorescence was used to demonstrate that each of these antisera contained antibodies which reacted with the egg and embryo cytoplasm as well as with the zona pellucida. Using immunohistological tests it was concluded that absorption of each antiserum with mouse serum rendered it specific to the zona when tested on zona coated eggs and embryos (34).

Experiments using the ovary specific antisera have been extended to look at the effects of antibodies on the egg-sperm interactions and its effect on fertilization. Treatment of the zona coated eggs of hamster and rabbit with the absorbed antisera have shown egg agglutination. Similarly treated hamster, mouse and rabbit ova have been demonstrated to form a precipitate in the outer region of the zona. This precipitate was observed to change the light scattering property of the zona surface and is thought to be due to the formation of a precipitate resulting from antibodies reacting with complementary antigens in the zona. The precipitate is seen within a few minutes after incubating the egg in the antiserum and was never observed in the zona of eggs treated with control antiserum. The precipitating effects of the antibody could be neutralized by absorption with ovary or zona coated eggs but not with any somatic tissues tested. The precipitation was always observed near the outer surface of the zona in each of the three species of mammals tested, suggesting that the active sites of the antigens are concentrated at the external surfaces (8). Further evidence for this external localization is that when the inner surface of the zona is exposed to the antibody following mechanical rupture of the egg, no precipitate is formed on the inner surface (35). This localization of the antigenic sites at the outer surface of the zona appears to coincide with the lectin binding sites that have been described

(25). The heterogeneity of the antigen being localized on the outer region of the zona fits well with the chemical and physical heterogeneity that has been observed in the zona of several mammals (19, 36).

Transmission and scanning electron microscopy have been used to demonstrate that the interaction of the zona precipitating antibody (ZPA) with the fibrous network of the hamster zona pellucida does not occlude interstitial pores. Transmission electron micrographs showed a distinct, dense precipitate on the surface of antibody treated eggs while treatment with control media, preimmunization serum, and ZPA absorbed with ovary, did not show much of a precipitate. Possibilities that were offered to account for the lack of pore occlusion, were that the ZPA might block or mask sperm receptors and/or enzyme substrate(s) during sperm penetration or mechanically inhibit zona dissolution during implantation (23).

Seemingly more significant effects of the zona-precipitating antibody are those that the antibody has on attachment and penetration of sperm through the zona. When hamster eggs were exposed to the antibody, washed, and inseminated in vitro with capacitated sperm, only a few sperm attached to the zona, although many were observed to collide with the precipitated surface. Those at the zona surface were easily removed by washing the inseminated eggs with saline so they were not firmly bound (Fig. 2). The possibility that the sperm might be effected by the antibody was not thought likely since the eggs were thoroughly washed following antibody treatment and sperm motility was not affected by the antibody. The sperm was also able to attach to and penetrate antibody-treated zona free eggs. It has been concluded from these studies that precipitation of the zona surface by specific antibodies interferes with sperm-egg interaction in vitro by rendering attachment sites on the zona unavailable to the sperm. As in the studies with the plant lectins, the interference of sperm binding might be explained in terms of a mechanical block due to

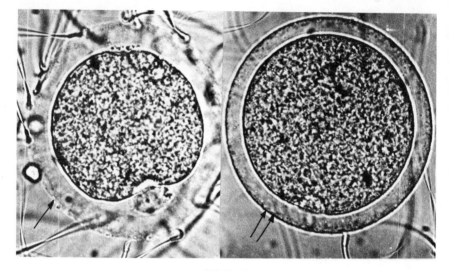

FIGURE 2

Hamster eggs treated with rabbit preimmune (left) and anti-hamster ovary (right) serum and then inseminated with capacitated hamster sperm. Note the precipitated zona (arrows) and lack of sperm attachment in the anti-ovary antiserum treated egg as compared to the preimmune serum treated egg.

the precipitate or to a reaction between the antibody and specific attachment sites for the sperm (18).

Experiments have been done to show that not only does antibody produced in rabbits against hamster ovary block fertilization of hamster eggs in vitro by binding to the surface of the zona pellucida and rendering it inpenetrable to spermatozoa but that the antibody was also effective in blocking fertilization in vivo. If the antibody was injected intraperitonealy into females, they were rendered infertile for approximately three oestrous cycles. It was concluded that this temporary sterility was due to binding of antibody to the zonae pellucidae of oviductal and ovarian oocytes, since observations of eggs recovered from the ovaries or oviducts of the immunized females showed light scattering in the zona as was seen in antibody binding in vitro (24).

Similar studies have also demonstrated that antibodies against mouse ovaries blocked fertilization of mouse eggs both in vitro and in vivo (37).

To date, the criteria that have been generally used for determination of specific sperm sites on the surface of the egg have been those of inhibition of attachment and/or inhibition of fertilization. Although it is generally accepted that the sperm binding sites on the zona pellucida contain protein (7, 37) it is not clear whether the protein per se is the important moiety of the sperm binding site or whether the saccharide moieties of the peptide-saccharide complex of the zona pellucida are responsible. In experiments using plant lectins and zona precipitating antibodies which inhibit sperm binding and fertilization (7, 8), it is still not possible to completely distinguish between stearic hindrance due to cross linkage of surface molecules or actual blockage of the sperm binding sites. Univalent Fab antibody fragments might give additional insight into the blockage of specific surface molecules. Although Fab fragments would reduce cross-linking of surface molecules, this still would not completely rule out stearic effects if sperm did not bind. Further studies might also make use of phytoagglutinins to separate and purify receptor sites using column affinity chromatography. Methods which might also be employed to determine specific sperm binding sites on the surface of the zona are those which have been used to look at sperm receptor sites on the plasma membrane of sea urchin eggs. These experiments were done by partially characterizing proteins on the egg plasma membrane and then titrating them against sperm. Decrease in fertilization which corresponded to an increase in egg surface protein was taken to be evidence that the sperm recognition sites are blocked by specific egg receptor sites (39).

The antigenic properties of the proposed sperm binding sites of the zona pellucida offer a means for the use of antibodies in understanding

early events in fertilization as well as the attractive and exciting
possibility for their use in contraception.

References

1. Odor, D.L., J. Biophys. Biochem. Cytol., 7:567, 1960.

2. Dickman, Z. and DeFeo, V.J., J. Reprod. Fert., 13:3, 1967.

3. Kang, Y., J. Anat., 139:535, 1974.

4. Austin, C.R., The Mammalian Egg. Blackwell, Oxford, 1961.

5. Hanada, A. and Chang, M.C., Biol. Reprod., 6:300, 1972.

6. Hartman, J.F., Gwatkin, R.B.L. and Hutchison, C.F., Proc. natn.
    Acad. Sci. USA., 69:2767, 1972.

7. Oikawa, T., Nicolson, G.L. and Yanagimachi, R., Exp. Cell Res.
    83:239, 1974.

8. Shivers, C.A., in Immunological Approaches to Fertility Control,
    edited by E. Diczfalusy, p. 223, Bogtrykkeriet Forum, Copenhagen, 1970.

9. Barros, C. and Yanagimachi, R., J. exp. Zool., 180:251, 1972.

10. Mintz, B., Science, 138:594, 1962.

11. Modlinski, J.A., J. Embry. Exp. Morph., 23:539, 1970.

12. Gwatkin, R.B.L., J. Reprod. Fertil., 6:325, 1963.

13. Austin, C.R. and Lovelock J., Expl. Cell Res., 15:260, 1958.

14. Sellens, M.H. and Jenkinson, E.J., J. Repr. Fert., 42:153, 1975.

15. Enders, A.C., in The Biology of the Blastocyst, edited by R.J. Blandau,
    p. 71-94, University of Chicago Press, Chicago, 1971.

16. Hastings, II, R.A., Enders, A.C. and Schlafke, S., Biol. Reprod.,
    7:288, 1972.

17. Fox, L.L. and Shivers, C.A., Fertil. Steril., 26:599, 1975.

18. Shivers, C.A. and Dudkiewicz, A.B., in Physiology and Genetics of
    Reproduction, edited by E.M. Coutinho and F. Fuchs, p. 81, Plenum
    Press, New York, 1974.

19. Pikó, L., in Fertilization, edited by C.B. Metz and A. Monroy, p. 325, Academic Press, New York, 1969.

20. Gould, K., Zaneveld, L.J.D., Srivastava, P.N., and Williams, W.L., Proc. Soc. Exp. Biol. Med. 136:6, 1971.

21. Inoue, M. and Wolf, D.P., Biol. Repr., 10:512, 1974.

22. Nicolson, G.L., Yanagimachi, R. and Yanagimachi, H., J. Cell Biol., 66:263, 1975.

23. Dudkiewicz, A.B., Shivers, C.W. and Williams, W.L., Biol. Repr. (in press), 1976.

24. Oikawa, T. and Yanagimachi, R., J. Reprod. Fert., 45:487, 1975.

25. Ownby, C.L. and Shivers, C.A., Bio. Repr., 6:310, 1972.

26. Hartman, J.F. and Gwatkin, R.B.L., Nature, Lond., 234:479, 1971.

27. Barros, C. and Yanagimachi, R., Nature, Lond., 233:268, 1971.

28. Bedford, J.M., in Physiology and Genetics of Reproduction, edited by E.M. Coutinho and F. Fuchs, p. 69, Plenum Press, New York, 1974.

29. Oikawa, T., Nicolson, G.L. and Yanagimachi, R., J. Reprod. Fert., 43: 133, 1975.

30. Oikawa, T., Yanagimachi, R., and Nicolson, G.L., Nature, 241:256, 1973.

31. Metz, C.B., Fed. Proc., 32:2057, 1973.

32. Sacco, A.G. and Shivers, C.A., J. Reprod. Fertil., 32:403, 1973.

33. Porter, G., Highfill, D. and Winovich, R., Int. J. Fertil., 15:171, 1970.

34. Glass, L.E. and Hanson, J.E., Fert. Steril., 25(6):484, 1974.

35. Garavagno, A., Rosada, J., Barros, C., and Shivers, C.A., J. exp. Zool., 189:37, 1974.

36. Guraya, S.S., Int. Rev. Cyt., 37:121, 1974.

37. Jilek, F. and Pavlok, A., J. Reprod. Fert., 42:377, 1975.

38. Hartman, J.F. and Hutchison, C.F., J. Reprod. Fert., 36:49, 1974.

39. Lennarz, W. (personal communication).

INDEX